Growing Professionally

Readings from NCTM Publications for Grades K–8

Edited by

Jennifer M. Bay-Williams • Karen Karp

University of Louisville • Louisville, Kentucky

Library of Congress Cataloging-in-Publication Data

Growing professionally: Readings from NCTM publications for grades K–8 /
edited by Jennifer M. Bay-Williams, Karen Karp.
    p. cm.
 ISBN 978-0-87353-605-9
  1. Mathematics teachers—Training of—United States. 2. Mathematics—
    Vocational guidance—United States. I. Bay-Williams, Jennifer M.
II. Karp, Karen.
QA10.5.G76 2008
372.7023—dc22

                                  2007047063

The National Council of Teachers of Mathematics is a public voice of
mathematics education, providing vision, leadership, and professional development
to support teachers in ensuring equitable mathematics learning of the highest
quality for all students.

Printed in the United States of America

This book is dedicated to the many teachers, teacher leaders, and teacher educators who devote endless hours, talent, and energy to the ongoing professional development of mathematics teachers.

We hope that you see some of the impact of your work on teachers; certainly the butterfly effect you create in the learning of mathematics is worthwhile, far-reaching, and significant.

DEDICATION

Contents

This book presents powerful articles over a range of topics that can support your own learning as well as those prekindergarten through eighth-grade educators with whom you offer professional development. Our process of selecting articles involved requesting suggestions of "favorite NCTM articles" from members of NCTM and two leadership affiliates of NCTM, the Association of Mathematics Teacher Educators (AMTE) and the National Council of Supervisors of Mathematics (NCSM). Additionally, we reviewed responders' recommended readings from such print resources as the NCTM Navigations series and mathematics methods books. And finally, we read untold numbers of articles, seeking those that could support the wide range of topics that are important for growing as a professional educator, from discourse to diversity to classroom practice to instructional tasks. We sought articles that would be successful across the continuum from aspiring teachers to master teachers. We hope that these articles become "tools for thought" so that preservice and in-service teachers can advance the conversations about improving practice and student learning.

This book is organized into three parts. Part 1, Ideas for Using Articles in Professional Development, presents suggestions for facilitators and activities that use articles in various ways with teachers in professional development settings. Part 2, Strategies for Growing as a Professional, includes articles about different forms of professional development. Collectively these articles offer a wide range of ideas for engaging teachers in ways that will support their growth. Part 3, Topics for Growing as a Professional, includes articles across seven themes. Within each theme are favorite articles that have been used in many different settings with success: articles that effectively illuminate an aspect of practice, articles that promote good discussion in group settings, and articles rich with exemplars to help us continue to refine and improve our practice.

The editors wish to extend our very sincere thank-you to the many individuals who contributed time, expertise, insight, and support at the myriad stages of creation of this collection.

How did we find the articles in this book? We started by setting up a database and inviting input from colleagues in the profession. The following people contributed suggestions, many of them offering more than one article. We are grateful for their thoughtful responses. Their recommendations totaled more than 100 articles, unfortunately many more than we could include. Many of these contributors offered explanations of how they used the recommended article, and their elaborations were enormously helpful to us in narrowing our selections to the collection presented here and listed in the Suggested Reading at the end of this book.

Keith Adolphson	Todd A. Grundmeier	Mari Muri
Honi J. Bamberger	A. T. Hayashi	Gretchen Murphy
Barbara Boschmans	Dennis Hembree	Tanna Nicely
Karen Brannon	Deborah A. Hill	Nancy Paugh
M. Lynn Breyfogle	R. Hill	Blake Peterson
JoAnn Cady	William D. Jamski	Kathy Rieke
Karen Droga Campe	Jane H. Jones	Josephine Rodriguez
Karen Cannon	Ginny Keen	Rheta Rubenstein
Kathryn Chval	Michael Kestner	Jacqueline Sack
Alison Claus	Beverly Kimes	Judith Sallee
Wendy Pelletier Cleaves	Bernadine Krawczyk	Edna O. Schack
Lee Anne Coester	Angela Krebs	Cynthia Schimek
David Coffey	John Lannin	Anne M. Seitsinger
Jacqueline Coomes	Hollylynne Stohl Lee	Heidi Shepard
Stan Dick	Suzie Legg	Karen Sherman
Carmella Ettaro	Larry Lesser	Margaret Smith
David Feikes	Eleanor Linn	Martha K. Smith
Francis (Skip) Fennell	Mary Lostetter	Sherilyn Stratton
Jeanette Fernandez	Johnny W. Lott	Phyllis Tam
Kathy Fick	Nancy Low	P. Mark Taylor
Carol Fisher	Carol A. Marinas	Chuck Thompson
Ann Fishman	Salvatore Marino	Sue Tombes
Melissa Freiberg	Heather Martindill	Lynn Trell
Susan Friel	James F. Marty	Juliana Utley
Elizabeth Gamino	Sharon McCrone	Pamela Wells
Mike Gilbert	Linda Metnetsky	Jane Wilburne
Eric Gold	Trudy Mitchell	Connie H. Yarema

We feel fortunate to have worked with the NCTM Publications staff. Prior to his retirement, Harry Tunis helped us set up the database and solicit article recommendations. Beth Skipper offered suggestions from her perspective as a journal editor. We extend very special thanks to Ken Krehbiel and Nancy Busse, who offered support, insights, and quality input, as well as to project manager Ann Butterfield and designers Randy White, Glenn Fink, and Jazminia Griffith.

(continued)

Acknowledgments

We each would also like to thank our families, who have offered support and have sacrificed evening and weekend time with us.

To my husband, Mitch, and children, MacKenna and Nicolas
—Jenny

To my family, Bob Ronau, Matthew Ronau, Tammy Ronau, Jessica Ronau, Zane Ronau, Joshua Ronau, Misty Ronau, Matthew Karp, Christine Marsal, Jeffrey Karp, and Pamela Morgan
—Karen

Mapping to NCTM Content and Process Standards

Titles of Articles	Number and Operations	Algebra	Geometry	Measurement	Data Analysis & Probability	Problem Solving	Reasoning and Proof	Communication	Connections	Representation
Part 2: Articles about Growing Professionally										
Mathematical Thinking—Helping Prospective and Practicing Teachers Focus	X							X		X
Using Students' Work as a Lens on Algebraic Thinking		X					X			X
Focusing Conversation to Promote Teacher Thinking	X							X		
Ideas for Establishing Lesson- Study Communities								X		
Using Cases to Integrate Assessment and Instruction	X									
Planning Strategies for Students with Special Needs: A Professional Development Activity		X						X		
Expanding Teachers' Understanding of Geometric Definition: The Case of the Trapezoid			X							
Using Teacher Produced Videotapes of Student Interviews							X			X
Part 3: Articles for Use in Professional Development										
Lifelong Learning										
Four Teacher-Friendly Postulates for Thriving in a Sea of Change										
Never Say Anything a Kid Can Say!						X		X		
Redefining Success in Mathematics Teaching and Learning		X				X	X	X		X
Signposts for Teaching Mathematics through Problem Solving						X	X	X		
Windows and Mirrors										
Helping English-Language Learners Develop Computational Fluency	X							X		
Mathematical Notations and Procedures of Recent Immigrant Students	X	X						X		X
Building Responsibility for Learning in Students with Special Needs	X									
Differentiating the Curriculum for Elementary Gifted Mathematics Students				X		X	X			
Meaningful Mathematical Tasks										
Selecting and Creating Mathematical Tasks: From Research to Practice	X	X	X	X	X	X	X		X	
Turning Traditional Textbook Problems into Open-Ended Problems	X	X	X	X	X	X				
The Role of the Textbook in Supporting Curriculum and Learning Principles				X		X		X	X	

(Continued)

Mapping to NCTM Content and Process Standards—*Continued*

Titles of Articles	Number and Operations	Algebra	Geometry	Measurement	Data Analysis & Probability	Problem Solving	Reasoning and Proof	Communication	Connections	Representation
Classroom Discourse										
Assessment and Accountability: Strategies for Inquiry-Style Discussions	X									
Strategies for Advancing Children's Mathematical Thinking						X	X	X	X	X
Questioning Our Patterns of Questioning		X						X		
Discourse That Promotes Conceptual Understanding	X							X		
Challenging Students										
Sometimes Less Is More						X				
Isn't That Interesting!	X					X	X	X		
Is a Rectangle a Square? Developing Mathematical Vocabulary and Conceptual Understanding			X				X	X		X
Using Counterintuitive Problems to Promote Student Discussion	X	X		X	X	X	X	X		
Deepening Understanding										
Relational Understanding and Instrumental Understanding							X		X	
Learning Strategies for Addition and Subtraction Facts: The Road to Fluency and the License to Think	X							X	X	
Using Language and Visualization to Teach Place Value										
Meaning and Skill—Maintaining the Balance	X									
Multiplying Fractions	X									
Developing Algebraic Reasoning through Generalization		X					X			X
Three Balloons for Two Dollars: Developing Proportional Reasoning	X	X				X	X			
Misconceptions										
Balancing Act: The Truth Behind the Equals Sign		X						X	X	X
Why Children Have Difficulty Mastering the Basic Number Combinations	X									X
The Harmful Effects of Algorithms in Grades 1–4	X									
Mean and Median: Are They Really So Easy?					X					X

Introduction

"What can I do to help my students succeed in learning mathematics?"

Introduction

TEACHERS consistently ask the question "What can I do to help my students succeed in learning mathematics?" In the process of rethinking what we do as teachers, we engage in an agenda for improvement. This agenda includes seeking knowledge, building collaboration, and developing an ethic of improvement. This agenda propels the professional development that teachers seek. As Liping Ma advises, "Do not forget yourself as a teacher of yourself" (Herrera 2002, p. 16).

One can engage in learning about teaching in many ways. Lesson study, workshops, action research, case studies, study groups, and immersion are just some of the strategies that are successful for improving teaching practice and, hence, students' learning of mathematics. For each strategy, the professional development is grounded in practice; is data driven, long lasting, and responsive to teachers' and students' needs; and engages participants in reflection, analysis, discussion, and application. Reading and discussing articles, either as its own form of professional growth or within the context of one of the forms of professional development listed above, can support teacher reflection and teacher learning.

The process of reading and discussing articles has the following advantages:

1. Accommodating differences. Every time a teacher shares an article with a group of fellow teachers, different aspects of the article "speak to" that teacher.

2. Learning independently and at one's own pace. Articles afford flexibility in how quickly or slowly they are read, and allow a teacher to read and reread passages that he or she finds useful.

3. Encouraging depersonalized conversations of professional practice and thereby offering high-quality opportunities for engaging in discussions about improving performance. Discussing the actions of a teacher in a case study or a vignette in an article can be less intimidating and more inviting to members of a group than sharing their own experiences. Articles can be used as a catalyst for conversations about practice that over time can support the development of a safe learning community in which teachers can talk about their own teaching.

4. Supporting teachers in analyzing their own work. Powerful, insightful learning occurs when teachers discuss an aspect of intense interest in a lesson they have just taught or observed (e.g., dividing fractions) and then have an opportunity to read an article that is closely related to their own teaching dilemmas.

5. Enriching the follow-up to a professional development session. Just as students need time to reflect on a mathematics lesson, participants need extended time to digest the ideas of a professional development session. Reading articles can reinforce or extend the ideas that were the focus of the session.

(Continued on page 4)

(Continued from page 3)

6. Helping focus the endeavor of a study group or community of learners. Reading articles can help teachers identify and focus on pertinent aspects of their learning. For example, reading articles on facilitating student discourse can help teachers identify specific targets to work on collaboratively over the course of a year.

7. Identifying successful high-quality tasks, questioning strategies, teaching techniques, and resources. Although a given article may focus on accommodating a lesson for students with special needs, for example, teachers can glean tangible strategies from its exemplars to incorporate in their practice.

8. Spreading the word. Articles can easily be shared. When articles that are used in a course or workshop are found to have an important message, a teacher can share that message with colleagues, a principal, or even parents. Giving and receiving articles are informal ways that professionals learn from one another.

Part

1

● ● ●

Ideas for Using Articles in Professional Development

Overview

Here we offer a few ideas of how to incorporate these articles in a professional development session or university classroom, although many other creative techniques can be used to engage small groups in discussion. You will also find specific suggestions for articles embedded in the articles themselves.

—Editors

Suggestions

Use the strategies, tasks, or vignettes prior to reading the article

When an article includes a mathematical problem, first ask teachers to engage in the task themselves, just as the article describes students' being engaged in the task in a classroom. Next, model the process of sharing a solution with a group of peers. Additionally, encourage teachers to share strategies they anticipate using and barriers they anticipate encountering when they employ this task with their students. Similarly, asking participants to compare vignettes that have been lifted out of an article prior to reading it is a way to raise curiosity and awareness before to reading the full article.

Allow for Opportunities to Discuss the Article Collaboratively

Because we each take away different things when we read something, sharing our "ahas" can help others glean more from the article. Opportunities for such sharing can be offered in simple ways, such as by providing ten minutes of time for exchange of ideas, as well as in more elaborate, entertaining ways. The following activities have been used successfully with collaborative groups of teachers in professional development settings.

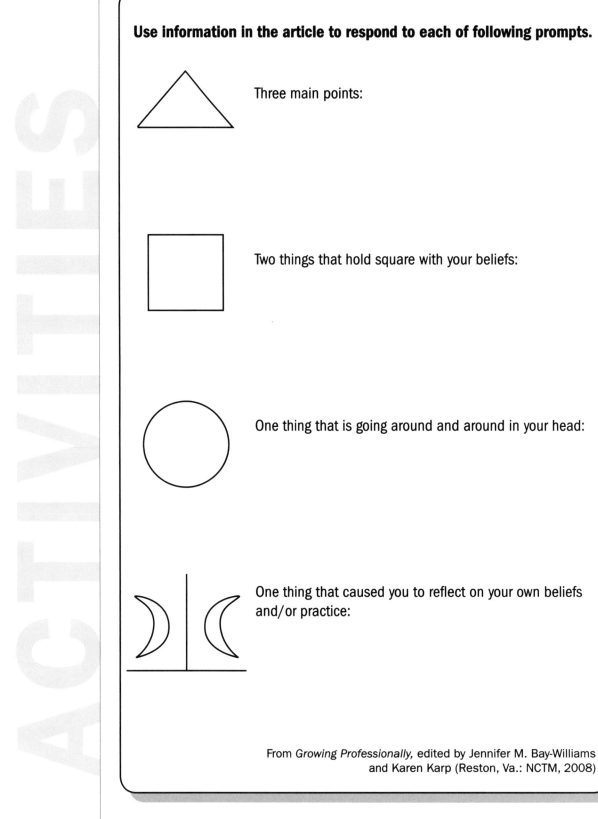

Use information in the article to respond to each of following prompts.

Three main points:

Two things that hold square with your beliefs:

One thing that is going around and around in your head:

One thing that caused you to reflect on your own beliefs and/or practice:

From *Growing Professionally,* edited by Jennifer M. Bay-Williams and Karen Karp (Reston, Va.: NCTM, 2008)

This coding later can be used in sharing, with each symbol being the focus of a round of discussion. Or, a poster can be prepared with each symbol, and groups can rotate to each station to record their group's comments for that symbol.

Rapporting (Not Reporting)

Ask teachers to read the article and relate the author's perspectives to their own. Give the teachers stacks of sticky notes, and ask them to write notes to the authors with the intent of delving deeper into the author's thinking or actions. Prompts for this activity might include—

- Relating one's own experiences that challenge or support the ideas in the article;
- Relating one's own experiences of learning school mathematics as compared with learning mathematics in the ways suggested in the article; and
- Asking the author further details, including next steps in implementing the ideas put forth in the article.

Answering these types of questions by jotting notes as one reads is a different approach to note-taking, resulting in a rapport of sorts with the author or authors of an article. The sticky notes can be shared after reading or can be sorted and organized by the group to further discuss the big ideas of the article.

From *Growing Professionally,* edited by Jennifer M. Bay-Williams and Karen Karp (Reston, Va.: NCTM, 2008)

Coding the Text

Adapted from the 4As Text Protocol, National School Reform Faculty (NSRF 2007)

Use the following markings as you read the text:

! = Important points

✓ = Things that you agree with

? = Question related to statement or idea presented

↔ = Connections with something else

This coding can later be used in a sharing session, with each symbol being the focus of a round of discussion. Alternatively, a poster can be prepared for each symbol, and groups can rotate to each station to record the group members' comments relating to that symbol.

From *Growing Professionally*, edited by Jennifer M. Bay-Williams and Karen Karp (Reston, Va.: NCTM, 2008)

Significant SenTENces

Ask participants to review the highlighting in the article they read, and with their group members to select ten (or a number of your choice) main sentences that are the essential points in improving students' performance.

Record (in short phrases if possible) each sentence on a sticky note or note card. Once each group has recorded its ten phrases, intergroup sharing can occur in two ways. In one approach, group members post all their selected sentences, then sort them according to repeated or related ideas. Alternatively, groups can trade sets and report similarities and differences in the ideas they prioritized in the article.

From *Growing Professionally*, edited by Jennifer M. Bay-Williams and Karen Karp (Reston, Va.: NCTM, 2008)

Share-Listen-Respond
Adapted from NSRF (2007)

Ask teachers to identify one important passage in the article.

In turn, each participant has 2 minutes to share the passage he or she selected and the reasons for picking it. Each group member then has 1 minute to extend, clarify, counter, or question the passage. After each person has had a turn, the originator of the passage gets a 1-minute rebuttal to offer reactions and reveal her or his current thinking in relation to what was heard from colleagues.

In a group of four participants, this activity takes 6 minutes per person. If each person has a turn, the discussion will take 25 minutes. If less time is available, groups of three can be used; or time allotments can be cut in half; or just one person can be chosen to take the lead on an article, with roles to be rotated on the next article.

From *Growing Professionally,* edited by Jennifer M. Bay-Williams and Karen Karp (Reston, Va.: NCTM, 2008)

Gallery Walk

To prepare for this discussion, prepare a large poster for each group, each with its own header. If using the "How Does the Article Shape Up?" activity described on page 8, for example, one poster would be labeled "Main Points"; another, "Round and Round"; and so on. Poster headers can be anything you like, including provocative questions.

Give members of each team same-color markers to keep with them as they rotate. At their first station, have them record the ideas that are shared among their group members. After about 5 minutes, have groups rotate to the next station.

At the new station, group members use their colored marker to (1) place a star next to ideas they agree with, (2) place a question mark next to ideas they are not necessarily sure of, and (3) add their own ideas.

Continue this process until the group returns to its original station. Give members 5–10 mintues to read and and prepare a summary of the ideas presented on their poster.

From *Growing Professionally,* edited by Jennifer M. Bay-Williams and Karen Karp (Reston, Va.: NCTM, 2008)

To lead people, walk beside them....

As for the best leaders,
the people do not notice their existence.

The next best, the people honor and praise.

The next, the people fear; and the next,
the people hate....

When the best leader's work is done the people
say, "We did it ourselves!"

—*Lao-Tsu*

Part 2

Articles about Growing Professionally

Synopses of Articles

Analyzing Students' Work Samples

Analyzing student work samples is an effective tool for growing profession-ally. Two of the reprinted articles emphasize the process of analyzing students' work, one from *Teaching Children Mathematics* (TCM) and one from *Mathematics Teaching in the Middle School* (MTMS). Both articles foster insights into effectively using students' work to promote teachers' learning. "Mathematical Thinking: Helping Prospective and Practicing Teachers Focus," by Jacobs and Philipp (2004), offers several examples of elementary school students' work and related discussion questions. The article outlines the central aspects of the process of using work samples to learn more about children's thinking. "Using Students' Work as a Lens on Algebraic Thinking," by Driscoll and Moyer (2001), also presents a model for analyzing students' work, in this example, related to algebra. Using a sample analysis of the "crossing the river" problem, they guide the reader through the process.

Coaching

Coaching is the focus of Huinker and Freckmann's (2004) "Focusing Conver-sations to Promote Teacher Thinking." This article, originally appearing in the Supporting Teacher Learning department of *Teaching Children Mathematics,* presents two scenarios focused on the importance of mentors' questioning skills. Actual dialogue between a professor and an in-service teacher serves as a model for helping teachers examine their decisions and engage their students.

Using Lesson Study

"Ideas for Establishing Lesson-Study Communities," by Takahashi and Yoshida (2004), describes the ground-up process for organizing a group of professions in lesson study. Using a research-based model, the authors provide a six-step process for colleagues to create, present, and reflect on a specific lesson. The lesson-study approach emphasizes that the participants are active learners in the process of improving their instruction and growing as professionals.

Using Cases

The exploration of lessons learned from using cases is the central feature of a professional development experience in "Using Cases to Integrate Assessment and Instruction," by Wilcox and Lanier (1999). The authors link assessment and instruction into a seamless process through the use of cases that incorporate meaningful work tasks (in this instance, fractions), concept maps, videos, the analysis of students' work, and the creation of personalized cases for study.

Conducting a Workshop Session

Brodesky, Gross, McTigue, and Tierney (2004), in "Planning Strategies for Students with Special Needs: A Professional Development Activity," offer a useful tool for focusing teachers' attention on what goes on inside classrooms. The discussion involves data and graphs but is generalizable to any mathematics lesson. The authors outline a workshop session built on six steps to identify potential barriers and brainstorm appropriate accessibility strategies.

Using Mathematics Tasks

Groth (2006) examines the challenges that geometric concepts pose for teachers in "Expanding Teachers' Understanding of Geometric Definition: The Case of the Trapezoid." By posing the provocative statement that "the trapezoid is one of the most controversial shapes in all of geometry," he pushes teachers to look at a variety of "proper" definitions of that shape. With a series of examples and nonexamples, the author discusses how to elicit thinking while maintaining a level of cognitive conflict and lack of certainty. Through experiencing the same dissonance their students experience, teachers confront fundamental questions about their practice and their need to consistently develop content knowledge.

Using Video Vignettes

Transforming classroom practices through the analysis of students' problem-solving strategies is the objective of "Using Teacher-Produced Videotapes of Student Interviews as Discussion Catalysts," by Jacobs, Ambrose, Clement, and Brown (2006). Teachers are engaged in learning through watching a video that depicts a teacher posing a problem, a student solving the problem, and the resulting discussions of the child's strategies. By having teachers observe other teachers at work using a stop-action approach, this professional development experience focuses on teachers' abilities to identify opportunities for learning.

Mathematical Thinking: Helping Prospective and Practicing Teachers Focus

Victoria R. Jacobs and Randolph A. Philipp

How did Misha solve the problems in **figure 1**, and what mathematical understandings do her strategies reflect? Misha's written work provides a rich context in which prospective and practicing teachers can discuss issues of mathematics, teaching, and learning. These discussions might include conversations about the similarities and differences between Misha's two strategies, what she must have understood about place value to generate these strategies, and what type of instruction likely preceded and should follow this problem-solving effort.

Teacher educators have found student work such as Misha's to be a useful discussion catalyst during professional development, and using student work in this way has been linked to positive changes in teaching practices and student learning (Little 2004). Student work provides an authentic context in which prospective and practicing teachers can explore how children think about mathematics and how they can use children's thinking in their instructional decision making. By learning to learn from student work, prospective and practicing teachers can develop the necessary dispositions and skills to continually learn throughout their careers from the written work that students in their classrooms generate.

Despite these benefits, little guidance exists for how teacher educators should use student-work examples. The benefits of using student work are not inherent in the written work itself but instead evolve from the discussions surrounding it. This article provides a questioning framework to help teacher educators enhance conversations about student work and focus attention on children's mathematical thinking and its role in instruction. We have found the following three categories of questions to be valuable discussion catalysts across multiple student-work examples:

Teaching Children Mathematics 11 (November 2004): 194–201

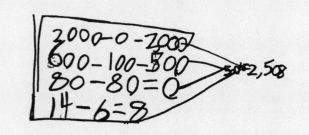

Fig. 1a. Ms. S. has 2694 bunnies. Ms. C. has 186. How many more bunnies does Ms. S. have than Ms. C.?

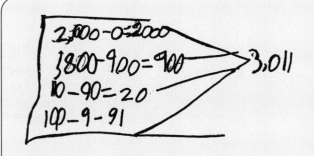

Fig. 1b. Ms. S. has 4010 bunnies. Ms. C. has 999. How many more bunnies does Ms. S. have than Ms. C.?

Fig. 1. Second grader Misha's solutions for two versions of a comparison problem

- Questions to prepare teachers to understand the child's thinking
- Questions to encourage teachers to explore the child's thinking in depth
- Questions to help teachers identify instructional "next steps" to extend the child's thinking

We present this framework to support the facilitation efforts of teacher educators in a variety of instructional settings with prospective and practicing teachers. Although these two audiences differ, both benefit from the use of student work. Therefore, when we describe the use of our framework with *teachers,* we envision audiences of either prospective or practicing teachers. Individual teachers who want to investigate the nuances in their students' work can also use our framework as a reflection tool; however, this article focuses on the use of the framework by teacher educators. We begin by explaining each of the questioning categories in the context of Misha's written work and then explore the questioning opportunities made possible by another example of written work in which the student, Heidi, generated an incorrect answer.

Misha's Nontraditional Strategies for Subtraction

As **figure 1** shows, Misha solved each comparison problem by breaking it into four partial subtraction problems in which she treated the digits, starting with the thousands, in terms of their values. She began solving the first problem by finding the difference between the quantities indicated by the thousands place (2000 – 0) and the hundreds place (600 – 100). The action that Misha performed in the tens place indicated that she had thought ahead and had realized that she would need to use one of the tens when working with the ones. Therefore, she computed 80 – 80 (instead of 90 – 80) and used the extra ten with the ones, computing 14 – 6 (instead of 4 – 6). To arrive at her final answer, she added her four differences.

For the second problem, Misha's actions again indicated that she began with the thousands, but she thought ahead and realized that she would need to decompose the 4010 so that she could use some of those thousands in subsequent computations. Specifically, she decomposed the 4010 into 2000, 1800, 110, and 100. This distribution allowed her to straightforwardly complete her subtractions within each place value, computing 2000 – 0, 1800 – 900, 110 – 90, and 100 – 9. As she did when solving the first problem, Misha added the four differences to arrive at her final answer. Misha's nontraditional strategies are interesting not only because they differ from the way in which most teachers would solve the problem but also because they reflect strong place-value understanding. The following sections apply our framework to identify questions to help teachers focus on Misha's mathematical thinking.

Preparing to Understand the Child's Thinking

Before sharing a student-work example, we encourage teachers to think about the mathematical problem from their own perspectives. This sequence gives teachers an opportunity to first engage with the mathematics and then, perhaps, extend their own understanding of that mathematics before trying to understand others' strategies.

How could you solve this problem using two different strategies? By asking teachers to solve the problem, we encourage them to make sense of the mathematics and to begin with strategies that are most comfortable for them. Requesting that teachers use a second strategy encourages them to *think again.* For example, with Misha's work, many teachers might rotely apply the traditional subtraction algorithm to subtract 186 from 2694 or 999 from 4010. This strategy is appropriate, but when asked to think again, these teachers are encouraged to move beyond the standard algorithm to consider conceptual ways to find the difference between two numbers. For example, in the first problem, they might count up from 186 to 694 and then add 2000; in the second problem, they might change 999 to 1000, mentally compute the difference between 1000 and 4010, and add 1.

Sometimes teachers have questions about why a second strategy is valuable when a problem has already been solved successfully and efficiently. In response, we share our belief that the purpose for solving mathematical problems is not only to find correct answers but also to develop powerful ways of reasoning that can be applied to similar, or even very different, problems. By asking teachers to solve problems in more than one way, we can encourage them to think flexibly, explore multiple ways of reasoning, and consider the connections among strategies. We also gain knowledge about how they think about numbers and operations, and we can use this information in our instructional decision making.

Once teachers have generated multiple strategies, asking them to share these strategies is beneficial. Sharing gives teachers opportunities to appreciate how other teachers solve problems in ways different from their own. Engaging with multiple adult strategies provides a starting point for understanding children's strategies because children so often think about problems in ways that adults might not expect, an idea that Misha's second solution nicely exemplifies. After exploring multiple strategies, teachers are better poised to think about children's creative thinking.

How might a child solve this problem? We again highlight the importance of considering different strategies by asking teachers to predict strategies that a child would use. Sometimes we ask for a variety of children's strategies, whereas other times we ask teachers to identify strategies that children with a particular understanding (or lack of understanding) might use. We do not expect teachers to predict the exact strategy in a student-work example. Instead, we want teachers to grapple with the mathematics and to consider how children commonly make sense of the mathematical situation. Research on children's mathematical thinking has identified patterns in children's strategies, and we can highlight these patterns by asking teachers to predict children's strategies. For example, children and adults often view mathematical situations differently but in predictable ways, as with the following problem: "Juan has 18 stickers. How many more stickers does he need to buy to have 34 stickers?" Whereas most adults approach this problem by subtracting, most young children explicitly follow the action (buying more stickers), thereby adding on from the lesser number to find the answer (Carpenter et al. 1999).

Although we have found that practicing teachers are generally better able than prospective teachers to envision children's strategies, we are convinced that prospective teachers benefit from trying. Asking teachers to consider how a child may solve the problem encourages them to step outside their own shoes and into a child's shoes to consider the mathematics embedded in the problem. After reviewing their own understanding of the mathematics and children's hypothetical strategies, teachers are better poised to engage with a specific child's thinking.

Exploring the Child's Thinking

Our second category of questions helps teachers explore the details of a child's strategy. We encourage teachers to consider the child's perspective so that they focus on what the child *understands* rather than on what he or she *does not yet understand*. This difference is subtle but essential for teachers who use children's thinking to inform their instructional decision making. We offer three questions to support teachers in focusing first on understanding the child's strategy before stepping back and considering the relationship between the child's thinking and the underlying mathematics.

How did the child solve this problem? To help teachers make sense of a child's strategy that may be different from their own, we begin discussions with a general question such as "How did Misha solve this problem?" Teachers may need time to examine and discuss strategies such as Misha's, not only because the strategies may be nontraditional but also because the child's way of recording strategies may be novel. In addition, even when written work is explicit, teachers may initially be unfamiliar with articulating a child's strategy in detail. They might focus on the operation ("She subtracted") or the tool ("She used paper and pencil") instead of describing, step by step, how the child solved the problem. Follow-up questions to encourage description of the entire strategy can be helpful. For example, we might ask, "What did Misha do first?" (and then next and so on) or "Can you describe each of the steps that Misha went through to solve this problem?"

Why might the child have done . . . (insert a specific aspect of the child's strategy)? Once teachers have generally described the child's strategy, asking specific questions about that strategy can be useful to make the details of the mathematics and the child's thinking explicit. For example, we could ask, "Why might Misha have started with the thousands instead of the ones?" or "Why might Misha have written 80 – 80 in her first strategy?" or "Why might Misha have used 110 in her second strategy?"

What is the mathematics embedded in this strategy? After understanding a child's strategy, teachers can reexamine the strategy to further explore the underlying mathematics. We want teachers to consider not only *how* a strategy works but also *why* it works and to what sorts of problems it would generalize. For example, we may ask teachers, "What makes Misha's strategies appropriate for each of these problems?" or "How do you think the numbers in the problems might have affected Misha's thinking?" To help teachers explore whether Misha's strategies are appropriate for solving other problems, we might specifically ask, "Can you write a problem that could not be solved using Misha's strategies?" The goal of these types of questions would be to generate a discussion about the generality of Misha's reasoning and whether it could be used to solve any subtraction problem.

Exploring the mathematics of a strategy can also include identifying the mathematical understandings that a child needs to use a particular strategy. For example, we could ask, "What place-value understanding did Misha need to meaningfully use her strategies?" This question might invite teachers to more broadly discuss the definition of *place-value understanding* and how we can know when children have that understanding. In Misha's case, place-value understanding was reflected in her ability to decompose numbers, manipulate them, and then recombine the results. Therefore, student-work examples can serve as entry points that motivate teachers to engage in discussions about broader mathematical issues. After teachers have spent time exploring the child's strategy and the mathematics underlying it, we ask teachers to think about how they can use this understanding to plan subsequent instruction.

Extending the Child's Thinking

Our last category of questions focuses on helping teachers use what they know about the child's mathematical thinking and the related mathematics to consider how they might further the child's understanding. Three questions in particular are useful for engaging teachers in discussions about how to extend a child's thinking.

What questions could you ask to help the child reflect on the strategy? Children need opportunities to articulate and reflect on their strategies as a way to deepen their knowledge and clarify their thinking for teachers. Many of the questions that we asked teachers to help them clarify their understanding of the strategy would be effective questions for teachers to consider asking children. For example, after asking teachers why they think Misha computed 80 − 80 in her first strategy or why she used 110 in her second strategy, we would encourage teachers to ask Misha the same questions. We might also ask teachers, "What questions could you ask Misha to help her think about her strategy in novel ways?" For example, a teacher could ask Misha whether she could identify the 186 or 2694 in her first strategy.

What questions might encourage the child to consider a more efficient strategy? Teachers often raise questions about the value of strategies that appear less efficient than standard algorithms. In response, we have facilitated discussions about efficiency and accuracy, including the idea that increased efficiency can take many forms, depending on the child, the problem, and the strategy. Although a single most efficient strategy for all children on all problems does not exist, children still need opportunities to improve their efficiency, and teacher questioning can help. The goal is not for teachers to tell children more efficient strategies but for teachers to pose questions that help children reflect on their existing strategies and consider improvements. For example, teachers might pose questions to help children simplify elaborate drawings or use groups of tens (instead of ones) to build or count quantities. Teachers can also encourage children to generate new strategies that are more efficient than their original strategies. In Misha's case, we might ask teachers, "What question could you pose to help Misha generate more efficient strategies that take advantage of number relationships, such as the fact that 999 is only 1 away from 1000?" Our goal would be to help teachers generate questions to pose to Misha, such as "Can you solve this problem without paper and pencil?" or "What do you know about 999 that could help you solve this problem?" These questions are reasonable for Misha because we already know that she has strong place-value understanding and an ability to decompose and recombine numbers. After teachers have identified questions to help the child generate a more efficient strategy or reflect on an existing strategy, we broaden the discussion to consider different but related tasks.

On the basis of the child's existing understandings, what task might you pose next? We encourage teachers to consider logical next steps that are based on their exploration of the child's work. We ask them to identify specific follow-up tasks, to suggest hypotheses about how the child might engage with those tasks, and to consider what they and the child might learn from the experience. To generate follow-up tasks, teachers may use the structure of the original problem and vary the numbers, or they may write new tasks about different but related mathematical concepts. Having teachers identify a *specific* next task they might pose is important because doing so gives teachers opportunities to translate their general understandings into instructional tasks. Giving teachers opportunities to reflect on and articulate the reasoning by which they chose these tasks is also essential.

Teachers, especially prospective teachers, often struggle when asked to generate follow-up tasks, but the use of multi-problem student work such as Misha's can help. Because we have Misha's work for more than one problem, we can ask teachers to compare and contrast what they learned from analyzing her work on these two problems. From Misha's first solution, we learned how she thought about regrouping when it was needed only once, but we did not know what she would do when multiple regroupings were needed. From her second solution, we learned that she could flexibly regroup to accommodate multiple regroupings. Examining a series of related problems that the same child has solved can help teachers better understand how to design follow-up problems so that each problem provides additional information about a child's thinking.

To explore new tasks to extend Misha's thinking, we often ask teachers to consider what they already know about Misha's understandings and then present the general question "What task might you pose next?" We also ask teachers to design tasks for specific goals. For example, we could ask, "What problem might you pose to encourage Misha to use her number sense to solve problems efficiently?" or "What strategy would you be curious to see whether Misha could explain?" In asking this latter question, our goal would be to encourage teachers to consider how other nontraditional strategies compared with Misha's strategies and what she could learn by exploring those comparisons. For example, a fictitious student may solve the problem $6215 - 398$ by beginning with $5000 - 0 = 0$ and $1000 - 300 = 700$. A teacher could show Misha these first two steps and ask her to finish solving the problem and explain her reasoning. Alternatively, a teacher might ask Misha to compare the reasoning in these first two steps with the reasoning that she used on her second problem. Asking Misha to engage with these types of tasks could help a teacher better understand how Misha looks ahead when decomposing numbers during subtraction. By asking teachers to generate these types of tasks, we create opportunities to raise mathematical, teaching, and learning issues.

Summary of Questioning Framework

Figure 2 summarizes the questioning framework that we applied to Misha's written work. This framework is designed to identify questions that can serve as starting points for discussions among prospective or practicing teachers. Using all three categories of questions with every student-work example is not necessary. Instead, we regularly select and modify questions on the basis of the goals that we hold for a particular instructional session, our knowledge about the audience, and the mathematics embedded in the specific example of student work under consideration. The following section contains an additional example of how to apply this framework to student work, in this case, work with an incorrect answer.

Heidi's Incorrect Solution

Heidi correctly used the standard subtraction algorithm to subtract 199 from 4002 as **figure 3** shows; however, she made a computational error when using the standard addition algorithm to add 199 to 3803. We conjecture that she added $9 + 1$ (in the tens column), realized that the answer was 10, and carried 1 to the hundreds column, but wrote 9 instead of 0 in the tens column of her answer. Heidi's incorrect answer masked the fact that the correct answer is the same as the initial number. Her strategy is interest-

---

Questions to prepare teachers to understand the child's thinking:
- How could you solve this problem using two different strategies?
- How might a child solve this problem?

Questions to encourage teachers to explore the child's thinking in depth:
- How did the child solve this problem?
- Why might the child have done . . . (insert a specific aspect of the child's strategy)?
- What is the mathematics embedded in this strategy?

Questions to help teachers identify instructional "next steps" to extend the child's thinking:
- What questions could you ask to help the child reflect on the strategy?
- What questions might encourage the child to consider a more efficient strategy?
- On the basis of the child's existing understandings, what task might you pose next?

Fig. 2. Discussion questions for student-work examples

ing, therefore, not only because of her computational error but also because of her failure to recognize that she was asked to subtract then add the same number and that doing so leaves that original number unchanged. Attending to these types of number relationships is foundational for algebraic understanding.

Each student-work example provides different questioning opportunities, and Heidi's work is different from Misha's in part because her strategy resulted in an incorrect answer. Teacher educators can use incorrect solutions to raise issues of how teachers can best support children. Heidi's written work, for example, can facilitate a discussion about the advantages and disadvantages of helping her correct her error versus encouraging her to use the algebraic thinking that the problem was designed to promote.

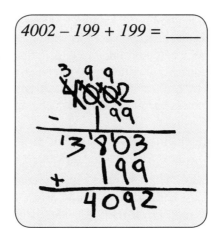

Fig. 3. Fifth grader Heidi's solution to an algebraic problem

Preparing to understand the child's thinking

Common approaches to this problem include computing from left to right, as Heidi did, or recognizing that computation is unnecessary because $-199 + 199 = 0$. Teachers need opportunities to consider the strategies they would use to solve this problem and the strategies children are likely to use. Teachers also need opportunities to consider how number selection and number order can influence strategies. To explore these issues, we pose questions such as "Why might the teacher have chosen the numbers 4002 and 199 in this problem?" or "How would what you learn about a child's understanding differ if the problem was $4002 - 200 + 200$ or $4002 + 199 - 199$ instead of $4002 - 199 + 199$?" We could also ask teachers how they might construct a problem to initially introduce this type of algebraic reasoning to students. After investigating the mathematics and children's hypothetical strategies, teachers should be better poised to consider Heidi's error in the context of this problem and the type of algebraic thinking that it was designed to promote.

Exploring the child's thinking

Heidi used standard algorithms for both addition and subtraction, and her approach is valid. One cannot determine whether Heidi's error is careless or conceptual, however, because her written work, like most written work, is an incomplete reflection of her understanding. To help teachers consider the limitations of written work, we might ask, "What do you understand about Heidi's thinking about addition and subtraction?" and "What else would you like to know?" To help them generate possible questions to gather further information, we could ask, "What specific question(s) could you ask Heidi to determine whether she made an isolated computational error or whether she was more generally confused about whole-number addition?" Similarly, Heidi's written work does not indicate whether she understands how the standard algorithms work. To address this issue, we ask teachers questions such as "Heidi wrote 1's above the 3, above the 8, and above the 0. What do you think those 1's mean to Heidi, and how would you like her to think about them?" or "What would you ask Heidi if your goal was to determine whether she understands the standard algorithms or whether she is simply applying them in a rote manner?"

In addition to helping teachers explore Heidi's actual strategy, we would want to help teachers compare this strategy with other possible strategies. In this case, Heidi missed an opportunity to look at the number relationships before beginning her computations. We might ask, "What understanding would Heidi have needed so that she could solve this problem without performing any addition or subtraction?" or "Why do these numbers encourage (or not encourage) algebraic thinking?"

Extending the child's thinking

When determining how to extend a child's thinking on the basis of an incorrect strategy, teachers must consider whether they will focus on correcting the error or helping the child more generally reason about the problem. Using Heidi's work, we might encourage teachers to consider the possibility of ignoring, at least temporarily, the computational error to instead promote algebraic thinking. Specifically, we may ask teachers, "What problems could orient Heidi toward looking at the whole problem before immediately computing from left to right?" The goal would be to have teachers generate other number sentences that could entice Heidi (and other children) to simplify calculations by taking advantage of the relationships between numbers (for example, $75 + 99 + 25 = n$; $6000 + 105 - 105 = n$; $247 + 324 = 246 + n$).

We could also help teachers consider how to support Heidi in correcting her error by asking, "How could you help Heidi recognize her error without directly pointing out her mistake?" Our goal in posing this question would be to generate suggestions such as asking Heidi to compare her work with that of another student or to solve the problem using a different strategy so that she could compare her own answers. Teachers might also consider the benefits of having Heidi read the original number sentence aloud with her (incorrect) answer. Because reading the entire problem can help children recognize number relationships that they otherwise might miss, Heidi may see that $-199 + 199 = 0$, realize that the answer should have been 4002, and, therefore, look for errors in her original computation.

Final Comments

Teacher educators can use student-work examples to ground conversations in the specific and important details of children's thinking, thereby helping prospective and practicing teachers develop deeper understandings about mathematics, teaching, and learning. Because the value in using student work is not inherent in the written examples themselves but instead evolves from the related discussions, we present our questioning framework to support teacher educators in facilitating these discussions. The *specific* questions that teacher educators might pose depend on the particular features of the student work under consideration. Important features include the problem, the answer, the strategy, and whether the strategy is presented alone or with other related work from the same child or different children.

Because each student-work example involves different mathematics and different instructional opportunities, teacher educators must carefully consider their goals when selecting examples. Student-work examples can be borrowed from many sources. For example, we have found figures illustrating children's strategies in most issues of *Teaching Children Mathematics,* especially in the "Problem Solvers" department. When working with practicing teachers, we also encourage them to bring student-work examples from their classrooms. Almost all the student work contains something of interest, and using participants' student work can increase the authenticity of discussions. Video examples provide another source of student work, and although this article has focused on students' written work, we have found that all the questions from our framework can be applied to video examples as well.

Whenever educators look at only one slice of a child's thinking, they lose the context necessary for a deep understanding of that child. For example, when we look at a child's written work, we may not be able to determine in what order it was written, how well the child could articulate the strategy, what role the teacher played during problem solving, and so on. These limitations do not detract from the value of using student work with prospective and practicing teachers, however, because the primary goal is not to understand the particular child but to provide a common context around which teachers may engage in rich discussions about mathematics, teaching, and learning.

REFERENCES

Carpenter, Thomas P., Elizabeth Fennema, Megan L. Franke, Linda Levi, and Susan Empson. *Children's Mathematics: Cognitively Guided Instruction.* Portsmouth, N.H.: Heinemann, 1999.

Little, Judith Warren. "'Looking at Student Work in the United States: Countervailing Impulses in Professional Development." In *International Handbook on the Continuing Professional Development of Teachers,* edited by Christopher Day and Judith Sachs, pp. 94–118. Buckingham, England: Open University Press, 2004.

This article is based in part on work supported by grants from the National Science Foundation (REC-9979902 and ESI-9911679). The views expressed are those of the authors and do not necessarily reflect the views of the National Science Foundation.

Using Students' Work as a Lens on Algebraic Thinking

Mark Driscoll and John Moyer

USING students' work in teachers' professional development programs has become popular for two reasons:

1. Teachers must consider teaching and learning issues jointly. By putting its focus on learners, the standards movement in mathematics education has increased the demand on professional development to help teachers understand issues of both teaching and learning and to help them interweave their knowledge of learning with their knowledge of teaching. In analyzing students' work together, teachers have numerous opportunities to focus on learning issues and how those concerns link with pedagogy.

2. Teachers must examine their own beliefs about the purpose, methods, and goals of mathematics education. Making inferences about students' mathematical abilities from their written work requires making decisions using information from outside sources and individual teacher's knowledge, beliefs, assumptions, and mind-sets—all of which affect the interpretation of the information.

We must distinguish different purposes for looking at students' work (Driscoll and Bryant 1998). For example, the purpose of understanding students' thinking differs from the purpose of evaluating students' achievement, which in turn differs from the purpose of improving instruction. The main focus of this article is on how teachers can use students' work to understand students' thinking. A secondary focus is how such knowledge can be used to improve instruction. This article does not discuss issues related to evaluating students' achievement.

Understanding Students' Algebraic Thinking

In the Linked Learning in Mathematics Project (LLMP), a program for middle school and high school teachers in Milwaukee, students' work serves as one tool to help teachers gain a deeper understanding of algebraic thinking. LLMP follows from two preceding projects: Leadership for Urban Mathematics Reform (LUMR) and Assessment Communities of Teachers (ACT). All three were funded by the National Science Foundation as teacher-enhancement projects. In this article, we describe how we have learned to use students' work to stay consistent with the purpose of understanding students' algebraic thinking. A more comprehensive account of our approach to algebraic thinking appears in *Fostering Algebraic Thinking: A Guide for Teachers in Grades 6 through 10* (Driscoll 1999).

We do not attempt to establish a comprehensive definition of algebraic thinking. Instead, our working definition encompasses several patterns of thinking noted in productive algebraic thinkers. The capacity for algebraic thinking includes the ability to think about *functions* and how they work and about the impact of a system's *structure* on calculations. These two aspects of algebraic thinking are facilitated by certain habits of mind:

- **Doing and undoing.** Effective algebraic thinking sometimes involves reversibility, that is, the ability to undo mathematical processes, as well as to do them. This ability means that the mathematician not only uses a process to get to a goal but

Mathematics Teaching in the Middle School 6 (January 2001): 282–87 also understands the process well enough to work backward from the answer to the starting point. For example, in a traditional algebra setting, algebraic thinkers are able not only to solve an equation, such as $9x^2 - 16 = 0$, but also to answer the question "What is an equation with solutions of 4/3 and –4/3?"

- **Building rules to represent functions.** Essential in algebraic thinking is the capacity to recognize patterns and organize data to represent situations in which input is related to output by well-defined function rules. For example, the following is a function rule that is based on computation: "Choose an input number, multiply it by 4, and subtract 3." This habit of mind is a natural complement to doing and undoing, in that the ability to understand how a function rule works in reverse generally makes the process more accessible and useful.
- **Abstracting from computation.** This ability enables mathematicians to think about computations independently of particular numbers used. One evident characteristic of algebra is that it is abstract. Thinking algebraically involves being able to think about computations apart from the particular numbers to which they are tied in arithmetic, that is, abstracting system regularities from computation. For example, students abstract from computation when they regroup numbers into pairs that equal 101 to make the following computation simpler: "Compute: $1 + 2 + 3 + \ldots + 100$."

A Lens on Algebraic Thinking

The following two conditions are required for students' work to become an effective professional development lens on algebraic thinking:

1. Criteria for choosing activities and selections of students' work, and
2. A process for analyzing and discussing students' work.

The following sections describe how LLMP has dealt with these conditions.

Criteria for choosing activities and students' work

LLMP looks for activities that encourage honing one or more of the habits of mind described previously. Because the project serves teachers from sixth through tenth grade, we also look for activities that allow for multiple approaches. In addition, the activities should permit extensions to related kinds of algebraic thinking. One such activity is Crossing the River (see **fig. 1**).

Crossing the River

Eight adults and two children need to cross a river. A small boat is available that can hold one adult, or one or two children. (So, the three possibilities are: 1 adult in the boat; 1 child in the boat; or 2 children in the boat.) Everyone can row the boat. How many one-way trips does it take for them all to cross the river? Can you describe how to work it out for 2 children and any number of adults? How does your rule work out for 100 adults?

What happens to the rule if there are different numbers of children? For example: 8 adults and 3 children? 8 adults and 4 children? Write a rule for finding the number of trips needed for *A* adults and *C* children.

One group of adults and children took 27 trips. How many adults and children were in the group? Is there more than one solution?

Fig. 1. The Crossing the River problem

The selection of students' work should be representative of the possible approaches to the activity. For a professional development session, the selection should be limited to four to six pieces but should include pieces that represent several different approaches to the problem and a few pieces in which the students' thinking is not absolutely clear. The responses of two students, Barbara and Denzel, from among those used for one of our sessions are shown in **figures 2** and **3**. Note that in answering the first question, Barbara uses both Spanish and English, saying in effect, "17 go, 16 return = 33 one-way trips."

Process for analyzing students' work

Using a model called *structured exploration* from the LUMR project, LLMP teachers first immersed themselves in the investigation. Working in small groups on the Crossing the River problem, they developed solution strategies, which they then presented to the full group. In the second step of the structured-exploration process, teachers worked again in small groups to analyze students' work. They compared the similarities and differences in algebraic thinking among the students, using the guidelines for analyzing algebraic thinking (see **fig. 4**) to help them focus on the algebraic habit of mind, "building rules to represent functions."

Sample Analysis

The following analysis captures the teachers' discussions of Barbara's and Denzel's work, using each of the parameters in the guidelines.

Indicator A in the guidelines is "systematically searches for a rule." When taken together, Barbara's responses to parts 1 and 2 of the problem indicate that she is systematically searching for a rule. In the diagram that Barbara drew in response to part 1,

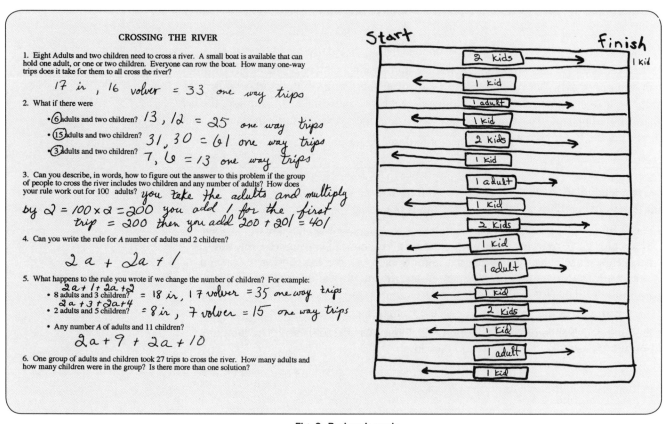

Fig. 2. Barbara's work

right and left arrows represent the direction of the trips needed to get the first 4 adults across in the 8-adult, 2-child situation. Note that the second part of Barbara's diagram, which is not shown, is identical except for a final additional trip in which the "2 kids" move from left to right. Barbara's words, "17 *ir*" (to go) "16 *volver*" (to return), indicate that she has mentally organized the trips into two groups, those in one direction and those in the other. In part 2, Barbara has drawn circles around the numbers 6, 15, and 3. This tactic probably indicates that she is searching for a rule to connect the varying number of adults with the corresponding numbers of "going" and "returning" trips.

Denzel's combined work on parts 1 and 2 shows that he is paying attention to a repeated group of 4 trips. We can infer from Denzel's statement in part 1, "this keeps going . . . ," that he is searching for a rule based on repetitions of same-sized groups of trips. This approach is in contrast with Barbara's, in which she searched for a rule based on two groups ("to go" and "to return") with varying sizes. The composition of Denzel's groups, however, is difficult to ascertain from part 1 alone. Although Denzel's answer to part 1, 33 trips, is correct, the enumeration of the trips is somewhat flawed. Step 2 actually involves 2 trips; child 10 gets off after going to the "other end" and adult "1 gets on and goes to the other side." Furthermore, Denzel does not reveal until part 2 that

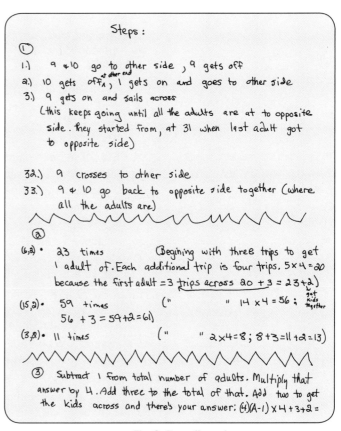

Fig. 3. Denzel's work

he is thinking of an initial group of 3 trips to get the first adult across, followed by 7 groups of 4 trips each to get the next 7 adults across, and ending with 2 final trips to get both children across.

Indicator B in the guidelines is "Attempts to describe a rule," which Barbara does not do until part 3. Unlike Denzel's first attempt, Barbara's response has the feel of a generalized conjecture (see indicator C).

In part 2, we see Denzel's first attempt to describe the rule verbally. At first glance, his answers of 23, 59, and 11 appear incorrect. However, a careful reading of his verbal description reveals that these answers are meant to be the number of trips needed to get 6, 15, and 3 adults across, but not both children. In each case, Denzel parenthetically adds 2 trips to "get kids to the other side," thus obtaining the correct answers of 25, 61, and 13 trips.

Indicator C, "Conjectures a generalized rule," is shown in Barbara's work on part 3, in which she verbally conjectures a rule for finding the number of trips needed in the 100-adult, 2-child situation. In fact, her description has the feel of a generalized conjecture, which she illustrates using the 100-adult example. We cannot tell whether Barbara conjectures the "multiply by 2" rule from her number pattern in part 2 or from the diagram of the situation itself in part 1, which shows two right arrows and two left arrows for each adult.

Part 3 of Denzel's work shows that he has generalized the specific descriptions of the rule given in part 2. His rule is given in both verbal and symbolic forms. Note, too, that Denzel consolidated all the computations into a single string in part 3, in contrast with the operation fragments that he wrote in part 2 (e.g., $56 + 3 = 59 + 2 = 61$).

> **Overarching question: What do these students know about building rules to represent functions? In looking for evidence, you may notice some of the following:**
>
> A. Systematically searches for a rule. Pays attention to how things are changing, perhaps to what gets repeated. (Ask yourself: what is the student paying attention to?)
>
>   E.g., "I'll draw out all the steps, then count how many trips in each direction."
>
>   E.g., "I've been comparing all the numbers I generated that give the number of trips needed for two adults, and they are all odd."
>
> B. Attempts to describe a rule.
>
>   E.g., "You get one adult over, then repeat the steps five times."
>
>   E.g., "To find out how many trips are needed for one more adult, add 4." (This describes a NOW-NEXT rule.)
>
> C. Conjectures a generalized rule. (Ask yourself: how does the student think things are working here?)
>
>   E.g., "With 2 children, it doesn't matter how many adults you have."
>
>   E.g., "From the pattern in the table it looks like you can find the number of trips by multiplying the number of adults by 4 and adding 1."
>
> D. Represents this rule in other forms:
>
>   Table ___   Graph ___   Equation ___   Expression ___
>   E.g., $4A + 1 = T$ E.g., $2a + 2a + 1$
>
> E. Links are made among the different representations to show how they connect. (Ask yourself: does the student seem to see connections between the different representations used?)
>
>   E.g., "The 4 in the formula means if the number of adults gets bigger by 1, the number of trips gets bigger by 4."

Fig. 4. Guidelines for analyzing student work

Indicator D in the guidelines is "Represents this rule in other forms." Barbara's work on part 4 shows that she uses the letter a consistently to represent the number of adults. Her expression $2a + 2a + 1$ is an algebraic representation of the verbal rule that she wrote in part 3. Part 5 shows how Barbara modifies her $2a + 2a + 1$ rule to fit situations with different numbers of children. This work implies that Barbara has conjectured a rule that extends her first rule. In particular, the regular form of the expressions in part 5 shows that Barbara is able to incorporate both types of repetition, trips for adults and children, into her rule, effectively handling a two- variable situation with just one variable. In other words, the expressions in part 5 indicate Barbara's conjecture that for each child beyond the first two, the rule must be modified by adding 1 to $2a$ and 1 to $2a + 1$. We cannot be sure whether Barbara realizes that the extra 1's represent 1 extra trip "to go" and 1 extra trip "to return" or whether they are just a result of a numeric pattern that she analyzed. Finally, although she may recognize a pattern for creating the different rules, Barbara does not seem able to represent the different rules as one single rule with two variables, for example, $2a + (k - 2) + 2a + (k - 1)$.

Part 3 of Denzel's work shows that he can write the rule in both verbal and algebraic forms. From his verbal description, we can infer that Denzel probably uses $A - 1$ to represent "subtract 1 from [the] total number of adults." That is, for Denzel, the expression $A - 1$ probably means "perform the operation of subtraction and get an answer" instead of "the number of adults who still need to cross the river after the first has done so." Note that the "(4)" at the beginning of the string "$(4)(A - 1) \times 4 + 3 + 2$" apparently indicates that Denzel is answering part 4 of the activity.

Indicator E directs teachers to look for connections among the different representations. Barbara's work on part 4 indicates that she has probably written her verbal rule as an algebraic expression, but she does not explicitly describe the connection between the two. In addition, we cannot tell whether Barbara's conception of the repetitions in the problem leads to the algebraic expression $2a + 2a + 1$ or whether the expression is simply the result of the number pattern that she noticed in part 2.

Denzel's work on part 2 shows that he has linked the arithmetic computations with the situation itself. His conceptualization eventually leads to the expression $(A - 1) \times 4 + 3 + 2$, which is algebraically, but not conceptually, equivalent to Barbara's. Parts 3 and 4 show that Denzel has linked the verbal rule with the symbolic form. His consolidation of the multiple steps in part 2 into a single string shows that he has linked the arithmetic computations with the generalized rule.

When asked, "What does this student seem to know about building rules to represent functions?" the teachers concluded that Barbara seems to be able to explicitly build rules that represent patterns and to use algebraic symbolism to represent the rules. Whether Barbara connected the numeric pattern in part 2 with the diagram of the situation itself is unclear. She may not explicitly realize, for example, that the underlying reason for multiplying the number of adults by 2 is that two right-pointing arrows are drawn for each adult. Barbara may simply notice that the numbers in part 2, that is, 12, 30, and 6, are twice the circled numbers of adults. Future interactions with Barbara should encourage her to connect numeric patterns and rules with actual situations.

Unlike Barbara, Denzel explicitly describes the connection between a numeric pattern for computing the number of trips and the situation itself. Denzel apparently interprets $A - 1$ as a subtraction problem that produces a number rather than a variable expression representing a number of adults. This interpretation shows that Denzel is able to link arithmetic computation with generalized rules; however, it may reveal a weakness in his understanding of algebraic symbolism. Future interactions with Denzel should encourage further and more powerful uses of algebraic symbols to represent quantities in realistic situations.

Improving Instruction

The ultimate goal of professional development is to improve instruction. Once teachers recognize the levels of algebraic thinking in their students, they need to plan how to help their students raise those levels. Planning appropriate algebra-based activities is not enough; appropriate implementation is also required. In particular, using effective questioning techniques that encourage the development of algebraic habits of mind is important. We worked with the LLMP teachers to identify the types of questions that they might use to help students of varying abilities develop their thinking further. **Figure 5** lists some of the questions that teachers can ask to help students sharpen their algebraic thinking. The LLMP teachers learned to pose such questions, selecting them by analyzing their students' existing algebraic capacity and identifying realistic expectations for growth.

Conclusion

Students' work can be a catalyst for effective professional development for teachers, but its use should be consistent with defined purpose. In LLMP, our purpose is to help teachers gain a deeper understanding of algebraic thinking. Rubrics and other tools for evaluating achievement are not appropriate for that purpose; instead, we fashioned our own set of criteria on the basis of a framework of algebraic habits of mind.

Our experience suggests that students' work can be the basis for designing similar professional development structures to focus on other kinds of mathematical

- Can you find a rule or relationship here?
- How does the rule work, and how is it helpful?
- Why does the rule work the way that it does?
- How are things changing?
- Is information given that lets me predict what is going to happen?
- Does my rule work for all cases?
- What steps am I doing over and over?
- Can I write down a mechanical rule that will do this job once and for all?
- How can I describe the steps without using specific inputs?
- When I do the same thing with different numbers, what still holds true? What changes?
- Now that I have an equation, how do the numbers (parameters) in the equation relate to the problem context?

Fig. 5. Sample questions for building rules to represent functions

thinking. For a particular content area, such as geometry, measurement, or statistics, two questions need to be asked: (1) What indicators characterize productive thinking in this area? and (2) What conceptual challenges typically arise to keep students from thinking productively in this area? Various resources, such as Owens (1993), summarize researchers' answers to these questions. From the answers can come analogues to our set of algebraic-thinking guidelines and indicators of common pitfalls in students' thinking. These guidelines and indicators are the raw material for criteria to allow students' work to act as a lens on mathematical thinking.

BIBLIOGRAPHY

Cai, Jinfa, John Moyer, and Constance Laughlin. "Algorithms for Solving Nonroutine Mathematical Problems." In *The Teaching and Learning of Algorithms in School Mathematics,* 1998 Yearbook of the National Council of Teachers of Mathematics (NCTM), edited by Lorna Morrow, pp. 218–29. Reston, Va.: NCTM, 1998.

Driscoll, Mark. *Fostering Algebraic Thinking: A Guide for Teachers in Grades 6 through 10.* Portsmouth, N.H.: Heinemann, 1999.

Driscoll, Mark, and Deborah Bryant. *Learning about Assessment, Learning through Assessment.* Washington, D.C.: National Academy Press, 1998.

Driscoll, Mark, Sydney Foster, and John Moyer. "Projects: Linked Learning in Mathematics Project." *Mathematics Teacher* 92 (January 1999): 72–73.

Driscoll, Mark, John Moyer, and Judith Zawojewski. "Helping Teachers Implement Algebra for All in Milwaukee Public Schools." *Mathematics Education Leadership* 2 (April 1998): 3–12.

Kelemanik, Grace, Susan Janssen, Barbara Miller, and Kristen Ransick. *Structured Exploration: New Perspectives on Mathematics Professional Development.* Newton, Mass.: Education Development Center, 1997.

Owens, Douglas T., ed. *Research Ideas for the Classroom: Middle Grades Mathematics.* New York: Macmillan, 1993.

Focusing Conversations to Promote Teacher Thinking

DeAnn Huinker and Janis L. Freckmann

TAMARA, a preservice teacher, taught a lesson on fractions to a class of fourth-grade students in her field-experience practicum. The instructor of her mathematics methods course observed the lesson. As they met in the hallway to debrief the teaching experience, the instructor began by asking, "How do you think your lesson went?" Tamara replied, "Great. I got through my entire lesson plan." Tamara's comment and the subsequent discussion revealed that she was focused on her actions as the teacher but not on the dynamics of instruction (Cohen and Ball 2001).

Similar conversations often have left us pondering our effectiveness as professional support providers, whether in the role of a mathematics teacher educator or a mentor, teacher leader, or staff developer. What questions should we be asking preservice teachers, novice teachers, and more experienced teachers so they think more deeply about these interactions? How can we frame these conversations as a time to think together? How can we support preservice and novice teachers in beginning a lifelong inquiry into their own mathematical teaching practices? How can we support experienced teachers as they shift toward reform-oriented practices in mathematics?

The instructional triangle in **figure 1** presents instruction as interactions among teachers, students, and mathematics (adapted from Cohen and Ball 1999 and National Research Council 2001). In our conversations with both preservice teachers and in-service teachers, focusing on how teachers and students attend, listen, and respond to one another while interacting with the mathematics is important. The purpose of this article is to engage professional support providers in thinking about the questions they ask that promote teacher thinking, deeper discourse, and reflection on practice. We begin with two scenarios based on debriefing classroom observations and then present a structure for formulating questions.

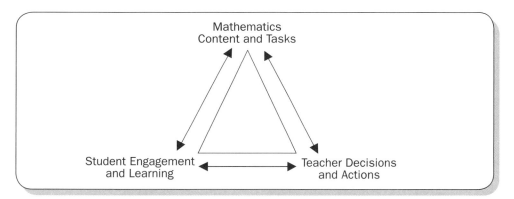

Fig. 1. Instruction as interactions among teachers, students, and mathematics

Teaching Children Mathematics 10 (March 2004): 352–57

Scenario 1: Debriefing a Lesson on Comparing Fractions

Alice, a preservice teacher, was working in a third-grade classroom for her field-experience practicum. One of the authors had an opportunity to observe Alice teaching a lesson on comparing fractions using paper strips. Alice engaged students in small-group work and then closed the lesson with a large-group discussion. The following debriefing conversation with Alice took place after she finished teaching the lesson.

Professor. Hi, Alice. It seems like your lesson on comparing fractions went pretty well. How do you think the lesson went?

Alice. Pretty good. The kids liked working with the fraction strips. I had never used them before.

Professor. The students really got into the class discussion. Morgan was showing everyone how she knew 3/8 was smaller than 3/6. I saw her lining the fraction strips up as she was explaining her thinking to Mark. Do you think Mark understood?

Alice. Mark has trouble with fractions. He still doesn't get it. He couldn't even fold the strips or mark them correctly.

Professor. As you were walking around the class monitoring small groups, were the fraction strips a useful tool for students?

Alice. I think so. The students could visualize the fractions more easily.

Professor. Do you think the students will need more practice comparing fractions before moving on?

Alice. No. I think they're ready to move on.

The conversation began with a general evaluation of the lesson; then the professor asked Alice to comment on the learning of a specific student. Next, the professor asked her to remark on the use of the fraction strips and consider subsequent instructional steps. Although this thinking is good for Alice, she was not pressed to think deeply about her teaching, student learning, or the mathematics. The conversation about the lesson was superficial. The support provider—in this case, the professor—posed only closed questions and therefore received only vague responses. The support provider also did most of the talking and most of the thinking in the conversation.

This conversation occurred a few years ago and is reflective of our past conversations with preservice and in-service teachers. We were asking questions of teachers but the conversations did not seem to go anywhere. We were not focused on helping teachers examine the interactions in their classrooms among the mathematics content, students' understanding, and the instructional decisions that the teachers made. We now know much more about questioning that will move a teacher to think more deeply. Our work as support providers is to press teachers such as Alice to be clear on their mathematical goals for students and how those goals relate to the task, and then to support teachers as they reflect on their monitoring of student understanding.

Scenario 2: Debriefing a Lesson on Adding Fractions

This next scenario illustrates how thinking shifted from the professor to the teacher. Notice how the line of questioning progresses and how the questions are structured. Laura teaches fourth grade and was a participant in a professional development program for in-service teachers. The following conversation with Laura took place after she finished teaching a lesson on adding fractions.

Professor. Hi, Laura. Seems like the lesson went pretty well. Why don't you begin by summarizing the mathematical ideas your students were learning today?

Laura. Well, one of the ideas was helping students see relationships among common fractions and the whole and then use those relationships to add the fractions together. The students were really into sharing their reasoning and explaining how they used fraction relationships by visualizing or using the fraction strips.

Professor. The class certainly seemed very willing to share ideas with each other. Let's talk about Travis. Take a moment and then describe the depth of understanding Travis has about fractions and where he is in regard to your goals for his learning.

Laura. Well, Travis can quickly add fractions with unlike denominators. He can picture common fractions in relation to a whole. He also has an understanding of equivalency. Yesterday he was explaining how he knew 1/2 was the same as 2/4 and 3/6. He actually brought out paper and was folding them into fraction strips. I think the students did that activity last year in third grade. He certainly was able to explain how he knew the fractions were the same amount in relation to the same whole.

Professor. I am sure there are other students like Travis. Now think a bit about Margaritte. Elaborate on the fraction ideas she is struggling with and describe your interventions.

Laura. Hmmm . . . Margaritte is a bit tougher to figure out. She wants to just add the numerators and denominators together. I know that this is a common error as students learn to add fractions. I thought about just working with Margaritte individually, but then decided it would benefit the whole class to engage in a discussion and let them convince each other whether or not it ever works to add fractions this way.

In this scenario, the conversation began with an articulation of the mathematical content goals for students. This mathematical content focus was sustained as the professor prompted Laura to talk about Travis's understanding of equivalency in relation to the mathematical goals. Laura made a strategic decision as she reflected on Margaritte's error and used it as a prompt to engage the whole class in a discussion and debate about adding fractions. The professor's questioning promoted a deeper and richer conversation and the teacher did most of the talking, thinking, and reflecting on the lesson. As support providers, we must ensure that teachers articulate students' needs and justify their decisions and actions based on the mathematical goals for student learning.

Focusing Conversations Using the Instructional Triangle

A major difference between the two scenarios is in the line of questioning. In the first scenario with Alice, the questioning focused on teacher management of the lesson and lacked specificity of the mathematical content and student understanding of that content. In the second scenario, the line of questioning with Laura focused on the instructional triangle. This enabled Laura to articulate and elaborate on the mathematical goals for the selected tasks and on her decisions and actions throughout the lesson in monitoring and promoting student learning. The following sections of this article examine each component of the instructional triangle in **figure 1** and the interactions among the components as a framework for focusing conversations with teachers.

Mathematics content and tasks

Teachers must be able to articulate the mathematics content that students will learn and to identify the aspects of the mathematical task that are going to help students learn this content. The support provider's questioning builds teachers' capacity to be more intentional and articulate about the mathematical goals for the lesson. This enables teachers to make more effective decisions in the midst of teaching a lesson because they have a clear picture of the mathematics content that students are learning. In the first scenario, the conversation focused on the activity, not on the mathematics. The professor did not ask Alice to verbalize the mathematical goals, so it was unclear whether she could envision the expectations for student learning. In the second scenario, the conversation began with an articulation of the mathematical goals and the connection to the task. This enabled Laura to be attuned to Margaritte's error and to use it as a teaching opportunity.

Student engagement and learning

Monitoring of student learning must focus on how mathematical ideas unfold in reaching the goals of the lesson. Teachers must be able to identify where students are in their understanding and where they are heading mathematically. This includes being more precise about students' struggles with the mathematical ideas and about which ideas students understand well. Only then can teachers make informed decisions about lesson modifications that provide additional support as well as those that push students along in their learning. Laura was able to specify levels of student learning in relationship to the mathematics content, whereas Alice was vague in expressing the level of students' understanding. The work of support providers is to press teachers' thinking to articulate how students are growing in their mathematical knowledge. For example, the professor could have asked Alice to compare the level of understanding between Mark and Morgan.

Teacher decisions and actions

Teachers must be able to articulate the mathematical goals for student learning and use them as the basis for making decisions before, during, and following a lesson. For example, Laura's decision of how to use Margaritte's error surfaced because of previous conversations about using student errors as opportunities to deepen the mathematical knowledge of all students. This illustrates the support provider's important role of raising the level of consciousness in the instructional decisions that teachers make.

Structuring Questions

Another difference between the two scenarios is in how the support provider structured the questions. A well-structured question is an invaluable tool in our repertoire to promote teacher learning. Costa and Garmston (2002) have extensively studied and delineated the structuring of questions that promote thinking. These questions keep the focus on inquiry into practice and produce more learning than do statements or closed questions (Garmston 2000). By choosing our words carefully and using intentionally designed questions, we can engage and transform another person's thinking and perspective. Well-structured questions engage individual teachers or groups of teachers in thoughtfully planning, reflecting on instruction, or analyzing situations. Well-structured questions to promote thinking include three essential parts: (1) an invitation to think, (2) a cognitive process, and (3) a specific topic. The order of the three parts may vary, but all three are necessary to formulate an effective question to promote thinking, such as the following: "As you review the mathematical goals for students' understanding of fractions, describe the assessment tasks you are considering." The invitation "as you review" invites the teacher to engage in thinking. The cognitive process is to "describe" assessment tasks and the specific topic is "understanding of fractions."

Invitation to think

The invitation invites complex thinking and reflection without making judgments. It is spoken in an approachable voice and uses language that is tentative and explorative to signal inquiry. It also uses plural word forms to signal the expectation of multiple responses, rather than a singular correct answer (Costa and Garmston 2002), as in the following example: "As you reflect on the lesson, what hunches do you have to explain the students' confusion about how to write a number sentence for the fraction word problem?" **Figure 2** lists additional examples of invitational stems. You might notice that our list does not include "why" or "what" questions because these presuppose a single correct response.

Cognitive process

Questions elicit different levels of complexity of thinking according to the verbs they use. The verbs elicit a specific level of thinking to prompt analysis and synthesis as well as application and evaluation. Asking questions at knowledge and comprehension levels also is appropriate at times. For example, asking teachers to describe student-generated subtraction strategies is at a lower level of thinking than is asking teachers to compare students' strategies. Asking teachers to evaluate students' subtraction strategies for fluency is at an even higher level of complexity. **Figure 3** lists verbs to use in formulating questions that shape how teachers think about a particular topic.

Specific topic

The third part of structuring questions is the selection of a specific topic. To promote deep teacher thinking, the topic should focus on the interactions in the instructional triangle among mathematical content and tasks, student engagement and learning, and teacher decisions and actions. NCTM (2000) highlights the complexity of effective mathematics teaching: It involves engaging and observing students in mathematical tasks, having clear mathematical goals for student learning, listening carefully to student ideas and explanations, and using the information to make instructional decisions. To achieve high levels of instructional effectiveness, teachers must develop a reflective disposition to carefully and intensely examine their practice (Schon 1987).

Closing Comments

Whether you are a mathematics teacher educator at a university or college, a district staff developer, or a classroom teacher with support responsibilities, thinking very carefully about the questions you pose to other teachers is important. We avoid questions that can be answered with a one-word response because they do not prompt deeper thinking and may even prevent the discussion from occurring. For example, we no longer ask, "How did your lesson go?" or "How are you doing with the new curriculum materials?" because the common response of "Fine" does not engage teachers in deeper thinking and reflection. We are more likely to say, "As you think about your mathematical goals for today's lesson, describe how you structured your lesson to impact student understanding" or "Given what you know about students' difficulties with fractions, summarize ways the new materials are supporting their learning." These latter instructions prompt teachers to talk further, elaborate, make connections, and even ask questions of themselves. As NCTM (2000) notes in the Teaching Principle, "Engaging in reflective practice and continuous self-improvement are actions good teachers take every day" (p. 18). One of our greatest tools as professional support providers is the formulation of well-structured questions to engage teacher reflection and conversation on instructional decisions made in the classroom. Our responsibility is to promote teacher thinking about the relationships among what teachers do, what students are learning, and the mathematics content.

REFERENCES

Cohen, David K., and Deborah Loewenberg Ball. "Making Change: Instruction and Its Improvement."
Phi Delta Kappan 83 (September 2001): 73–77.

As you think about…

As you consider…

Given what you know about…

In what ways…

In regard to the decisions you made…

In your planning…

From your previous work with students…

Take a minute…

When you think about…

Fig. 2. Invitational stems that use plural forms and exploratory language to invite reflection

Observe
Decide
Notice
Identify
Remember
Compare
Contrast
Predict
Interpret
Explain
Evaluate
Conclude
Summarize
Speculate
Envision
Relate
Differ
Consider
Distinguish
Describe

Fig. 3. Verbs that elicit specific cognitive processes to engage thinking

——. "Instruction, Capacity, and Improvement." CPRE Research Report No. RR-043. Philadelphia: University of Pennsylvania, Consortium for Policy Research in Education, 1999.

Costa, Arthur L., and Robert J. Garmston. *Cognitive Coaching: A Foundation for Renaissance Schools.* Second ed. Norwood, Mass.: Christopher-Gordon, 2002.

Garmston, Robert J. "Glad You Asked." *Journal of Staff Development* 21 (winter 2000): 73–75.

National Council of Teachers of Mathematics (NCTM). *Principles and Standards for School Mathematics.* Reston, Va.: NCTM, 2000.

National Research Council. *Adding It Up: Helping Children Learn Mathematics.* Washington, D.C.: National Academy Press, 2001.

Schon, Donald A. *Educating the Reflective Practitioner.* San Francisco, Calif.: Jossey-Bass, 1987.

Ideas for Establishing
Lesson-Study Communities

Akihiko Takahashi and Makoto Yoshida

MANY educators in the United States have recently become interested in lesson study, a professional development approach popular in Japan, as a promising source of ideas for improving education (Stigler and Hiebert 1999). Numerous schools and school districts have attempted to use lesson study to improve their teaching practice and student learning (Council for Basic Education 2000; Germain-McCarthy 2001; Lewis 2002; Research for Better Schools 2002; Stepanek 2001; Weeks 2001).

Teachers at one such school, Paterson Public School in Paterson, New Jersey, have been conducting lesson study since 1999. Cynthia Sanchez, a sixth-grade teacher, shared some of her experiences in *Currents* (2002), the Research for Better Schools newsletter:

While preparing the lessons, the group and I were very thoughtful. We looked at everything from how to introduce a new lesson to anticipated student responses, the use of the blackboard, manipulatives, and student engagement. This made me realize that there is more to teaching math than just opening a textbook and working on problems, or "spoon feeding" formulas just to get quick answers. (p. 5)

The concept of lesson study originated in Japan, where it is widely viewed as the foremost method of professional development for teachers (Fernandez et al. 2001; Lewis 2000; Lewis and Tsuchida 1998; Shimahara 1999; Stigler and Hiebert 1999; Yoshida 1999). Lesson study is an important feature of the Japanese educational system and has enabled Japanese elementary school teachers to improve their classroom instruction

This article provided the basis for a closing discussion about next steps at a professional development workshop for postsecondary faculty that included faculty from both the mathematics and the education departments. Several of our mathematics faculty members have since participated in a Lesson Study Project, learning more about our own mathematical knowledge and pedagogy.

—Edna O. Schack.

Teaching Children Mathematics 10 (May 2004): 39–48

(Lewis and Tsuchida 1998; Stigler and Hiebert 1999; Takahashi 2000; Yoshida 1999). In fact, Japanese mathematics instruction has transformed from teacher-directed instruction to child-centered instruction during the past fifteen years (Lewis and Tsuchida 1998; Yoshida 1999). The ability to make this change has widely been attributed to the efforts of lesson study.

What Is Lesson Study?

During lesson study, teachers work collaboratively to—

- formulate long-term goals for student learning and development;
- plan, conduct, and observe a "research lesson" designed to bring these long-term goals to life, as well as to teach particular academic content;
- carefully observe student learning, engagement, and behavior during the lesson; and
- discuss and revise the lesson and the approach to instruction based on these observations (Lewis 2002).

The research lesson is taught in a regular classroom, and participants observe as the lesson unfolds in the actual teaching-learning context. Debriefing following the lesson develops around the student-learning data collected during the observation. Through the lesson-study process, participants are given opportunities to reflect on the teaching process as well as on student learning (Murata and Takahashi 2002; Yoshida 1999). **Figure 1** shows a typical model of school-based lesson study. A lesson-planning group develops a research lesson and implements it in a classroom. All the members of the lesson-study group observe the lesson and collect data, then engage in debriefing the lesson. As a result, the lesson is sometimes revised and implemented again in other classrooms. This is called a lesson-study cycle. Other teachers at the school often observe these lessons. When the school decides to open its research lesson to the public, groups from outside the school such as teachers, educators, and university professors have an opportunity to attend this "lesson-study open house." At this event, all the participants can observe the research lessons and engage in discussions of those lessons in order to think about improving teaching and learning. This system contributes to the development of new ideas for teaching and learning as well as images of good teaching practices in the classroom.

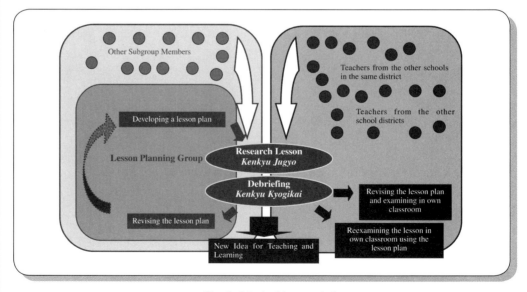

Fig. 1. A typical lesson study

This article draws on our experiences as practitioners, educators, and researchers of lesson study in the United States and Japan. We describe what lesson study is and why it is significant, and discuss how teachers can begin effective lesson-study activities at their own schools in order to improve teaching and learning.

Why Is Lesson Study Powerful?

Lesson study has played an important role in professional development in Japan since the beginning of Japanese public education more than a hundred years ago. One reason for this popularity might be that lesson study gives Japanese teachers opportunities to—

- make sense of educational ideas within their practice;
- change their perspectives about teaching and learning;
- learn to see their practice from the child's perspective; and
- enjoy collaborative support among colleagues.

For example, one Japanese teacher said the following about lesson study:

> It is hard to incorporate new instructional ideas and materials in classrooms unless we see how they actually look. In lesson study, we see what goes on in the lesson more objectively, and that helps us understand the important ideas without being overly concerned about other issues in our own classrooms. (Murata and Takahashi 2002)

Why is lesson study so appealing to so many researchers and educators in the United States? We think it is because lesson study has certain characteristics that set it apart from typical professional development programs.

First, lesson study gives teachers the opportunity to see teaching and learning as it takes place in the classroom. Lesson study provides the context for teachers to focus their discussions on planning, implementation, observation, and reflection on classroom practice. By looking at actual classroom practice, teachers are able to develop a common understanding or image of what good teaching practice entails, which in turn helps students understand what they are learning.

Another unique characteristic of lesson study is that it keeps students at the heart of the professional development activity. Lesson study provides an opportunity for teachers to carefully observe students during the learning process and discuss actual classroom practice.

A third characteristic of lesson study is that it is teacher-led professional development. Through lesson study, teachers are actively involved in the process of instructional change and curriculum development. Paterson Public School No. 2 in Paterson, New Jersey, a pre-K to grade 8 urban school that serves a population of mostly Latino and African-American students, has been implementing lesson study since 1999. Principal Lynn Liptak shared her thoughts about the differences between lesson study and traditional professional development in the United States in Lewis's *Lesson Study: A Handbook of Teacher-Led Instructional Change* (2002). Liptak explained that lesson study is teacher-led professional development in which all the participants reciprocally learn from one another's experiences. In addition, the collaboration through lesson study helps reduce isolation among teachers and develops a common understanding of how to systematically and consistently improve instruction and learning in the school. For example, teachers can establish a common expectation for student learning in the classroom and provide consistent and coherent instruction in the entire school. Moreover, lesson study is a form of research that allows teachers to take a central role as investigators of their own classroom practices and to become lifelong, autonomous thinkers and researchers of teaching and learning in the classroom.

Another important characteristic of lesson study is that it has played a significant role in improving curricula, textbooks, and teaching and learning materials in Japan. In fact, most Japanese mathematics textbook publishers employ classroom teachers who are deeply involved in lesson study as authors, and their materials are often based on classroom teaching and learning through lesson study. As a result, the content of student textbooks and teacher guides is focused, connected, and coherent in order to help students construct an understanding of the mathematics they are learning (Schmidt, Houang, and Cogan 2002).

How Can You Begin Lesson Study?

Create an informal study group

Because lesson study is a form of teacher-led professional development, any teacher can begin lesson study by starting to collaborate with other teachers. Effective models of lesson study in Japan often begin as grassroots movements by enthusiastic teachers rather than as top-down formations (see Yoshida [1999] and Lewis [2002]). For example, lesson study at Paterson Public School No. 2 started as a voluntary study group in 1999. Soon, members of the group were able to spread the idea of lesson study at the school and convinced other classroom mathematics teachers to join them. Now, the school is conducting school-based lesson study and trying to provide consistent and coherent mathematics education to its students. Informal study groups that focus on improving mathematics teaching and learning can be a step toward developing a lesson-study group. If you are not already part of such a group, you might initiate the practice by sharing what happened in your mathematics class with your colleagues during a grade-level meeting or prep time. You do not have to begin lesson study with all the teachers at your school. Forming a comfortable, collaborative group is the most desirable step toward beginning successful lesson study.

Experience lesson study

The idea of lesson study is simple: collaborating with fellow teachers to plan, observe, and reflect on lessons. Developing effective lesson study, however, can be a complex process (Lewis 2002). In order to be effective, lesson study must become a cultural activity, woven into the fabric of teachers' everyday teaching experiences. Teachers cannot learn effective lesson study simply by reading about it. They must experience it firsthand by participating in it on a long-term basis. Stigler and Hiebert (1998) and Gallimore (1996) claim that the cultural script can influence observable instructional pattern. The written recommendations, demonstrations, and one-shot workshops that have characterized U.S. educational reform cannot easily fix or improve the cultural script.

Practicing lesson study makes it possible to learn such subtle yet important things as how a lesson plan for lesson study is different from a traditional lesson plan, why such a detailed lesson plan is necessary, what kinds of data must be collected during observation in order to conduct meaningful discussions, and how to carry out effective debriefings.

Identify your research goal or theme

All the members of the lesson-study group determine the lesson-study goal or theme and the subject to study. For example, the Chicago Lesson Study Group, a small group launched in the fall of 2002, decided to investigate how to improve the teaching and learning of measurement in the elementary and middle-school grades. Members of the group chose this theme because standardized test scores showed that measurement was the weakest area in mathematics for their students and because measurement was the most difficult topic for them to teach. This theme emerged from a discussion about what topics teachers found difficult to teach. Paterson Public School No. 2 chose "fostering student problem-solving and responsibility for learning" as its lesson-study goal by identifying students' weaknesses in mathematics throughout the K–8 program. At the beginning of the 2003–04 school year, however, teachers realized that they needed to expose students to many different solutions to a problem in order to help them develop good problem-solving skills. The students needed to be able to share their ideas and record

what was shared to retain their learning. After identifying this professional growth target, the teachers refined their 2003–04 lesson-study goal as "to encourage, record, and share student thinking." This goal was written not only for the students but also for the teachers. The teachers believed that they were lacking such skills themselves. By developing this goal, the school is trying to cope with inadequate skills among students and teachers in the hope of providing consistent education throughout the school.

Decide on a topic to investigate

Students are the center of classroom teaching; therefore, identifying which topics students have difficulty with is the obvious place to start in deciding on a topic. Another way to choose the topic is to determine the most difficult or uncomfortable topic to teach among the group members. Recent changes in standards or curriculum, or the time of year to conduct the research lesson, might become issues to consider when choosing the topic. Regardless of how you choose the topic, do not choose an isolated topic for lesson study. Choose topics that are important in the curriculum, and think about the topic as a unit of lessons instead of a single lesson. The word "lesson" sometimes creates a misunderstanding of the lesson-study process. Teachers do work to develop a research lesson, but it must be connected to other lessons in the unit in order to maximize the learning results.

Investigate a variety of materials

Even after identifying a lesson-study theme, it still is too early to develop a lesson plan. Some groundwork is necessary. For example, if a group decides to explore how to teach measurement of the area of a rectangle to fourth-grade students, the group must know how this topic relates to other topics in the same grade, what prior knowledge students should have, and how this topic can help students learn new mathematics concepts in future classes. Moreover, teachers must know what kind of instructional materials various textbooks use to teach this topic to students, and what the research suggests (if anything) about methods for teaching the topic. Good understanding of the content and the relationships among topics is very important in order to carry out effective lesson study. This investigation is called *Kyouzai-kenkyuu* in Japanese. It means studying—

- a variety of teaching and learning materials, such as curricula, textbooks, worksheets, and manipulatives (for example, investigating how the topic you chose relates to the sequence in the curriculum, what prior knowledge is necessary for teaching this topic, and how various instructional materials present new concepts of the topic);
- a variety of teaching methods;

- the process of student learning, including students' typical misunderstandings, mistakes, and anticipated solutions to problems, as well as how teachers can react to them;
- the state of students' learning (what they know or are able to do); and
- research related to the topic.

Japanese teachers often begin this process by examining and comparing teacher's guides published by different textbook companies.

Develop a research lesson and write a lesson plan

Japanese teachers usually make a simple unit plan before developing the research lesson. They first determine the main mathematical concepts they need to help their students understand the topic. They also consider how many lessons they can afford to teach in the unit, how the students can learn new mathematical concepts by recalling their prior knowledge, what are the most important or difficult concepts the students must learn, how the research lesson that introduces important mathematical concepts fits into the whole unit plan, and so on. The Japanese teachers generally believe that one lesson cannot guarantee that all students will acquire the mathematical concepts they need to understand; therefore, they believe that a good instructional plan for the unit is important. Once they have chosen the research lesson topic, the teachers begin developing the lesson and write a very detailed lesson plan.

Japanese teachers use many different types of lesson plans. Although no single universal form is available, every lesson plan is expected to provide lesson-study participants with such information as—

- why the lesson-planning group decided to use a certain problem for the lesson;
- why the group chose a particular manipulative; and
- why the group used particular wording for the important questions.

To answer these questions, a typical lesson plan includes the title of the lesson, the lesson goal, the relationship to the standards or curriculum, information about the lesson (such as the background and rationale), the expected learning process, and evaluation points to determine whether students are learning. Teachers in a novice lesson-study group might want to begin writing their lesson plans using the provided lesson-plan format.

Conduct a research lesson and a debriefing

Respecting the natural atmosphere of the class is always a priority during a research lesson; therefore, a research lesson ideally should be held in the instructor's regular classroom. If the regular classroom cannot hold enough participants, however, the instructor might teach the research lesson in a larger classroom. Out of respect for maintaining the natural environment, neither the members of the lesson-planning group nor participants should give any advice, coaching, or comments to students because (1) it distracts from the natural interaction between the students and the instructor; (2) it affects the data that the lesson-planning group is collecting; and (3) in the regular class, having many teachers helping the students is not common.

The main goals of observing a research lesson are to understand student thinking and learning processes, collect the data to back up those points, and determine how students received the plan of the lesson so the observers understand what the teachers intended to teach. To collect the most useful data, observers must adhere to the following guidelines:

- Collect data with the lesson goal in mind.
- Use the lesson plan, seating chart, and worksheets to record observations.
- Document student learning processes, including the many ideas for solving the problem, common misunderstandings the students had, and how and when their understanding changed.

A debriefing is usually held immediately after the research lesson. Holding the debriefing in the classroom in which the research lesson was held might be a good idea because participants can see the blackboard writing and the materials that the students used during the lesson. In addition, teachers should bring all the resources (such as textbooks, teacher's manuals, and manipulatives) that they used to develop the lesson, as well as data collected from the lesson (such as observation notes, students' worksheets and notebooks, and notes from pilot lessons).

Before the debriefing, several people should be assigned to conduct the debriefing session—usually a facilitator, a recorder (note taker), and a final commentator. The facilitator, who typically is one of the more experienced lesson-study practitioners at the school, keeps the discussions focused during the debriefing. The note taker keeps minutes of the meeting and is responsible for writing a summary of the important things discussed.

The debriefing session usually begins with an instructor's short comments on his or her teaching. The instructor addresses how the lesson went, what difficult decisions he or she made during the lesson, and what he or she would like to discuss with participants. Next, a member of the lesson-planning group explains the lesson plan. The instructor's comments and the lesson-plan explanation are meant to set the focus and tone of the discussion. Therefore, the lesson-planning group must carefully think about these comments in order to lead the discussion toward the predetermined goals. The facilitator must also know the goals so that he or she may direct and guide the discussion appropriately.

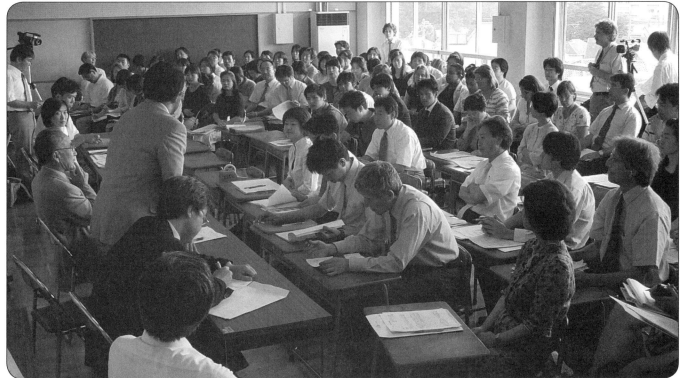

Next, data collected by the lesson-planning group may be discussed in relation to the focus of the discussion previously identified. The main purpose of the discussion is to find out how the students understand and learn the topic through the research lesson. Discussing student learning based on the evidence collected during the observation is important.

Afterward, the discussion is open to all the participants, usually beginning with a focus on the topics of discussion, then gradually opening up to a more general discussion. Discussion is always focused on the lesson, however, not on other topics such as how the school conducts lesson study, how teachers create time to do lesson study, and so on. Remembering that the skill of the facilitator greatly affects the quality of the discussion is important.

At the end of the session, a final commentator (*Koshi*) is given an opportunity to summarize the session. In the United States, the final commentator is usually invited from outside the group or school. Sometimes the commentator is the person conducting lesson study with the group as an outside advisor, or is someone who does not know much about the group's lesson study but is able to contribute his or her pedagogical and content knowledge. Some important qualities that this person should have are—

- the ability to read the audience and provide appropriate comments that help participants learn or want to learn;
- the ability to point out something that no one in the audience noticed but is important to learning about the topic; and
- an attitude that he or she is also a learner through the lesson study and an appreciation of the teachers' efforts.

More information about guidelines for lesson observations and debriefings is available in *Lesson Study: A Handbook of Teacher-Led Instructional Change* (Lewis 2002) and the spring/summer 2002 issue of *Currents* (Research for Better Schools 2002).

Write a summary of a research lesson

Although the research lesson and its debriefing are finished, the lesson-study activity should not end at this point. The lesson-planning team should meet again to reflect on the whole lesson-study process and summarize it in writing. We have already mentioned why a detailed lesson plan that can be a record of the lesson activity is important. Storing or distributing the lesson plan by itself, however, is not effective in developing a professional community of teachers. Accompanying the lesson plan with teachers' written reflections and samples of students' work to complete the summary of the research lesson is a good idea. Research-lesson summaries in Japan vary in their content, but they usually include what group members have learned in the course of planning, conducting, and discussing the lesson; the notes from the debriefing; and sample student work. The summaries also may include a word-by-word transcription of the segment of the lesson that shows the student-teacher interaction and highlights the learning of the topic. This helps the reader of the report construct a vivid image of what happened during the lesson and helps the writer recall the image.

At the end of the school year, lesson-study groups in Japan often gather all the research-lesson reports to compile a lesson-study report. In school-based lesson study, each sub-group that developed a research lesson brings its research-lesson reports to the research committee to be compiled as a lesson-study report. The report usually includes the lesson-study goal, the rationale for setting the goal, reflections about lesson study at the school, a summary of achievements, and a list of the investigative/research tasks. These lesson-study reports usually are stored at each lesson-study group site as well as at the board of education and education centers. In addition, some lesson-study groups seek to publish their reports. In Japan, teachers publish more than educational researchers, and many of the research-lesson reports developed through lesson study are available at large bookstores.

Conclusion

Research, as well as many educational associations in the United States, suggests that mathematics classes should shift from traditional teacher-led instruction to student-centered instruction. As a result, many schools and teachers are working hard to change their classroom teaching to maximize student learning. Many educational reports published in recent years encourage collaboration among teachers. Professional development through lesson study provides many qualities of the professional development approaches that have been suggested to improve classroom practice and learning. It is collaborative and concrete, and it has student learning and understanding as its center. It is continuous and teacher-led. The lesson-study approach permits teachers to be involved in professional development as active learners, as they expect their students to be involved in their own learning.

The following Web sites contain information on future research lesson events and lesson study:

- Global Education Resources :
  www.globaledresources.com
- Lesson Study Group at Mills College:
  www.lessonresearch.net
- Lesson Study Research Group at Teacher's College, Columbia University:
  www.tc.columbia.edu/lessonstudy
- Research for Better Schools:
  www.rbs.org/index.shtml

Beginning lesson study and embarking on the road to improving teaching is within the reach of any teacher or group of teachers with enthusiasm and commitment to the profession.

REFERENCES

Council for Basic Education. "The Eye of the Storm: Improving Teaching Practices to Achieve Higher Standards." Paper presented at the Wingspread Conference, Racine, Wisc., September 24–27, 2000.

Fernandez, Clea, Sonal Chokshi, Joanna Cannon, and Makoto Yoshida. "Learning about Lesson Study in the United States." In *New and Old Voices on Japanese Education,* edited by Edward Beauchamp. New York: M. E. Sharpe, 2001.

Gallimore, Ronald. "Classrooms Are Just Another Cultural Activity." In *Research on Classroom Ecologies: Implication for Inclusion of Children with Learning Disabilities,* edited by Deborah L. Speece and Barbara K. Keogh, pp. 229–50. Mahwah, N.J.: Lawrence Erlbaum Associates, 1996.

Germain-McCarthy, Yvelyne. *Bringing the NCTM Standards to Life: Exemplary Practice for Middle Schools.* Larchmont, N.Y.: Eye on Education, 2001.

Lewis, Catherine. "Lesson Study: The Core of Japanese Professional Development." Paper presented at the American Educational Research Association, April 2000.

———. *Lesson Study: A Handbook of Teacher-Led Instructional Change.* Philadelphia, Pa.: Research for Better Schools, 2002.

Lewis, Catherine, and Ineko Tsuchida. "A Lesson Is Like a Swiftly Flowing River: How Research Lessons Improve Japanese Education." *American Educator* (winter 1998): 14–17, 50–52.

Murata, Aki, and Akihiko Takahashi. "Vehicle to Connect Theory, Research, and Practice: How Teacher Thinking Changes in District-Level Lesson Study in Japan." Paper presented at the twenty-fourth annual meeting of the North American chapter of the International Group of the Psychology of Mathematics Education, Columbus, Ohio, 2002.

Research for Better Schools. "What Is Lesson Study?" *Currents* 5 (spring/summer 2002).

Schmidt, William, Richard Houang, and Leland Cogan. "A Coherent Curriculum: The Case of Mathematics." *American Educator* (summer 2002): 1–17.

Shimahara, Nobuo K. "Japanese Initiatives in Teacher Development." *Kyoiku Daigaku Gakko Kyoiku Sentaa Kiyo* 14 (1999): 29–40.

Stepanek, Jennifer. "A New View of Professional Development." *Northwest Teacher* 2 (2001): 2–5.

Stigler, James, and James Hiebert. "Teaching Is a Cultural Activity." *American Educator* (winter 1998): 1–10.

———. *The Teaching Gap: Best Ideas from the World's Teachers for Improving Education in the Classroom.* New York: Free Press, 1999.

Takahashi, Akihiko. "Current Trends and Issues in Lesson Study in Japan and the United States." *Journal of Japan Society of Mathematical Education* 82 (2000): 15–21 (in Japanese).

Weeks, Denise J. "Creating Happy Memories." *Northwest Teacher* 2 (2001): 6–11.

Yoshida, Makoto. "Lesson Study: A Case Study of a Japanese Approach to Improving Instruction through School-Based Teacher Development." PhD diss., University of Chicago, 1999.

Using Cases to Integrate Assessment and Instruction

Sandra K. Wilcox and Perry E. Lanier

ASSESSMENT may be the most neglected area in the initial preparation and continuing professional development of mathematics teachers. Stiggins (1989) reports that when programs do attend to assessment, the focus is almost always on paper-and-pencil tests. Seldom do teachers have opportunities to consider how to construct and use high-quality performance assessments or to evaluate curriculum-embedded assessments critically. Little attention is given to using multiple forms of assessment—including observing and talking with students—to develop a more complex picture of student learning. And because instruction and assessment typically are separated, teachers do not learn how they might use assessment to shape their own instructional decisions.

In most traditional practice, teachers use assessment primarily to assign grades at the end of a unit of instruction and to rank order their students. They rely heavily on students' written work at completing routine exercises and solving stylized word problems. As Lesh and Lamon argue (1992), students are *tested,* their efforts are *measured,* and they are *evaluated*. As a teacher with whom we worked once said, in self-criticism, "I cover something, test, grade, and move on."

This practice contrasts sharply with visions of assessment reflected in NCTM's document on teaching standards (1991). There assessment is portrayed as ongoing activity in which teachers gather information about student learning in multiple ways: listening, observing, talking with students, posing questions, and examining students' written work. According to the *Standards* document, "What teachers learn from this should be a primary source of information for planning and improving instruction in both the short and the long term" (NCTM 1991, 62). Teachers need help developing the knowledge and habits of mind to construct practices that embody the integration of assessment with instruction.

Linking Assessment with Instruction through Cases

For the past five years, we have been collaborating with middle school mathematics teachers to learn more about how teachers can use the assessment of student learning as an ongoing activity to shape instructional practices. From this collaboration we have produced a number of what Merseth (1992) calls decision-making cases (Wilcox and Lanier, in press). They are built around actual classroom events; present the raw data of the event just as it happened; and are crafted to engage groups of professionals in analysis, problem solving, reflection, and decision making. Each case can be minimally interrogated with a core set of questions. What do I see going on here? What do students seem to understand, and what is the evidence? What are students struggling to understand, and what is the evidence? What does this analysis of students' understandings suggest about where a teacher might go next in instruction? What would I do next?

We have used the cases in various professional-learning settings, including teacher education courses, in-service-workshop series, and seminars. Teacher-leaders have subsequently used them with colleagues in professional-development activities in their local settings. Here we describe the use of decision-making cases in a recent three-

Mathematics Teaching in the Middle School 4 (January 1999): 346–41

week graduate course titled "Alternative Assessment in Mathematics: The Link between Curriculum, Teaching, and Learning." We highlight two case investigations that were especially fruitful in moving beyond concerns about standardized tests and grading. And we offer some compelling evidence of changes in participants' beliefs about the purposes and methods of assessment.

Getting Started: Using Concept Maps to Represent Ideas about Assessment

We began our first session by gathering information about the participants, the ideas about assessment that they were bringing to the course, and the questions that they hoped the course would address. In the class were eight teachers of mathematics with teaching experience of five to more than twenty years. Mary and Lori taught in elementary classrooms; Dean and Laurie, in the middle grades; and Joanne, in an adult-education program. John was the science department chair and occasional mathematics teacher in a small rural high school; Mel was the K–12 mathematics coordinator of a large urban district. Chiang was an international student on leave from a teacher's college in his home country, where he was a mathematics educator.

To learn how the teachers thought about assessment, we asked each of them to draw a concept map with assessment as the center node (see one example in **fig. 1**). We asked the following questions as together we examined the displayed maps: What seems to be similar across the maps? What do I see on other maps that I did not include or did not think of?

In our discussions, we noted that common entries tended to link assessment to goals and objectives, to types of assessments, and to uses of assessment. A single map referred to grading. Dean was the only one to make explicit reference to teaching or teachers, linking assessment to the "evaluation" of teachers, methodology, and curriculum. No evidence was seen that the teachers were thinking about how assessment might be used to help them make instructional decisions. John had an interesting insight that although he used alternative forms of assessment, he was using them in just the same way that he used traditional assessments. That fact seemed to bother him. He thought that alternative assessments should contribute something else but was not sure what or how. Chiang raised several questions about reliability and validity and seemed highly skeptical that alternative assessments were as good as, or superior to, multiple-choice tests.

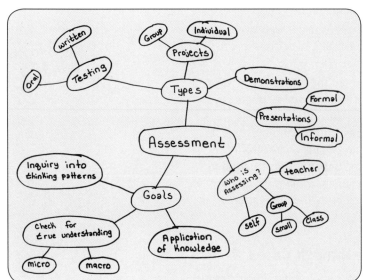

Fig. 1. John's concept map

Exploring a Case to Push the Conversation

One assignment presented a case-development task that course members were to carry out in small groups. In preparation, we examined a set of case materials that we fondly call "fractions of a square." The case is built around students' engagement with a task on fractions of a whole (**fig. 2**) and includes samples of students' written work, the transcript of an audio recording of one pair of students, and a video recording of several small groups as they worked together on the task.

Name _____   Date _____

Fractions of a Square

The aim of this assessment is to provide the opportunity for you to—
· analyze and reason about rational numbers.
· use spatial and numerical reasoning.

The large outer square represents 1 whole unit. It has been partitioned into pieces. Each piece is identified with a letter.

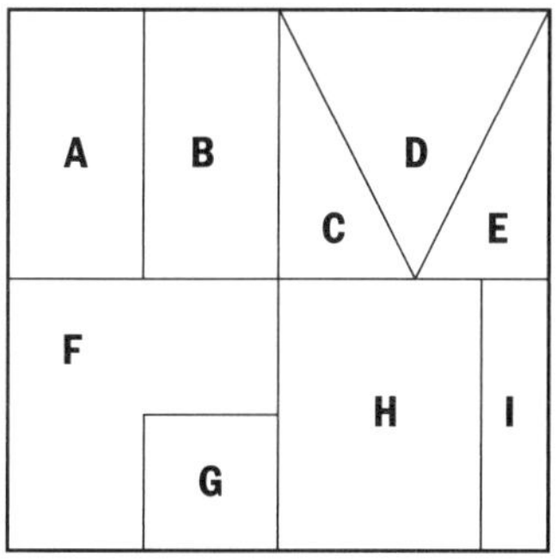

1. Decide what fraction of the whole square each piece is, and write it on the shape.

2. Explain how you know the fractional name for each of the following pieces.

    A

    C

    D

    F

3. Which piece or collection of pieces from this square on the first page will give you an amount close to—

    a) $\dfrac{1}{5}$?

    b) $\dfrac{2}{3}$?

4. a) Design your own fraction square using the square below. Your square must contain at least four differently sized fractional pieces other than 1/2. At least two of your fractional pieces have to be a different size than what is in the original square (used for problems 1, 2, and 3).

    b) Give the fractional names for each of your pieces.

Adapted from the Balanced Assessment for the Curriculum Project, funded by the National Science Foundation

Fig. 2. Fractions of a Square activity

Working on the task themselves

We began by having the teachers work the task themselves and then considered the following questions:

- Why would a teacher use a task like this one?
- What does the task have the potential to reveal about students' understanding of fractions?
- What do you think students are likely to do with this task?

Teachers shared their responses to the task. Several commented on their struggles to write an explanation for the second question. They were comfortable talking about their reasoning but challenged by putting it on paper for someone else to read and understand. They attributed this discomfort to a lack of experience in writing about their mathematical thinking. Dean's elaborate fraction square (**fig. 3**) elicited a nice discussion about the opportunity that the task afforded for more able and motivated students to construct complex squares while allowing all students to engage in the task. And it tested the teachers' own subject-matter knowledge as we discussed whether the three triangles in the upper left of the square were each one-sixth of the square and where the lines should be drawn to ensure that they were of equal area.

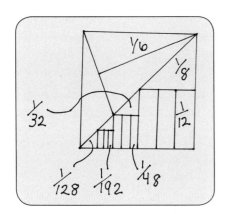

Fig. 3. Dean's fraction square

Examining students' written work

Next the teachers looked at some sixth graders' work on a task intended to assess their conceptual understanding of common fractions as fractions of a whole. The teacher was planning to introduce decimal forms next and had used this ungraded stocktaking task to ascertain whether the students were ready to move on. We posed the following questions:

- What do you think students understand, and what is the evidence?
- What do you think students are struggling to understand, and what is the evidence?

Three pieces of work—those of Matt, Linda, and Heather—were particularly intriguing to our teachers. They wondered why Matt (**fig. 4**) had used fraction strips to lay on the square. They knew that this method was wrong, but few could explain why. They noted that Linda (**fig. 5**) had seemed to lose track of the whole, but they did not notice the computation along the edge of the square. They puzzled about why Heather (**fig. 6**) had named F as 3/4 and H as 1/6, and then were truly stumped in trying to make sense of her fraction square for question 4. Rather than focus on what students were doing wrong, we pushed them to consider what they could say about what students *did* understand and what they would cite as *evidence* to support their claims.

Adding to students' written work

Throughout the examination of students' written work, teachers wanted to know more than they could learn from the papers alone. We supplied a transcript of an audio recording of Linda and her partner as they worked on the first question. Linda on occasion overrode the fraction name that her partner suggested. The teachers' interpretation was that Linda seemed to control the conversation, although they were wary of drawing firm conclusions about group dynamics without actually observing the interaction. The transcript drew their attention to the computation along the side of the square. Linda had "added" all the fraction pieces by adding numerators and adding denominators, a mistake all too familiar to teachers. But Linda's comment to her partner about their answer

Name Matt   Date

Fractions of a Square

The aim of this assessment is to provide the opportunity for you to—
- analyze and reason about rational numbers.
- use spatial and numerical reasoning.

The large outer square represents 1 whole unit. It has been partitioned into pieces. Each piece is identified with a letter.

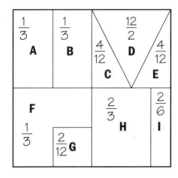

1. Decide what fraction of the whole square each piece is, and write it on the shape.

2. Explain how you know the fractional name for each of the following pieces.

   A — I took the fraction strips and I got 1/3

   C — I took the fraction strip and I got 4/12

   D — I took the fraction strip and I got 12/12

   F — I took the fraction strip and I got anything.

4. a) Design your own fraction square using the square below. Your square must contain at least four differently sized fractional pieces other than 1/2. At least two of your fractional pieces have to be a different size than what is in the original square (used for problems 1, 2, and 3).

   b) Give the fractional names for each of your pieces.

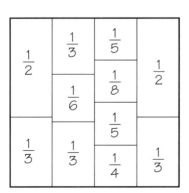

Adapted from the Balanced Assessment for the Curriculum Project, funded by the National Science Foundation

Fig. 4. Matt's work

Fractions of a Square

> The aim of this assessment is to provide the opportunity for you to—
> • analyze and reason about rational numbers.
> • use spatial and numerical reasoning.

The large outer square represents 1 whole unit. It has been partitioned into pieces.
Each piece is identified with a letter.

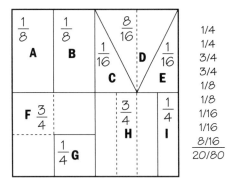

1. Decide what fraction of the whole square each piece is, and write it on the shape.

2. Explain how you know the fractional name for each of the following pieces.

   A   If you divided into 4th's and then divided each 4th in half, you would have 2, 1/8 in each
       square.

   C   If you divided into 4th's and then into 8th's and put an equilateral triangle, that would give
       you 1/16.

   D   This is the same as letter C.

   F   If you divide the whole square into 4th's and then divide each individual square into 4th's, 3
       of those 4th's would be .7

Fig. 5. Linda's work

of 20/80 was a real surprise: "If you add eighty plus twenty, that'll be a hundred, and a hundred equals a whole thing." This statement led to a discussion of how students draw on prior learnings, even partial understandings, to try to make sense of situations.

We played a video recording of Heather's group as it worked on the task. She and three classmates conversed at length about how to name F. Heather and Brad argued that it should be 3/16 because G was 1/16. Initially James seemed to agree but then said that F should be 3/4 because four boxes made up the square with G and F and that G was one

Name ___Heather___   Date _____

Fractions of a Square

> The aim of this assessment is to provide the opportunity for you to—
> • analyze and reason about rational numbers.
> • use spatial and numerical reasoning.

The large outer square represents 1 whole unit. It has been partitioned into pieces.
Each piece is identified with a letter.

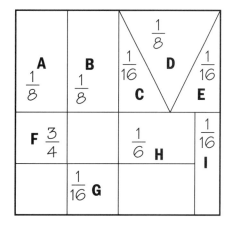

1. Decide what fraction of the whole square each piece is, and write it on the shape.

4. a) Design your own fraction square using the square below. Your square must contain at least four differently sized fractional pieces other than 1/2. At least two of your fractional pieces have to be a different size than what is in the original square (used for problems 1, 2, and 3).

   b) Give the fractional names for each of your pieces.

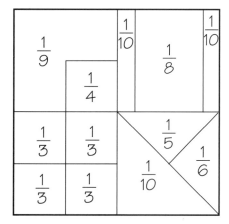

Fig. 6. Heather's work

box and F was three boxes. The fourth student said little. In the end, neither Heather nor Brad seemed convinced by their own arguments and acquiesced to James. Their discussion about H was harder to follow. The teachers thought that at times the group was using numerical reasoning, and at other times, visual-spatial reasoning. Heather appeared to divide H into two parts and then to describe the three pieces in the lower right as each one-third; but her comment as she pointed to H that "one-third and one-third, and that's one-sixth all together" was troublesome.

Deciding where to go next in instruction

Finally, we posed the question at the heart of our cases: If you were the teacher, how would your analysis of this information help you think about your next instructional moves? Although one teacher claimed that students had not "mastered" what the teacher had taught and that "remediation" was called for, the others were more reflective. Mary thought that the students needed "hands-on to feel and do" and that introducing tangrams might be helpful. This idea led to a thoughtful discussion about the use of manipulatives and about whether simply putting concrete materials in students' hands was a panacea. As Mel pointed out, these students had made and used fraction strips yet not all students knew when fraction strips were an appropriate representation or to what kinds of situations they applied. Laurie thought that she might have students cut up the drawing and stack the pieces to find those with the same amount as a way of proving a fraction name. Lori wondered whether a teacher might have students fold the fraction square, thereby making a connection to the earlier work they had done with folding fraction strips. Chiang seemed troubled, thinking that students should be able to apply a rule to name each fraction piece. He was uncomfortable with the "openness" and "subjectivity" that he thought ran through our conversation.

Standing back to reflect on the usefulness of cases

We posed the following question: In what ways might using case materials (a) expand your ideas of what it means to assess for students' understanding and (b) help you learn to use assessment to inform your practice? Laurie spoke first: "Why have I been giving a test and checking answers and not paying any attention to students' thought processes? I need to look at the kinds of tasks I give kids, revise some, discard others." Lori also commented on the nature of tasks. "Can you and students learn from working on a task?" she asked. Mary said that our discussion of the case made her want to go back and reexamine the objectives she had for her students and that she was beginning to question her notions of mastery and remediation. John pointed to a significant shift that the case discussion was fostering: "I ought to be trying to figure out what they *do* understand rather than trying to 'fix' their problems." Dean commented on the contributions to his learning that discussing case materials with a group of professionals had made and worried about the scant time for collaboration with colleagues.

Adding to our case

At our next class session, Chiang shared that he had given the fraction task to his sixth-grade daughter and fifth-grade son. He noted that both had correctly named each piece, although not with the same name. When questioned, his daughter said that she had used G as her unit of comparison and then named all pieces in terms of the number of sixteenths. His son said that he had named H by seeing it as C and D together. He described visually cutting off a corner of D and turning it to create a rectangle with C that was congruent to H. This powerful experience was pivotal for Chiang. He was amazed that his children had not applied a "rule" as he thought that they should, and he was surprised by the different ways in which they reasoned about the problem. Subsequently, he seemed to be much more open to thinking about a broader conception of the methods and purposes of assessment.

Creating Their Own Case

We then launched our teacher-students on their own case development. One set of materials came from a grade 3–4 task that involved looking for and describing patterns and generalizing about patterns by using algebraic sentences. The data set included copies of students' notebook work and video recordings and transcripts of several lessons during which students explained their approaches using the overhead projector. Mary, Lori, and John crafted a case around the use of notebooks in the elementary mathematics classrooms and what teachers can learn about student reasoning and sense making from students' notebook writing. They focused on notebooks, rather than the students' mathematical understanding, as an instructional assessment strategy.

Mel and Laurie examined case materials from a middle-grades preassessment data-handling task. The situation involved three people trying to make the closest estimate of when thirty seconds had elapsed. Each person had five turns at guessing, and their times were recorded with a stopwatch. The task asked students to select the best estimator and to give reasons for their choice. The case materials contained students' written work on the task and a transcript of the recording of the culminating whole-class conversation. Mel and Laurie built their case around the following set of questions: What can you determine about a student's understanding from the written work? What do you think these students know about data, and what counts as evidence of this knowledge? What more do you know about students' thinking about data from the transcript? How might a teacher augment or revise the task to further students' reasoning? They used these questions to investigate the data themselves and then suggested how a facilitator might use them. They paid considerable attention to the mathematics of the task, particularly to students' tendency to use the mean as indicating the best guesser to the exclusion of other relevant analysis, such as the spread of guesses.

Dean, Joanne, and Chiang worked with a set of materials from a middle-grades small-group probability task, a dice game with a set of scoring rules. The task asked whether the game was fair and why. The case materials included students' written work on the task and a video recording of the groups' presentations of their strategies and conclusions. These teachers dug deepest into what they thought the students understood about the mathematics embedded in the task. They asked us for research articles that might shed light on students' difficulties with probability. They wondered how commonly these particular students were asked to explain their strategies or their reasoning. They were especially struck by students' interpretations of what made a game fair, particularly when students' conclusions were at odds with their analysis of the possible outcomes.

In their final reflection paper for the course, several teachers wrote explicitly about the personal learning they took from the case investigations. Chiang's reflection follows:

> Just because they got the answer right, does it mean they know? This is precisely what case analysis and development reveal to me—that assessment based on one or limited sources of information is not sufficient.

Joanne wrote the following:

> While studying the Fractions of a Square task, I discovered new insights about the nuances and subtleties of equivalent fractions. Also, spatial reasoning and numerical reasoning now have a much more personal significance for me. . . . I want to go back and relearn geometry from a spatial perspective so I can be a better geometry teacher.

John's contribution included this statement:

> When we began our analysis of the case, I wondered what I was going to get out of it.
> . . . What I learned during those two weeks was that just going through the process of
> analyzing a case taught me what kinds of things could be counted as evidence, where
> to look for information, and what kinds of tasks allow you to best assess your students
> level of learning. I didn't see this as the focus when I began, but it was the most helpful
> aspect of the case.

An excerpt from Laurie's paper follows:

> I am going to get more tape recorders for group work. This will allow me not only to
> listen more closely to students' thinking aloud but to have a better idea of each student's
> contribution.

Revisiting Concept Maps: End-of-Course Self-Assessment

Typically, most assessment is controlled by the teacher. The teacher decides which
problems to pose, which questions to ask, and what kinds of responses to accept. What
a student is able to demonstrate about her or his learning over a unit of study is conditioned, in part, by what the teacher decides to assess and how. At our final class meeting,
we asked each teacher to construct a second concept map (see **fig. 7**) to reveal how and

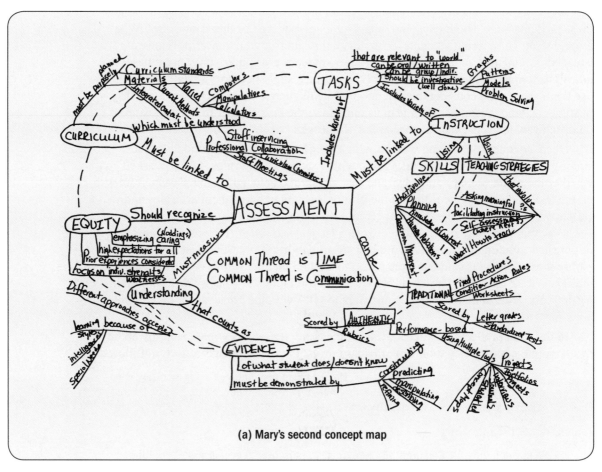

(a) Mary's second concept map

Fig. 7. Concept maps drawn at the end of the course

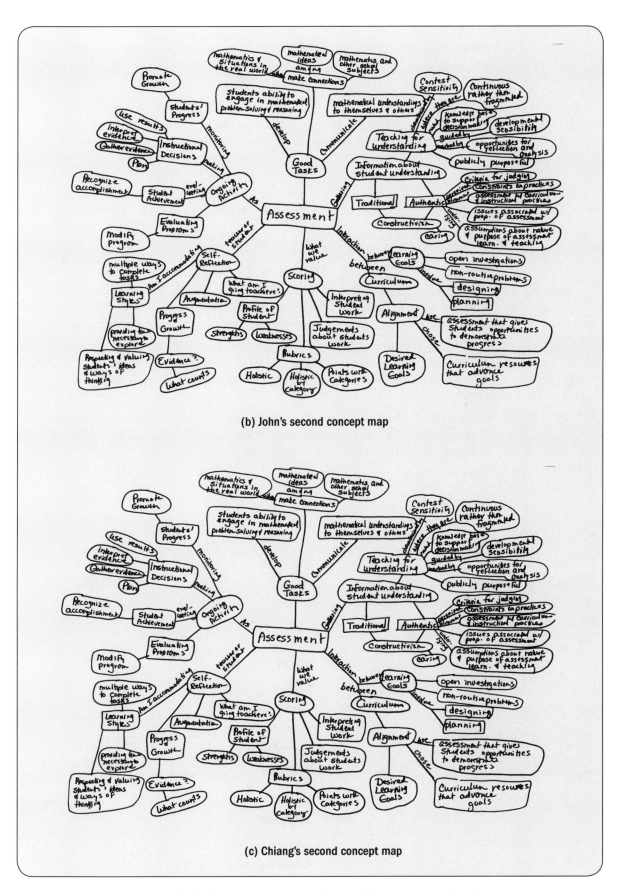

(b) John's second concept map

(c) Chiang's second concept map

Fig. 7. Concept maps drawn at the end of the course—*Continued*

in what ways their thinking about assessment had grown. These maps revealed much richer reflection, packed as they were with many levels of connected ideas developed from the workshop.

In his final reflection paper, Chiang wrote the following:

> My second concept map just amazed me how much I have progressed and grown in just three weeks about assessment. It reminds me of an eagle eyes viewing from above, analyzing the linkages of parts as whole and whole as parts.

We concur.

REFERENCES

Lesh, Richard, and Susan Lamon. "Trends, Goals and Priorities in Mathematics Education." In *Assessment of Authentic Performance in School Mathematics,* edited by Richard Lesh and Susan Lamon. Washington, D.C.: American Association for the Advancement of Science, 1992.

Merseth, Katherine. "Cases for Decision Making in Teacher Education." In *Case Methods in Teacher Education,* edited by Judith H. Shulman. New York: Teachers College Press, 1992.

National Council of Teachers of Mathematics (NCTM). *Professional Standards for Teaching Mathematics.* Reston, Va.: NCTM, 1991.

Stiggins, Richard. "Teacher Training in Assessment: Overcoming the Neglect." In *Teacher Training in Assessment,* edited by Steven L. Wise, pp. 27–40. Hillsdale, N.J.: Lawrence Erlbaum Associates, 1989.

Wilcox, Sandra K., and Perry E. Lanier, eds. *Using Assessment to Reshape Teaching: A Casebook for Mathematics Teachers and Teacher Educators, Curriculum and Staff Development Specialists.* Hillside, N. J.: Lawrence Erlbaum Associates, 2000.

Planning Strategies
for Students with Special Needs:
A Professional Development Activity

Amy R. Brodesky, Fred E. Gross, Anna S. McTigue, and Cornelia C. Tierney

I N TODAY'S mathematics classrooms, teachers are confronted with an increasing range of learners, including students with special needs. On the national level, 13.2 percent of students have identified disabilities. This translates to 6,195,113 students, a jump of 30 percent from 1990 to 2000 (National Center for Education Statistics 2001). The Individuals with Disabilities in Education Act of 1997 (IDEA) mandates that students with disabilities have access to the general education curriculum. This legislation has led to an increase in the number of students with disabilities who are included in regular education classes. Many classroom teachers feel overwhelmed by the challenges of responding to the learning needs of all their students. We often hear teachers say, "I want all my students to be successful in math, but I'm not sure what to do. I don't have training in special education and I don't have much support."

How, then, might a school or district begin to address these issues through professional development? The intent of this article is to share a workshop activity that mathematics coordinators, professional developers, and teacher leaders can use to help teachers plan accessibility strategies for teaching mathematics. The central premise of the workshop activity is based on the Equity Principle in *Principles and Standards for School Mathematics* (NCTM 2000):

> Equity does not mean that every student should receive identical instruction; instead, it demands that reasonable and appropriate accommodations be made as needed to promote access and attainment for all students. (p. 12)

The challenge for teachers lies in applying this principle to daily classroom practice. Having a top-ten list of accommodations and strategies for working with students with special needs in mathematics is not enough. To be effective, those strategies must be connected to teachers' specific mathematics curricula, to their students, and to their classroom situations. In order to make these essential connections, this workshop activity is designed for use with a mathematics lesson of the teachers' choice so that the discussions and strategies are responsive to their specific curriculum. The activity provides opportunities for regular educators and special educators to collaborate in planning strategies, so that their combined expertise strengthens the lesson. This emphasis on collaboration and making connections to mathematics content and classroom practice reflects the research on effective professional development (Daley and Bierema 2002; Smith 2001).

Overview of Workshop Activity

In this activity, teachers use an accessibility planning process to identify potential barriers in a lesson from their mathematics curriculum. They begin by doing the mathematics problem themselves and discussing the goals of the lesson. The process then adds another layer: the students. The teachers are given written "snapshots" of students with disabilities who have a variety of strengths and weaknesses that impact their learning of

Teaching Children Mathematics 11 (October 2004): 146–54

Planning Strategies for Students with Special Needs 61

mathematics. By considering the lesson in terms of its accessibility to these students, teachers identify potential barriers and brainstorm possible strategies. As a final step, teachers make a plan to implement specific strategies with their own students (see **figs. 1** and **2**).

A Closer Look

Step 1: Focus on the mathematics

The activity begins with regular educators and special educators working together on the lesson, which gives them a common experience to draw on throughout the session. By taking on the role of learners, they deepen their understandings of the mathematics and experience firsthand what their students must do to solve the problem. After completing the problem, the group discusses two key questions: "What are the mathematical goals?" and "What is most important for all students to learn?" In order to deepen the discussion, participants identify the mathematical priorities and write them in their own words. If these goals are not taken into consideration, participants run the risk of planning strategies that do not get to, or even conflict with, the essential mathematics.

In order to illustrate the workshop activity, we use a lesson from a fourth-grade unit, Changes over Time (Tierney et al. 1998). Students work with multiple representations of data on plant growth by matching graphs with written descriptions and numerical data (see **fig. 3**). This lesson appears toward the end of the unit and builds on students' prior experiences measuring real plants and graphing the changes in height over time. The mathematical priority is for students to connect three representations: graphical, qualitative, and quantitative. This involves interpreting each representation and translating among them.

Goals:

Participants will—

- use a process for identifying potential barriers in mathematics lessons for a range of students, including ones with disabilities;

- plan accessibility strategies to address barriers for students while maintaining the integrity of the lesson and mathematics;

- collaborate across disciplines and share expertise and perspectives; and

- leave with specific strategies to try with their students and a process to apply to future planning.

Audience:

General educators and special educators in grades 4–8

Time:

1 1/2–2 hours

Preparation:

- Select a lesson from your curriculum.

- Make copies of the "K.I.D.S." handout for each small group.

- Make a copy of the Accessibility Planner for each participant.

Fig. 1. Workshop activity overview

Step	Important Questions
1. Focus on mathematics	• What are the mathematics goals? What are the tasks? • What is most important for all students to learn?
2. Focus on students	• In the group of students that you were given, what are their strengths and weaknesses?
3. Identify potential barriers	• What is the match or mismatch between the lesson's mathematics content and tasks and the students' strengths and weaknesses?
4. Brainstorm accessibility	• What strategies would you use to meet students' needs and strategies enable them to reach the mathematics goals?
5. Share the strategies	• What strategies would you choose for your group's sample of students? Why? • What strategies would you use if you had all these students?
6. Plan follow-up actions	• What strategies would you implement with your students? • How could you collect evidence to see if the strategies were effective?

Fig. 2. Sequence of workshop activity

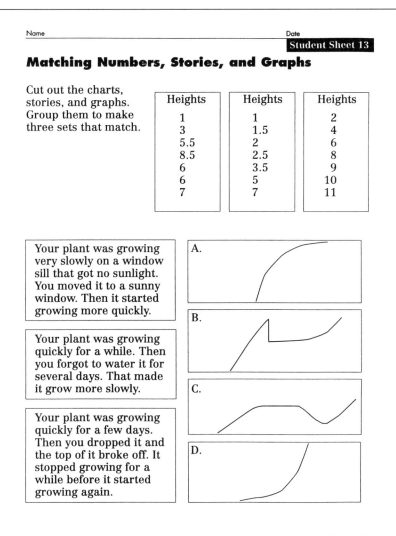

Name _____ Date _____

Student Sheet 13

Matching Numbers, Stories, and Graphs

Cut out the charts, stories, and graphs. Group them to make three sets that match.

Heights	Heights	Heights
1	1	2
3	1.5	4
5.5	2	6
8.5	2.5	8
6	3.5	9
6	5	10
7	7	11

Your plant was growing very slowly on a window sill that got no sunlight. You moved it to a sunny window. Then it started growing more quickly.

A.

Your plant was growing quickly for a while. Then you forgot to water it for several days. That made it grow more slowly.

B.

Your plant was growing quickly for a few days. Then you dropped it and the top of it broke off. It stopped growing for a while before it started growing again.

C.

D.

Fig. 3. Lesson from Changes over Time

Step 2: Focus on students

After finishing this discussion about the mathematics, the focus shifts to thinking about the students. Participants are given "written snapshots" of students (the "K.I.D.S."; see **fig. 4**) that illustrate a variety of common issues and behaviors that teachers are likely to encounter with their own students who have high-incidence disabilities, such as Learning Disabilities (LD) and Attention Deficit Disorders (ADD/ADHD). For example, Kevin struggles with reading, whereas Isabelle is easily distracted. Danny has difficulties with organization and with reading nonverbal cues, whereas Sarah tends to think concretely and struggles to see connections among mathematics concepts. The snapshots also highlight the students' strengths because these strengths provide an avenue for helping students bypass their weaknesses (Levine 2000). Too often, discussions about students with disabilities dwell on their diagnostic labels or on what they cannot do. In order to deepen the discussion and avoid this deficit model, the planning process focuses on both strengths and weaknesses.

At the workshop, each small group is given two of the "K.I.D.S." students from which to choose. Each group selects one student and fills out the student's strengths and weaknesses on the Accessibility Planner. The brief descriptions of the students are designed to be discussion starters and are not meant to be extensive learner profiles of disabilities. Each small group has a common student to focus on and teachers can share their own experiences with similar students.

Kevin

Kevin struggles with reading and feels overwhelmed when there is a lot of text on a page. He often gets confused by the wording of directions and therefore is unsure what he is being asked to do. Embarrassed by these difficulties, Kevin is hesitant to ask the teacher or his classmates for help. He is more comfortable when things are presented visually. Often, he draws pictures as a way to figure out the solutions to problems.

Isabelle

Isabelle gets easily distracted when the teacher is giving directions and has trouble sitting still during class discussions. Sometimes, she tunes out of the mathematics lesson and tries to sneak reading a book in her desk. Although she does not listen well to her classmates, she is quite articulate and likes to talk about her own ideas, particularly when she works in small groups. Isabelle tends to rush through mathematics problems, leaving out important parts and making careless errors.

Danny

Danny's desk is a mess. He can spend an entire mathematics lesson looking for his mathematics book or his homework. When he has a multistep problem to solve, he tends to lose track of the steps, get confused, and not finish in time. Danny often manages to get by because he has an excellent memory for mathematical facts and vocabulary. He sometimes misses social cues so his classmates do not like to work with him on small-group activities.

Sarah

Sarah has difficulty making connections with prior lessons, so each problem looks new to her. Often, she looks blankly at the page and waits for the teacher to help her. When Sarah knows what to do, she works slowly and carefully. Her papers are neat and well-organized. In class discussions, she is eager to participate but tends to talk about the steps she used to solve a problem without explaining why.

Fig. 4. The "K.I.D.S.": Examples of students from the Addressing Accessibility in Mathematics project

A Window on the Workshop

The goal of these sections is to give images of the workshop in action, and the dialogues are drawn from several workshops.

The student "Isabelle" strikes a chord with the teachers in one small group (see **fig. 5**):

- "I know Isabelle."
- "My 'Isabelle' always seems to be looking up at the ceiling when I'm talking. I can picture her weaknesses. What are her strengths?"
- "Her strengths are in language. She likes to talk and to read."

Another small group zeroes in on Sarah:

- "I have a few 'Sarahs' and it's a challenge."
- "The hardest part is that they don't see any connection between today's problem and what they did yesterday. It looks totally new."

Step 3: Identify barriers

By discussing the examples of students, participants have laid the groundwork for identifying potential barriers in the lesson's mathematics content and tasks. They now consider the question "What is the match or mismatch of the lesson with the student's strengths and weaknesses?" This is an opportunity to identify barriers that may get in the way of students being able to learn the mathematics, participate, and show their understanding. A task that is a barrier for one student may be a benefit for another; this is important to keep in mind. For example, the task of reading the descriptions of plant growth may be a barrier for Kevin, who has reading difficulties, while it is a benefit for Isabelle, who is an avid reader.

A Window on the Workshop

One small group identifies a barrier in the lesson's organizational tasks for Danny:

- "The 'Dannys' in my class would be thrown off because there are four graphs and three of everything else."
- "I can just see Danny's messy desk. He's bound to lose some of his pieces and not be able to finish. You know that those pieces are probably on the floor under his desk."

Another group focuses on the difficulties that Sarah may have with the mathematics content:

- "I think those graphs are going to be a problem for Sarah. They look pretty abstract without any numbers or labels. I think she'll have difficulty connecting them with the one she made for her own plant's growth."
- "It's not just the graphs. I don't think she'll know what to do with the numbers."

Strengths	Weaknesses
· Reading	· Difficulty focusing when teacher presents information and during class discussions
· Verbal communication	
· Eager to talk about her own ideas	· Trouble sitting still
	· Works quickly and carelessly
	· Weak attention to detail

Fig. 5. Isabelle's strengths and weaknesses

Step 4: Brainstorm accessibility strategies

Participants are now ready to plan accessibility strategies to address the potential barriers for the "K.I.D.S." students. The goal of these accommodations is to address the barriers by providing the scaffolding and support that students need to reach the mathematics goals. The caution for these strategies is that they can go too far and lose the integrity of the mathematics content and pedagogy or set expectations too low for students.

As a guiding principle for planning accommodations, we suggest that teachers begin by considering instructional practices, such as pairing students or using large visuals, before deciding to alter the curriculum materials. For example, a teacher could help make the text less overwhelming to Kevin by suggesting that he highlight or underline words that describe changes in growth, such as "growing quickly." He can then focus on connecting these important descriptions with the graphs because he prefers visual representations. This strategy may also benefit Isabelle, who tends to work quickly and carelessly, by helping her focus and slow down.

The process of planning accessibility strategies involves thinking proactively about the kinds of difficulties that students may encounter in the lesson. For example, teachers might prepare an organizer for Danny by making a handout that has a large 3-by-3 grid on which he can position the matched sets of stories, graphs, and charts. Teachers might prepare questions and hints to use with students such as Kevin and Sarah who are often unsure how to get started. These questions could be general ("What is the activity asking you to do?") or more specific to the lesson ("Where on graph B is the plant growing quickly?"). Some students may need strategies to help them understand the mathematics, whereas other students may need support to carry out the tasks in the activity. In teaching the lesson, teachers might choose to use the accessibility strategies from the start or to keep them in their "back pocket" to be taken out if students need them.

A Window on the Workshop

This small group's discussion about Kevin highlights the importance of grounding the strategies in an understanding of the student's strengths and weaknesses:

- "Because Kevin has reading difficulties, I could read the descriptions aloud to him or pair him up with another student who is a strong reader."
- "What if he also has trouble processing auditory information? Then that strategy wouldn't be effective."
- "Well, we know his strengths are in the visual area. How can we build on that to deal with the reading difficulties?"
- "We could suggest that he highlight important information in the story, such as 'very slowly.' He could then use the same color to highlight the parts of the graph that correspond to the words."

Other participants grapple with ways to help Sarah make connections in the mathematics (see **fig. 6**):

- "The unlabeled graphs will look so abstract to Sarah. I'd check to see if she knows what each axis represents. Then I'd ask her to find places on the graphs where the plant grew slowly and where it grew fast."
- "It seems likes she's pretty dependent on the teacher for help. So I might ask questions to guide her through connecting one story to a graph and then have her try to do the others on her own. I could pair her with another student, maybe Kevin."
- "What about the numbers? They look even more abstract than the graphs. If Sarah's struggling, I'd give her some graph paper and suggest that she graph one number sequence. Then she might be able to connect the sequences without graphing them."

Mathematical Goals: What are the priorities?

The priority is for students to connect three representations of data on plant growth. This involves interpreting graphs, written descriptions, and number sequences and translating among the representations.

Student's Strengths	Student's Weaknesses	Potential Lesson Barriers/ Student Difficulties	Accessibility Strategies
· Enthusiastic about participating in discussions · Works carefully and neatly · Organized	· Difficulty making connections to prior mathematics problems · Does not take initiative to ask teacher for help · Concrete in her explanations—talks about steps and not about reasons	· There are many choices of which graph, story, or number sequence to start with. She may not know how to get started and may not ask for help. · The graphs do not have grids, numbers, or labels. She may have difficulty connecting these abstract graphs with the ones she made for her own plant and for prior lessons. · She may have difficulty interpreting the number sequences and connecting them to the unlabeled graphs. Will she recognize that these are measurements like the ones she took of her plant?	· Check to see if she is able to get started, so she is not waiting around if she is stuck. · To help her make connections, start the lesson by having students talk about their own plants. Use Sarah's graph of her plant's growth as an example for the class and call on her to talk about it. She likes participating. · Ask questions to see what she understands about the unlabeled graphs: "Where is the plant growing quickly?" · To make the numbers more concrete, give her graph paper to graph one sequence. Have her try matching the other sequences without graphing. · Ask her to draw in the air with her hand to show what she thinks the graph for one of the stories would look like.

Fig. 6. Accessibility planner for Sarah

The focus of these discussions is on planning accommodations that are a good match to both the students' needs and the mathematics content of the lesson. The value of these discussions goes far beyond the specific accommodations planned because they give teachers opportunities to extend their own repertoires of strategies as they listen to one another. The challenge of reaching a wide range of students becomes less overwhelming when teachers share their perspectives and build on one another's expertise.

Step 5: Share the strategies

After brainstorming a list of strategies, each small group selects two or three suggestions to share with the whole group. Each group presents its chosen strategies and explains why they are a good match to their student's strengths and weaknesses. Teachers often find that the strategies they planned for one student would also benefit other students. For example, the organizer for Danny would also be helpful for Isabelle, who tends to be careless.

To address the reality of most classrooms, the workshop then poses the question "What if you had all these students?" Teachers often say, "I do have all these students, plus twenty-one more with their own needs." The expectation is for teachers to create not twenty-five individual lesson plans but one plan with accessibility built into it. One approach is to select three *focal students* to act as proxies for the range of diverse learners in the class. For example, a teacher could select a typical student, a struggling student with an Individualized Education Program or Plan (IEP), and a high-performing student and list their strengths and weaknesses. By planning with these focal students in mind, teachers can create one lesson plan that meets most of the needs of the range of students in the class. Teachers may need to make some additional adaptations for individual students, particularly those with more severe disabilities.

Although the accessibility strategies must be a good match for students' needs, they also must be reasonable for teachers to prepare and implement given their classroom constraints. Teachers can build on instructional practices that they already use and gradually add new strategies to their repertoires. "Best teaching practices," such as providing frequent and constructive feedback and using meaningful contexts, are effective approaches for students with disabilities (Bottge 1999; Miller and Mercer 1997). Teachers can build accessibility into their classroom environment through the use of strategic seating plans, organizational systems, and word walls, and by reducing visual and auditory distractions. For example, Isabelle might be able to better focus if her desk is near the teacher and if the classroom walls are not covered with a lot of distracting visuals. By building a supportive classroom culture in which respect for learner differences exists, teachers can help students such as Kevin feel comfortable asking for help and taking risks.

Step 6: Plan follow-up actions

In this final section, teachers select accessibility strategies to use with their own students when they teach the lesson or similar lessons. They make a follow-up plan to address the questions "What will you implement? How? With whom?" Teachers may choose different ways to implement the same strategy in order to fit their teaching styles and classroom cultures. One teacher may hand out organizers at the start of the lesson, whereas another may have copies available for students on an as-needed basis. An important question for implementing accessibility strategies is whether to use them with one student, some students, or the whole class. The answer to this question will vary depending on the types of barriers in the lessons and the needs of the student population. Some students with significant disabilities may require accommodations, such as enlarged texts, that are not appropriate for use with the whole class. Some accommodations may be beneficial for students who are struggling, regardless of whether they have disabilities. Other adaptations may benefit the whole class by clarifying directions or focusing attention on the important mathematics.

While planning, teachers must consider the questions "How will you determine whether the strategy is effective for students?" and "What would you hope to see?" It is helpful to identify ways to collect evidence of the strategy's impact on students' understanding of the mathematics by examining their work and by asking questions. After the workshop, the participants are expected to implement the strategies in their classrooms and then report to the group at a follow-up meeting. Scheduling a follow-up meeting gives teachers the impetus to use the strategies with their students. Research on effective professional development emphasizes the importance of making these connections so that the activity does not become a one-shot deal with little impact on classroom practice (Daley and Bierema 2002; Mott 2000; Smith 2001).

At the follow-up meeting, participants have the opportunity to reflect on their experiences using the strategies with students. By examining work from the students, teachers discuss the impact of the accommodations: "What evidence do you see of students' understanding of the mathematics? What difficulties do you see? What additional strategies might be needed?" These discussions help teachers refine or revise their strategies to further strengthen the connection between the accommodations, the students, and the mathematics. The follow-up meeting also serves as a much-needed opportunity for teachers to share some stories of success for students with disabilities in mathematics and to support one another in meeting the challenges.

Conclusions

In order to make mathematics more accessible to students with disabilities, teachers need opportunities to plan proactively and to expand their own repertoires of strategies. In this workshop activity, regular educators and special educators share expertise to identify potential barriers and brainstorm accommodations for a lesson from their mathematics curriculum. Teachers leave the workshop with specific strategies that they can try in their classrooms and on a broader level, with an accessibility planning process.

The components of this activity are applicable beyond a workshop setting and can be used by teachers for co-planning or individual planning. The Accessibility Planner is a useful tool that teachers can use to incorporate accessibility into their mathematics lessons. By selecting their own focal students and listing strengths and weaknesses, teachers can anticipate difficulties in upcoming lessons. They can then plan and implement strategies to address these barriers, helping students reach the mathematical goals. As teachers make accessibility an integral part of lesson planning and classroom practice, they are better able to meet the challenge of reaching *all* the "K.I.D.S."

Visit the "Addressing Accessibility in Mathematics" Web site at www.edc.org/accessmath to learn more about strategies and to download blank copies of the Accessibility Planner and other resources.

REFERENCES

Bottge, Brian A. "Effects of Contextualized Math Instruction on Problem Solving of Average and Below Average Achieving Students." *The Journal of Special Education* 33 (2) (1999): 81–92.

Daley, Barbara J., and Laura L. Bierema. "Visible Connections and Subtle Intersections of Continuing Professional Education and Human Resource Development." Paper presented at the Academy for Human Resource Development Preconference on Continuing ProfessionalEducation, Honolulu, Hawaii, 2002.

Levine, Mel. *Educational Care: A System for Understanding and Helping Children with Learning Problems at Home and in School.* Revised edition. Cambridge, Mass.: Educators Publishing Service, 2000.

Miller, Susan Peterson, and Cecil D. Mercer. "Educational Aspects of Mathematics Disabilities." *Journal of Learning Disabilities* 30 (1997): 47–56.

Mott, Vivian W. "The Development of Professional Expertise in the Workplace." In *Charting a Course for Continuing Professional Education: Reframing Professional Practice,* edited by Vivian W. Mott and Barbara Daley, pp. 23–31. San Francisco: Jossey-Bass Publishers, 2000.

National Center for Educational Statistics. nces.ed.gov. 2001.

National Council of Teachers of Mathematics (NCTM). *Principles and Standards for School Mathematics.* Reston, Va.: NCTM, 2000.

Smith, Margaret Schwan. *Practice-Based Professional Development for Teachers of Mathematics.* Reston, Va.: National Council of Teachers of Mathematics, 2001.

Tierney, Cornelia, Ricardo Nemirovsky, and Amy Shulman Weinberg. *Changes over Time: Investigations in Number, Data, and Space.* Menlo Park, Calif.: Dale Seymour Publications, 1998.

The preparation of this article was supported, in part, by a grant from the National Science Foundation. Any opinions expressed in this article are those of the authors and do not necessarily represent the views of the National Science Foundation.

Expanding Teachers' Understanding of Geometric Definition: The Case of the Trapezoid

Randall E. Groth

R ESEARCH findings show that teachers often need considerable help developing conceptual understanding of geometric definitions (Borasi 1996). Ball, Lubienski, and Mewborn (2001) argue that conceptual understanding is essential to the practice of teaching mathematics:

> Without such knowledge, teachers lack resources necessary for solving central problems in their work—for instance, using curriculum materials judiciously, choosing and using representations and tools, skillfully interpreting and responding to their students' work, and designing useful homework assignments (p. 433).

Their observation is especially pertinent to teaching geometry, because textbooks and curriculum materials sometimes differ in the conventions they use for defining geometric objects. If teachers do not understand the various conventions and the consequences of each, they will have limited ability to guide their students' learning.

Coming to understand the nature of mathematical definition is a complex matter. Borasi (1992) identified some cognitive hurdles that individuals must overcome in doing so, such as recognizing the difference between listing properties of a concept and defining it, creating definitions that exclude unwanted cases, and coming to grips with the idea that more than one definition for a mathematical concept can be considered acceptable. This article describes an instructional episode aimed at helping a group of preservice elementary school teachers overcome the last of these three hurdles. Specifically, it details my efforts to help them understand two commonly accepted definitions for *trapezoid* and the consequences of each.

Initial Class Discussion

"You know, the trapezoid is one of the most controversial shapes in all of geometry," I announced one day to my class of preservice elementary school teachers. The comment was met by more than a few looks of disbelief. The facial expressions of some class members seemed to convey such thoughts as "What could possibly be interesting about the trapezoid?" and "How can there be any room for controversy in mathematics?" Determined to convince them of my claim, I decided to have the class experience first-hand the kind of debates that an "innocent" shape such as a trapezoid can spark. The resulting discussion provided an opportunity for considering mathematics content issues alongside pedagogical issues, and it also fostered important insights about the preservice teachers' thinking that were valuable for designing follow-up instruction.

To further clarify my initial claim about the controversial nature of the trapezoid, I informed the class that more than one definition for the term *trapezoid* is commonly accepted. I did not tell them, however, what those two commonly accepted definitions are. I withheld the information that some textbooks define a trapezoid as a quadrilateral with *at least* one set of parallel sides, whereas other textbooks define it as a quadrilateral with *exactly* one set of parallel sides (Cathcart et al. 2003).

Teaching Children Mathematics 12 (March 2006): 376–80

After letting the class in on the fact that no definition for *trapezoid* is universally accepted, I had them work in groups to produce as many examples and nonexamples of trapezoids as they could. Using examples and nonexamples to illustrate geometric concepts is a useful approach to teaching children properties of shapes (Fuys and Liebov 1997), so I saw this occasion as an opportunity for the preservice teachers to learn a pedagogical strategy while refining and extending their own geometric content knowledge. Each of the six groups of preservice teachers in class produced several examples and nonexamples. To wrap up the small-group portion of our discussion, I asked each group to send someone to the chalkboard to draw one example of a trapezoid, one nonexample, and one shape for the class to debate about. A sketch representing the shapes that ended up on the board is shown in **figure 1**.

With the shapes posted on the board for everyone to see, my first question to the class was "Are there any shapes that should be removed from the 'examples' portion of the board?" One preservice teacher immediately stated that shape C, which appeared to be a parallelogram, should be moved into the nonexamples category. Another stated that shape F, which was a pentagon, should be moved there, as well. When I asked how many thought that shape C should be moved, several, but not all, of the preservice teachers raised their hands. When I asked how many thought shape F should be moved, the almost universal consensus was that it should be considered a nonexample. The group that put shape F on the board defended its decision by saying that the shape had a set of parallel sides. Others in class stated that it was not enough for a shape to have a set of parallel sides but that it also needed to be a quadrilateral. By the end of this exchange, everyone seemed to at least tentatively agree that a trapezoid is some sort of quadrilateral and that parallel sides play some role in defining it.

The next bit of class conversation took us back to considering shape C. I questioned why we should remove it from the "examples" category. After all, it appeared to be a quadrilateral with some parallel sides. I got rather vague responses to this new line of questioning at first. One preservice teacher said that one of the sides has to be

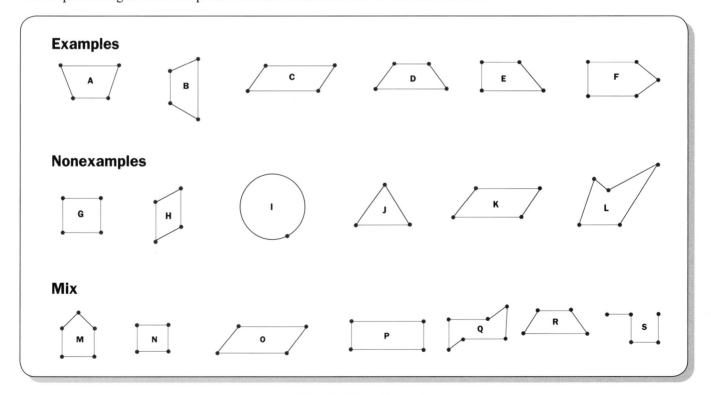

Fig. 1. Sketches of trapezoids

"tilted out" to make a shape a trapezoid. Another said that there should be only a certain number of parallel sides. Although these responses indicated some intuitive, informal thinking about the defining characteristics of a trapezoid, some work still remained to be done in moving toward a more formal, precise definition for *trapezoid*.

To begin to move toward a definition of *trapezoid* that we could agree on as a class, I pointed out that we had already reached some consensus about what makes a shape a trapezoid. Nearly everyone agreed that it should be some type of quadrilateral. I then said, "Putting our disagreement about shape C aside for the time being, suppose we allow it to be called a trapezoid. If I write 'A trapezoid is a quadrilateral with _____,' what needs to go in the blank?" This statement prompted one preservice teacher to reply that a trapezoid is "a quadrilateral with two sets of parallel sides." The murmur in the classroom indicated that this definition did not sit well with some of the other preservice teachers. I asked them to explain why they did not like making two sets of parallel sides a defining characteristic. One of them pointed out that under that definition, shape D could not be considered a trapezoid. This comment led to a suggestion by another class member that one definition for trapezoid could be "a quadrilateral with *at least* one set of parallel sides." At this point, we had one definition that almost everyone seemed to agree on.

To elicit thinking about another possible definition for *trapezoid,* I asked how our definition would change if we did not allow shape C in the "examples" category. This question elicited the other commonly accepted definition relatively quickly. A preservice teacher stated that in such a situation, a trapezoid would have to be defined as a quadrilateral with exactly one set of parallel sides. This second definition was put on the chalkboard alongside the first. We had now arrived, as a class, at the two commonly accepted definitions for *trapezoid*.

Some members of the class were not content to let the class discussion on this subject end simply with the two stated definitions. Questions immediately arose about the teaching implications of our discussion. One class member asked what teachers should do if a child has been taught both conflicting trapezoid definitions. This question resulted in a brief discussion of the importance of frequent discussions among teachers at various grade levels within a school district. Another class member asked whether one should teach children that two different trapezoid definitions are accepted. Others in class seemed to prefer the approach of exposing children to just one definition initially, perhaps building on that definition later on. Several in class, however, did not express opinions about the questions that were raised. Curious about how others in the class were thinking about the issues raised, I decided to elicit thoughts from everyone by using writing prompts.

Responses to Writing Prompts

About two weeks after our discussion about the trapezoid took place, I presented the following two writing prompts:

1. Suppose that the curriculum materials your district has adopted define a trapezoid as "a quadrilateral with exactly one set of parallel sides." A student tells you that he does not think your definition is right, because the teacher he had last year said a rectangle can be a trapezoid, and it has more than one set of parallel sides. What would you do in this situation?

2. Suppose you discover that the textbook used at the high school level in your school district defines a trapezoid as "a quadrilateral with exactly one set of parallel sides," whereas the curriculum materials used at the elementary and middle school levels define a trapezoid as "a quadrilateral with at least one set of parallel sides." What would you do?

These two questions were based on the concerns raised at the conclusion of the class discussion of the two different definitions of *trapezoid*.

Their responses to the writing prompts indicated that five of the nineteen preservice teachers in class still held to the idea that just one correct definition for trapezoid exists. Apparently, the class discussion that had occurred two weeks earlier was not enough to dislodge that belief. Four of those five preservice teachers apparently remembered parts of the earlier class discussion, because they suggested using examples and nonexamples to teach geometric concepts. They said that they would use the strategy to teach students how to "correctly" define a trapezoid. The fifth preservice teacher who held to the belief that a trapezoid can be defined in only one way said that she would go to the board of education for the school district to suggest that they buy different curriculum materials for the elementary school. Each of the five preservice teachers suggested some sort of intervention. However, the extent to which their proposed interventions would be effective is questionable, since they did not have the content knowledge needed to guide them in a fruitful direction.

Fourteen of the nineteen preservice teachers acknowledged the ambiguous nature of the definition for trapezoid in their writing prompts. They took several different approaches to dealing with the ambiguity. Six preservice teachers said that they would simply tell students that two different definitions for *trapezoid* are commonly accepted. Eight preservice teachers said that they would guide students to discover the two commonly accepted definitions on their own, either by researching textbooks and Web sites or by engaging in an examples-nonexamples activity similar to the one they had experienced two weeks earlier. Among the fourteen who acknowledged the ambiguity inherent in defining *trapezoid,* two mentioned discussing the discrepancy between definitions with teachers at other grade levels.

Interpreting the Responses

Perhaps the most surprising aspect of the responses to the writing prompts was that some of those who had participated in the class session described previously still held to the idea that just one "correct" definition for *trapezoid* exists. This fact raises two important questions for consideration: (1) Why did they hold to the idea of the existence of just one definition for trapezoid? and (2) What can be done to help them move beyond this idea?

Possible reasons for the responses

At first glance, the preservice teachers who stated that just one correct definition for *trapezoid* exists appear to have simply forgotten an essential point from the class activity done two weeks earlier. However, the question of why many of them remembered the strategy of using examples and counterexamples but not the mathematical content that had been discussed still remains.

One explanation for the apparent lack of retention of content is that some preservice teachers may have had a resilient concept image of trapezoid that interfered with their willingness to consider other definitions. A concept image is defined as follows:

> The total cognitive structure that is associated with the concept, which includes all the mental pictures and associated properties and processes. It is built up over the years through experiences of all kinds, changing as the individual meets new stimuli and matures. (Tall and Vinner 1981, p. 152)

An important point to note is that a concept image is built up over several years of schooling. If one has several years of experience encountering only one sort of trapezoid, then to build a concept image that accommodates more than one type can be difficult. In addition, school experiences with geometry often are not structured to allow individuals to move past simple recognition and naming of geometric shapes to tasks requiring analysis of the relationships among them (Fuys, Geddes, and Tischler 1988). The result is that some students leave their study of geometry in grades K–12 with impoverished concept images for geometric ideas.

Another explanation for the limited content knowledge displayed in some writing prompts is that some of the preservice teachers may have held Platonist beliefs about the subject of mathematics. Platonists generally hold that mathematics is not a product of human invention, but that objects in mathematics have an existence of their own (Dossey 1992). Such beliefs about the nature of mathematics could interfere with one's acceptance of the idea that mathematical objects, such as trapezoids, can be redefined by mathematicians for the sake of convenience. In practice, teachers' beliefs often do contain Platonist elements (Thompson 1992). Hence, by the time students reach preservice teacher-preparation courses, they are likely to have had a great deal of exposure to Platonist conceptions of mathematics.

Strategies to enhance content knowledge

From the preceding discussion, it can be inferred that building preservice teachers' geometric content knowledge involves helping them develop richer concept images and guiding them to question Platonistic assumptions about mathematics. Since concept images and beliefs about mathematics are developed long before teachers reach a university-level course, a necessary part of preservice teacher education is "inciting doubt and making the previously unproblematic problematic" (Cooney, Shealy, and Arvold 1998, p. 330). The challenge is to incite doubt and simultaneously provide an appropriate level of support.

One strategy for helping preservice teachers develop knowledge about the nature of definitions is to have them examine textbooks. In the example of the trapezoid, a helpful approach can be to ask them to look at one textbook that defines a trapezoid as a quadrilateral with exactly one set of parallel sides and another that defines it as a quadrilateral with at least one set of parallel sides. A textbook using the former definition would consider a reference to a square as a trapezoid to be incorrect, whereas one using the latter definition would consider it correct. By looking at the entire textbook, and not just the two definitions, teachers can gain a sense of how the differences in the definitions result in different consequences. This type of exercise can serve to challenge beliefs in the existence of one static, universal definition for mathematical objects while providing the experiences necessary for building richer concept images.

In-service teacher educators can go beyond examinations of textbooks to lead teachers to reflect on their classroom experiences. Experienced in-service teachers may have encountered in their practice similar situations in which definitions in geometry curriculum materials seem to conflict with one another. Such perceived conflicts can serve as fertile ground for discourse that builds content knowledge. During that discourse, existing beliefs and concept images can be discussed and re-examined. Important questions that an in-service teacher educator can pose during such discourse include these:

1. How did you deal with the conflicting definitions? Why?
2. What sort of impact did the conflicting definitions have on students?
3. When should students be exposed to the idea that geometric objects can be redefined to suit a particular system of reasoning? Why?

In addition, in-service teachers can be encouraged to examine the concept images and beliefs of students they teach by gathering data about their students' thinking through responses to writing prompts, individual interviews, and classroom teaching episodes.

Concluding Thoughts

The trapezoid lesson described in this article can be viewed as an important step, although not necessarily the final one, toward the development of content knowledge needed for teaching geometry. Conducting the trapezoid discussion and posing the writing prompts can generate important insights about preservice and in-service teachers' thinking. Those insights, in turn, can help teacher educators design instruction that is responsive to teachers' needs. The trapezoid problem, and other questions that elicit and build on teachers' thinking, can be instrumental in developing the rich knowledge of mathematics content needed for teaching.

REFERENCES

Ball, Deborah L., Sarah T. Lubienski, and Denise S. Mewborn. "Research on Teaching Mathematics: The Unsolved Problem of Teachers' Mathematical Knowledge." In *Handbook of Research on Teaching,* 4th ed., edited by Virginia Richardson, pp. 433–56. New York: Macmillan, 2001.

Borasi, Rafaella. *Learning Mathematics through Inquiry.* Portsmouth, N.H.: Heinemann, 1992.

———. *Reconceiving Mathematics Instruction: A Focus on Errors.* Norwood, N.J.: Ablex Publishing Corporation, 1996.

Cathcart, W. George, Yvonne M. Pothier, James H. Vance, and Nadine S. Bezuk. *Learning Mathematics in Elementary and Middle Schools.* 3rd ed. Upper Saddle River, N.J.: Pearson, 2003.

Cooney, Thomas J., Barry E. Shealy, and Bridget Arvold. "Conceptualizing Belief Structures of Preservice Secondary Mathematics Teachers." *Journal for Research in Mathematics Education* 29 (May 1998): 306–33.

Dossey, John A. "The Nature of Mathematics: Its Role and Its Influence." In *Handbook of Research on Mathematics Teaching and Learning,* edited by Douglas A. Grouws, pp. 39–48. Reston, Va.: National Council of Teachers of Mathematics, 1992.

Fuys, David, Dorothy Geddes, and Rosamond Tischler. *The van Hiele Model of Thinking in Geometry among Adolescents. Journal for Research in Mathematics Education* Monograph no. 3. Reston, Va.: National Council of Teachers of Mathematics, 1988.

Fuys, David, and Amy Liebov. "Concept Learning in Geometry." *Teaching Children Mathematics* 3 (January 1997): 248–51.

Tall, David, and Shlomo Vinner. "Concept Image and Concept Definition in Mathematics with Particular Reference to Limits and Continuity." *Educational Studies in Mathematics* 12 (2) (1981): 151–69.

Thompson, Alba. "Teachers' Beliefs and Conceptions: A Synthesis of the Research." In *Handbook of Research on Mathematics Teaching and Learning,* edited by Douglas A. Grouws, pp. 127–46. Reston, Va.: National Council of Teachers of Mathematics, 1992.

Using Teacher-Produced Videotapes of Student Interviews as Discussion Catalysts

Victoria R. Jacobs, Rebecca C. Ambrose,
Lisa Clement, and Dinah Brown

TEACHER. Let's pretend you have 7 stuffed animals and your baby sister has 5 stuffed animals. How many more stuffed animals do you have than [your baby sister] Carina?

Krystal. Seven more.

Teacher. How did you get that? Tell me about that.

Krystal. Because I have more than her.

Teacher. You do have more than her. How many do you have?

Krystal. I have 7 and she has 5.

Teacher. And how many more do you have than her?

Krystal. Seven more.

Teacher. And how many extras do you have than her?

Krystal. Seven.

Teacher. OK.

Krystal. Two more extras.

Teacher. Oh, you have 2 more extras. How did you figure that out?

Krystal (puts out one hand with 5 fingers raised and one with 2 fingers raised). Because 7 has 2 more and because, see, see, if this *(7 fingers raised)* wasn't 7, then that means there would be no 7.

Teacher. Did I see you think on your fingers that this is 5 *(holds up hand with 5 fingers raised)* and this is 7 *(adds the other hand with 2 fingers raised)*? *(Krystal agrees.)* So if your sister has 5 and you have 7, then you have 2 more. *(Krystal agrees.)* Because you were comparing your fingers. Good thinking! All right. Are you ready? *(Krystal nods.)* So how many more stuffed animals do you have than your sister?

Krystal. Seven.

Teacher. OK. Are you ready for your next problem?

This excerpt was taken from a longer problem-solving interview conducted by a first-grade teacher participating in a mathematics study group comprised of 18 teachers of grades K–6 and a university facilitator. The group meets monthly to explore children's mathematical thinking, and about two-thirds of the members have been working together for several years. The group's principle collaborative activity is discussing video clips. Each teacher conducts and videotapes interviews with at least three students per year, and once a year, he or she selects a video clip to share with the group. In this article, we provide a glimpse of these discussions, and we explore the use of teacher-produced, teacher-selected video clips as a catalyst to support teachers' growth.

Teaching Children Mathematics 12 (February 2006): 276–81

Why Use Problem-Solving Interviews?

In a problem-solving interview, an individual child works one on one with a teacher to solve a set of mathematical problems. The teacher poses a problem, the child works to solve it, and then they discuss the child's strategy. By carefully observing and listening during interviews, teachers have opportunities to learn about student thinking. They may also support or extend a child's thinking by asking follow-up questions. These conversations are similar to the interactions teachers have during classroom instruction when they constantly make "in the moment" decisions about what to do next in response to what students have said or done.

Many professional developers use problem-solving interviews to help teachers investigate and develop respect for how children think (Ginsburg, Jacobs, and Lopez 1998; NCTM 2000). We have found that conducting interviews can also help teachers develop expertise in eliciting and responding to children's thinking. When teachers share and discuss their interviewing experiences with one another, the benefits are even greater. In this article, we illustrate some of these benefits by sharing selected excerpts of the teachers' conversation surrounding the Krystal video clip.

Viewing the Krystal Video Clip

One month before the study group meeting, teachers were given Krystal's comparison problem to try with their own classes. Different sets of numbers were provided to accommodate the range of grade levels (K–6), and teachers were encouraged to adapt the problem context while keeping the problem structure constant. Trying this problem with their own students gave teachers a common starting point and gave breadth to the study-group discussion when teachers reflected on how a variety of children thought about the same problem.

To begin the study-group conversation, the teacher who had interviewed Krystal posed questions that she wanted the group to help her consider:

- Does Krystal understand the [comparison] concept?
- Does Krystal understand the language: "How many more do you have than _____?"
- What are ways in which children compare naturally? How can those natural experiences be used?

After viewing the videotape, all 18 teachers participated in the 85-minute discussion, focusing on both the specific interaction with Krystal and, more generally, how to help children learn the mathematical concept of comparison. The teachers intertwined their experiences from watching the video clip, trying Krystal's comparison problem in their classrooms, and reflecting on other classroom activities related to comparison (e.g., graphing).

Why Share Teacher-Produced Video Clips?

When teachers produce and share video clips, they have opportunities to (*a*) see other teachers at work, (*b*) pose and discuss hypotheses about children's mathematical thinking, (*c*) identify instructional strategies that build on how children think about mathematics, and (*d*) both support and act as critical friends to other teachers. In the next four sections, we share excerpts of the study group's conversation to illustrate the opportunities that arose while the teachers discussed the Krystal video clip.

To see other teachers at work

Given the isolated nature of teaching in the United States (Little 1990), teachers rarely have opportunities to observe one another at work. Discussing one another's video clips offers teachers a glimpse of their colleagues' professional styles. In the Krystal video clip, they see how the teacher uses knowledge of her students' lives to present a problem about Krystal and her baby sister, Carina. They see how the teacher is persistent in her questioning, rephrasing and revisiting her questions to understand and stimulate Krystal's thinking. Teachers enjoy seeing one another at work, and these experiences can help a group develop cohesion. Teachers are also likely to relate to videos of teachers they know because they work in the same geographical area and because the unedited videos generally include subtle details that contribute to their authenticity. For example, a phone may ring, another teacher may interrupt, or children may drop manipulatives. Finally, the teachers have mentioned that watching one another makes them feel better about their own efforts to change because, in contrast to published cases that often present polished exemplars, video clips produced by teachers typically depict teaching in the process of change.

To pose and discuss hypotheses about children's mathematical thinking

Discussing video clips with other teachers allows for the exploration of children's mathematical thinking from multiple perspectives. After watching the Krystal video clip, the teachers posed and discussed hypotheses about the details and possible interpretations of what had occurred in the interaction. For example, one teacher suggested that Krystal's strategy of comparing fingers worked well with 5 and 7 because the difference was visually obvious on her hands. The group then wondered how children's strategies might change if the comparison was less visually obvious, as in the case of 4 and 7. The group also considered how larger numbers might influence children's strategies. A second-grade teacher shared her experience:

When [my students] had the lower numbers, they built them and they compared. They made one row of the first number and one row of the second number, and they could see the 3 more or 2 more or whatever it was. But when I went to higher numbers, they would just add them together. . . . It's like they lost sight of the problem.

The group worked together to understand why children's thinking changed in the context of larger numbers. They also discussed how children naturally compare outside school and offered examples in which children use "more than" language in phrases such as "I have more than you" and "I'm bigger than you." They noted that children typically use the "more than" language without specifying quantities. For example, children may say, "She's bigger" or "He's taller," without identifying how much bigger or taller. Recognizing children's limited experiences with quantifying comparisons helped the teachers better understand Krystal's (and other children's) struggles.

Throughout the teachers' discussion, there were underlying assumptions that children solve problems in different ways and that children's thinking should be respected. Because the teachers shared these critical assumptions, they did not worry that the discussion would turn into a negative evaluation of children or teachers. Instead, they viewed video clips as venues for inquiry into children's thinking. For example, consider the following excerpt in which a first-grade teacher shared a surprising strategy that her student had used to compare 8 versus 11 bears:

Misha. One kid [Jose] had 11 bears. Another kid had 8. And [the child] said, "Jose had more. If you take away 3 from Jose, then they're equal."

Jana. That's cool! That's very cool.

Tami. A kid said that?

Maria. That surprises me. . . . It surprises me that—as we were just talking earlier about, it's natural for us to count on. It's never natural for us to count backward.

Kim. Depends on what the problem is, though.

Carla. Some kids think that way, and it always amazes me.

Misha. But I don't think he was thinking of numbers. He was just thinking visually. I think you could see it visually.

Maria. Because he just moved them over, and that was what he was taking away.

Jana. Kind of "extras" idea—like if you take those extras.

Misha. Right, whoever doesn't have a partner is an extra one.

Several teachers were surprised by this child's strategy because most children find counting backward less natural than counting on. They expressed their doubt and their reasons for this doubt, and then jointly built an explanation for how a child might generate this strategy. Throughout this inquiry, the teachers neither evaluated the merit of the strategy nor focused on what the child did not know or had not been taught. Instead, they tried to understand how the child was thinking and reasoning so that their instruction could be informed by children's perspectives.

To identify instructional strategies that build on children's thinking

Teachers rarely have the luxury of brainstorming with other teachers about how to interpret and respond to a particular interaction with a child. Video clips make the details of these interactions accessible for exploration. For example, the Krystal video clip showed a child who had some understanding of the comparison concept but did not understand the language of "how many more." Krystal twice incorrectly answered "7" to the question "How many more stuffed animals do you have than your sister?" However, when the question was rephrased to "How many extras do you have than her?" Krystal correctly answered "2."

On the basis of Krystal's performance, the teachers spent substantial time exploring the distinction between understanding the concept of comparison and understanding the language used in comparison problems. The conversation cycled between hypotheses about why children might struggle with comparison language and what teachers can do to engage children with the idea of comparison. Sometimes in their instructional suggestions, they advocated changing the language in the problem. For example, the teachers explored the possibility of asking, "Who has more?" followed by "Do you have a lot more or a little more?" In the excerpt below, the teachers discussed the benefits of introducing the idea and language of comparison with missing-addend problems, in which the action is explicit and the lower number is stated first.

Jana. Another one that works too is "How many more does your little sister need to get to have the same as you?"

Marta. So you add action.

Jana. You add the action, but you keep the terminology "how many more," and then that starts to link it for them, and they start to be able to do that more.

Shawn. But those are kind of two different things. . . .

Jana. Yes, it's changing the problem, but it helps them understand that terminology and what you're asking them, the more you do it.

Marta. What you're saying about how kids compare naturally—it seems like when I listen to my kids do that problem, they always say, "This one needs this many more to be the same as."

Maria. I've just started doing some missing [-addend] problems [like Jana suggested], and I put the lower number first. And so I would start with "Carina had 5 stuffed animals and you had 7." I think starting from the lower number first would have made her automatically want to count on.

Lili. Oh, good idea. . . .

Maria. And then I feel that has given my kids the language. I didn't have to say more than that. They already notice that from the counting on.

Jana. That's good. . . . And then this is the third [question posed by the presenting teacher]: "What do the kids naturally compare?" I always start this kind of problem with age. You have, "You're 6 and your buddy's 11. How many more years until you're as old as your buddy?" It makes more sense to them than anything [else], because they already know they have an older brother or sister, and a lot of them already know how many years older that brother is.

This excerpt illustrates how the teachers seamlessly intertwined their discussions of children's thinking and instructional implications. Jana suggested that using missing-addend problems with the language of "how many more" can help children engage with the concept and language of comparison. This suggestion reflects an understanding of how children think about different types of problems; missing-addend problems (with explicit action) are generally easier than comparison problems for young children (Carpenter et al. 1999). Marta endorsed this suggestion by drawing on her knowledge of how children typically talk about comparison. She argued that the language children naturally use is more consistent with the language in missing-addend problems than in comparison problems. The teachers then built on this suggestion by sharing implementation tips from their classroom experiences in using missing-addend problems. Maria shared the utility of starting with the smaller number, and Jana suggested comparing ages as an initial context. Throughout the discussion, the teachers listened to one another, built on one another's ideas, and collaboratively engaged in inquiry. A spirit of inquiry was important because it removed the focus from evaluating any particular episode to instead continually exploring children's mathematical thinking and its instructional implications.

To support and act as a critical friend to other teachers

Collaborative inquiry about their own video clips can promote a strong community in which teachers support and challenge one another. In the discussion of the Krystal video clip, the group helped the presenting teacher recognize the skillful components of her questioning. The teachers were also willing to move beyond the polite, positive conversation that is typical of teachers' interactions (Lord 1994). Teachers challenged one another by disagreeing and asking for justification of interpretations and suggestions. Because the teachers had already established a trusting environment, these challenges were nonthreatening. For example, several teachers commented on the importance of re-asking the original question: "So how many more stuffed animals do you have than your sister?" When the presenting teacher attributed this action to luck, her colleagues disagreed and expressed their admiration for her skillful questioning. They also wanted to better understand how she decided to ask that question. In the spirit of inquiry, they explored the possibility that she had posed the question because she had articulated Krystal's finger-based strategy instead of having Krystal describe it.

Carla. I don't think I would have [re-asked the original question]. Were you maybe not sure?

Presenting teacher. I think [Krystal] had done it, but I reinterpreted her hands.

Kim. So you were afraid you led her too much, and you just wanted to make sure?

Presenting teacher. Yes. And I also wasn't sure because I said "extras." [The original question] came out again because I don't know if she really had that language [of "how many more"].

The tone of this interchange was respectful, and the teachers were not evaluating the presenting teacher's decision. Instead, they were suggesting that the situation could be handled in multiple ways, and they wanted to explore the reasoning by which one might make a decision.

Teaching is full of ambiguous situations such as this, in which several teaching moves would be warranted. Providing teachers with opportunities to talk about how they make decisions has two positive effects. First, teachers increase their repertoire of possible responses for any given situation. Second, teachers recognize the futility of searching for a single "right" response and, instead, learn to make more informed decisions by considering the advantages and disadvantages of different responses. Video-clip discussions do not provide prescriptions for teachers, but they do support exploration of possible responses to complicated situations.

Final Thoughts

When teachers collaboratively investigate children's mathematical thinking, the conversations can be powerful. The study group's discussion about comparison problems raised issues similar to those reported in the research of Fuson and her colleagues (Fuson, Carroll, and Landis 1996). These teachers constructed this knowledge as part of a dynamic, personally meaningful process in which they were their own experts, collaboratively exploring their own questions about their own students. Realizing these benefits can be challenging. We conclude with suggestions for getting started.

Logistical issues

Teachers may struggle to fit interviewing into their schedules. Incentives such as providing stipends, course credit, or substitutes can help teachers make interviewing a priority. Once teachers have experience with interviewing, they often recognize the benefits and find ways to make time for interviewing in their regular routines.

Technological issues

Although most schools have some type of video camera, teachers may be unfamiliar with this equipment, and sound can sometimes be problematic. Time should be planned for helping teachers learn to work with the video equipment. Sound can be enhanced by using inexpensive external microphones.

Facilitation issues

We initially feared that certain video clips might not generate much conversation, but we have found that children's thinking is inherently interesting, and thus no matter what video clip a teacher shares, valuable discussion can ensue. We have also learned that these conversations about children's thinking can provide a foundation for improving teachers' decision making so that instruction builds on children's ideas. Using this lens, we offer the following prompts as starting points for discussing a video clip:

- What strategy did the child use?
- What did we learn about the child's understanding of the mathematical concept?
- What questions might we pose next to better understand or further the child's thinking?

Finally, we have found that teachers' video-clip selections have often been better than our own selections. Teachers have been willing to share problematic video clips that we, as facilitators, might have hesitated to show for fear of placing teachers in a potentially negative light. Therefore, although facilitators need to be sensitive to the imperfections in teachers' and children's efforts, we have found the use of teacher-produced, teacher-selected video clips to be a productive and respectful way to engage teachers in collaborative inquiry about the difficult issues involved in teaching and learning mathematics.

REFERENCES

Carpenter, Thomas, Elizabeth Fennema, Megan Franke, Linda Levi, and Susan Empson. *Children's Mathematics: Cognitively Guided Instruction*. Portsmouth, N.H.: Heinemann, 1999.

Fuson, Karen, William Carroll, and Judith Landis. "Levels in Conceptualizing and Solving Addition and Subtraction Compare Word Problems." *Cognition and Instruction* 14 (1996): 345–71.

Ginsburg, Herbert, Susan Jacobs, and Luz Stella Lopez. *The Teacher's Guide to Flexible Interviewing in the Classroom: Learning What Children Know about Math*. Boston, Ma.: Allyn & Bacon, 1998.

Little, Judith Warren. "The Persistence of Privacy: Autonomy and Initiative in Teachers' Professional Relations." *Teachers College Record* 91 (summer 1990): 509–36.

Lord, Brian. "Teachers' Professional Development: Critical Colleagueship and the Role of Professional Communities." In *The Future of Education: Perspectives on National Standards in Education*, edited by Nina Cobb, pp. 175–204. New York: College Entrance Examination Board, 1994.

National Council of Teachers of Mathematics (NCTM). *Principles and Standards for School Mathematics*. Reston, VA: NCTM, 2000.

Acknowledgments

The authors would like to thank the mathematics study group for their commitment to understanding children's mathematical thinking as a means to improving their instruction. We have regularly benefited from their rich discussions such as the one described in this article. This study group was supported, in part, by a Mathematics Education Trust (MET) grant from the National Council of Teachers of Mathematics. The views expressed in this article are solely the responsibility of the authors.

Editor's Note

MET supports the improvement of mathematics teaching and learning through the funding of grants, awards, and other projects by channeling the generosity of contributors into classroom-based efforts that benefit all students. For more information, visit www.nctm.org/about/met.—*Ed.*

Part 3

....

Articles for Use in Professional Development

COMMENTARY

Teaching is a lifelong learning process. An ancient Chinese proverb states, "The best time to plant a tree is twenty years ago; the second best time is today." Professionals exhibit a readiness for change, even for change that is so radical that it may cause disequilibrium. In our own journeys, we have discovered, for example, that often more than one way can be used to solve a particular problem and more than one answer is acceptable, that procedural proficiency is attainable when conceptual understanding underpins the procedures (see the section on "Deepening Understanding"), and that growth and change as a teacher require time and reflection. As Steven Leinwand, former director of mathematics education in the state of Connecticut, once said, "If you don't feel inadequate, you are probably not doing the job" (see Leinwand's article "Four Teacher-Friendly Postulates for Thriving in a Sea of Change," the first of the reprinted articles in this section).

"In an increasingly complex world, sometimes old questions require new answers."

Synopses of Articles

Steven Leinwand's classic piece titled "Four Teacher-Friendly Postulates for Thriving in a Sea of Change" was originally published in *Mathematics Teacher* in 1994 and was reprinted in that journal in 2007 as a President's Choice article, selected by former NCTM President Cathy Seeley. Each postulate is provocative and can generate an animated conversation in any group of preservice or in-service teachers. This piece is not only a favorite of many but an essential conversation starter in addressing change.

"Never Say Anything a Kid Can Say!" by Steven Reinhart (2000) is a popular, widely used article. It outlines the author's personal journey from teacher-directed to student- oriented instruction, identifying eleven techniques he used for arriving at his desired destination. These very tangible, specific strategies and the manner in which they are presented resonate with teachers and help them see how they can take steps toward shifting their own classroom practices.

Margaret Schwan Smith's (2000) article "Redefining Success in Mathematics Teaching and Learning" addresses the natural struggles experienced in any change process. Using a teacher's story as the centerpiece, Smith guides the reader through the story by a series of carefully placed reflective questions that would be excellent conversation starters in a professional develop-ment experience. The article contains is a particularly useful analytic grid that teachers can use to redefine their expectations for students, define their corresponding actions, and list their classroom-based indicators of success.

"Signposts for Teaching Mathematics through Problem Solving," by James Hiebert (2003) in NCTM's *Teaching Mathematics through Problem Solving: Prekindergarten–Grade 6,* underscores the consistent benchmarks that can be anticipated as teachers move students successfully into deeper levels of mathematical understanding. By problematizing the subject, the author demonstrates how teachers can guide students through predictable signposts to success.

Four Teacher-Friendly Postulates for Thriving in a Sea of Change

Steven Leinwand

MANY of us chose mathematics teaching because it was always so neat and clean. We felt an affinity toward teaching and learning mathematics because it was orderly and logical. Almost always, we arrived at only one numerical answer by using one right procedure that could be easily graded either right or wrong. We knew that with our beloved mathematics, we suffered none of the gray areas that plague the disciplines of language arts and social studies. And we knew that we would be rewarded for teaching mathematics the way we ourselves were taught. But, oh, how things have changed!

Let's face it: the NCTM's standards documents have made our professional lives much more challenging. Given how much the teaching of mathematics must change to serve a digitized world of calculators and computers and given the breadth of the recommendations of the standards documents (NCTM 1989, 1991), it is not surprising that many teachers of mathematics are frustrated and feel thoroughly challenged. To ease this inevitable frustration, I offer four perspective-building postulates for thriving in a sea of change.

Postulate 1: We are being asked to teach in distinctly different ways from how we were taught. Long-accepted truths state that most people parent as they were parented and most teachers teach as they were taught. We build on what is familiar because the familiar "feels right." However, to teach concepts, not just skills; to rely on cooperative groups; to work collaboratively with colleagues; and to assume the availability of calculators are parts of a very unfamiliar terrain. Neither previous generations of mathematics teachers nor our colleagues in other disciplines have had to face such a chasm between how they were taught and how they are being asked to teach. No wonder many of us feel disoriented and inadequate (see postulate 4).

Since teachers can't do what they haven't seen or experienced, we need to create tangible and accessible models of curricular and instructional reform. We need to increase opportunities for collegial classroom visits, and we need to increase our reliance on videotapes of what the distinctly different forms of pedagogy look like.

Postulate 2: The traditional curriculum was designed to meet societal needs that no longer exist. The bedrock upon which this entire reform movement rests is a clear understanding that society's needs and expectations for schools have shifted radically. No longer are schools expected to serve as society's primary sorting mechanisms. Instead, schools must become empowering machines. Schools cannot remain perpetuators of the bell curve, where only *some* were expected to survive and *even fewer* to truly thrive; education must be a springboard from which *all* must attain higher levels. For this reason behaviors and attitudes that were rewarded a short decade or two ago are now under such scrutiny.

In the face of such emotionally trying bombardments, two very different responses to the standards and to other aspects of the reform movement have become common. Some teachers have basically ignored the entire movement, believing that "this too will pass." Others understand that change is required but, sensing that they themselves are not really moving fast enough, feel guilty about not doing more sooner. Denial and guilt

> We are being asked to teach in distinctly different ways from how we were taught
>
> • • • • • • • •
>
> The traditional curriculum was designed to meet societal needs that no longer exist

Mathematics Teacher 87 (September 1994): 392–93; reprint *Mathematics Teacher* 100 (May 2007): 580–83

It is unreasonable to ask a professional to change much more than 10 percent a year, but it is unprofessional to change by much less than 10 percent a year

• • • • • • • • •

If you don't feel inadequate, you're probably not doing the job.

are entirely appropriate responses to the magnitude of the change swirling around us. However, neither response is particularly comforting and neither represents the level of professionalism we expect from ourselves.

For comfort and a professional safety net, I find it helpful to remember that ignoring the need for change in mathematics also ignores how radically different society's expectations for schools have become. And feeling guilty about what we've done in the past or about not changing fast enough masks acknowledging how effectively schools once met a set of needs that simply no longer exists.

Postulate 3: It is unreasonable to ask a professional to change much more than 10 percent a year, but it is unprofessional to change by much less than 10 percent a year. We can easily argue that the most disorienting element of our lives is the rate at which things are changing. Many researchers have written about people's ability to accommodate to the ever-increasing rate of change. In somewhat arbitrary, but certainly comforting, fashion, I have come to believe that about 10 percent a year is a reasonable rate of change to expect—large enough to represent real and significant change but small enough to be manageable.

One way to visualize change at this rate is to think about substituting one new unit each year, shifting four weeks of instruction to address a new topic, or doing something in a very different way, such as changing questioning techniques or introducing journals. Using this incremental approach will result in five years in changing nearly half of what we do today. Even the most radical proponent of reform should be satisfied with a change of this magnitude in our mathematics classes, and our most cautious and tradition-bound colleagues should be able to retain a real sense of control over such a rate of change.

Postulate 4: If you don't feel inadequate, you're probably not doing the job. Just think what we are asking each other to do: increase the use of technology; use manipulatives and pictures with far greater frequency; make regular use of group work; focus on problems, communication, applications, and interdisciplinary approaches; teach groups that are far more heterogeneous; increase attention to statistics, geometry, and discrete mathematics; assess students in ways that are far more authentic and complex—and do it all yesterday and in ways that boost achievement overnight! Feeling overwhelmed by this torrent of change is neither a weakness nor a lack of professionalism—it is an entirely rational response.

No one can do it all. Just as no physician is expected to be an expert in all aspects of medicine, no mathematics teacher in the 1990s can reasonably be expected to be an expert in all aspects of teaching mathematics. We must select a few areas of focus and balance the fear and worries we understandably have in some areas with the pride of accomplishment and success we find in other areas. We must accept the inevitability of a sense of inadequacy and use it to stimulate the ongoing growth and learning that characterize the true professional. Only then will we be sufficiently armed, intellectually and emotionally, to thrive in the exhilarating, exhausting, and often overwhelming sea of change.

BIBLIOGRAPHY

Leinwand, Steven J. "Sharing, Supporting, Risk Taking: First Steps to Instructional Reform." *Mathematics Teacher* 85 (September 1992): 466–70.

National Council of Teachers of Mathematics [NCTM]. *Curriculum and Evaluation Standards for School Mathematics.* Reston, Va.: [NCTM], 1989.

———. *Professional Standards for Teaching Mathematics.* Reston, Va.: [NCTM], 1991.

Never Say Anything a Kid Can Say!

Steven C. Reinhart

A FTER extensive planning, I presented what should have been a masterpiece lesson. I worked several examples on the overhead projector, answered every student's question in great detail, and explained the concept so clearly that surely my students understood. The next day, however, it became obvious that the students were totally confused. In my early years of teaching, this situation happened all too often. Even though observations by my principal clearly pointed out that I was very good at explaining mathematics to my students, knew my subject matter well, and really seemed to be a dedicated and caring teacher, something was wrong. My students were capable of learning much more than they displayed.

Implementing Change over Time

The low levels of achievement of many students caused me to question how I was teaching, and my search for a better approach began. Making a commitment to change 10 percent of my teaching each year, I began to collect and use materials and ideas gathered from supplements, workshops, professional journals, and university classes. Each year, my goal was simply to teach a single topic in a better way than I had the year before.

Before long, I noticed that the familiar teacher-centered, direct-instruction model often did not fit well with the more in-depth problems and tasks that I was using. The information that I had gathered also suggested teaching in nontraditional ways. It was not enough to teach better mathematics; I also had to teach mathematics better. Making changes in instruction proved difficult because I had to learn to teach in ways that I had never observed or experienced, challenging many of the old teaching paradigms. As I moved from traditional methods of instruction to a more student-centered, problem-based approach, many of my students enjoyed my classes more. They really seemed to like working together, discussing and sharing their ideas and solutions to the interesting, often contextual, problems that I posed. The small changes that I implemented each year began to show results. In five years, I had almost completely changed both *what* and *how* I was teaching.

The Fundamental Flaw

At some point during this metamorphosis, I concluded that a fundamental flaw existed in my teaching methods. When I was in front of the class demonstrating and explaining, I was learning a great deal, but many of my students were not! Eventually, I concluded that if my students were to ever really learn mathematics, *they* would have to do the explaining, and *I*, the listening. My definition of a good teacher has since changed from "one who explains things so well that students understand" to "one who gets students to explain things so well that they can be understood."

Getting middle school students to explain their thinking and become actively involved in classroom discussions can be a challenge. By nature, these students are self-conscious and insecure. This insecurity and the effects of negative peer pressure tend to discourage involvement. To get beyond these and other roadblocks, I have learned to ask the best possible questions and to apply strategies that require all students to participate. Adopting the goals and implementing the strategies and questioning techniques that

One idea for this article is to have teachers identify three ideas or strategies that they will use in their classroom. During sharing time, all participants will be interested to hear the three ideas or strategies their peers have selected.

—Karen Droga Campe

Mathematics Teaching in the Middle School 5 (April 2000): 478–83

follow have helped me develop and improve my questioning skills. At the same time, these goals and strategies help me create a classroom atmosphere in which students are actively engaged in learning mathematics and feel comfortable in sharing and discussing ideas, asking questions, and taking risks.

Questioning Strategies That Work for Me

Although good teachers plan detailed lessons that focus on the mathematical content, few take the time to plan to use specific questioning techniques on a regular basis. Improving questioning skills is difficult and takes time, practice, and planning. Strategies that work once will work again and again. Making a list of good ideas and strategies that work, revisiting the list regularly, and planning to practice selected techniques in daily lessons will make a difference.

Create a plan. The following is a list of reminders that I have accumulated from the many outstanding teachers with whom I have worked over several years. I revisit this list often. None of these ideas is new, and I can claim none, except the first one, as my own. Although implementing any single suggestion from this list may not result in major change, used together, these suggestions can help transform a classroom. Attempting to change too much too fast may result in frustration and failure. Changing a little at a time by selecting, practicing, and refining one or two strategies or skills before moving on to others can result in continual, incremental growth. Implementing one or two techniques at a time also makes it easier for students to accept and adjust to the new expectations and standards being established.

1. Never say anything a kid can say! This one goal keeps me focused. Although I do not think that I have ever met this goal completely in any one day or even in a given class period, it has forced me to develop and improve my questioning skills. It also sends a message to students that their participation is essential. Every time I am tempted to tell students something, I try to ask a question instead.

2. Ask good questions. Good questions require more than recalling a fact or reproducing a skill. By asking good questions, I encourage students to think about, and reflect on, the mathematics they are learning. A student should be able to learn from answering my question, and I should be able to learn something about what the student knows or does not know from her or his response. Quite simply, I ask good questions to get students to think and to inform me about what they know. The best questions are open-ended, those for which more than one way to solve the problem or more than one acceptable response may be possible.

3. Use more process questions than product questions. Product questions—those that require short answers or a yes or no response or those that rely almost completely on memory—provide little information about what a student knows. To find out what a student understands, I ask process questions that require the student to reflect, analyze, and explain his or her thinking and reasoning. Process questions require students to think at much higher levels.

4. Replace lectures with sets of questions. When tempted to present information in the form of a lecture, I remind myself of this definition of a lecture: "The transfer of information from the notes of the lecturer to the notes of the student

The best questions are open-ended

without passing through the minds of either." If I am still tempted, I ask myself the humbling question "What percent of my students will actually be listening to me?"

5. Be patient. Wait time is very important. Although some students always seem to have their hands raised immediately, most need more time to process their thoughts. If I always call on one of the first students who volunteers, I am cheating those who need more time to think about, and process a response to, my question. Even very capable students can begin to doubt their abilities, and many eventually stop thinking about my questions altogether. Increasing wait time to five seconds or longer can result in more and better responses.

Good discussions take time; at first, I was uncomfortable in taking so much time to discuss a single question or problem. The urge to simply tell my students and move on for the sake of expedience was considerable. Eventually, I began to see the value in what I now refer to as a "less is more" philosophy. I now believe that all students learn more when I pose a high-quality problem and give them the necessary time to investigate, process their thoughts, and reflect on and defend their findings.

Share with students reasons for asking questions. Students should understand that all their statements are valuable to me, even if they are incorrect or show misconceptions. I explain that I ask them questions because I am continuously evaluating what the class knows or does not know. Their comments help me make decisions and plan the next activities.

Teach for success. If students are to value my questions and be involved in discussions, I cannot use questions to embarrass or punish. Such questions accomplish little and can make it more difficult to create an atmosphere in which students feel comfortable sharing ideas and taking risks. If a student is struggling to respond, I move on to another student quickly. As I listen to student conversations and observe their work, I also identify those who have good ideas or comments to share. Asking a shy, quiet student a question when I know that he or she has a good response is a great strategy for building confidence and self-esteem. Frequently, I alert the student ahead of time: "That's a great idea. I'd really like you to share that with the class in a few minutes."

Be nonjudgmental about a response or comment. This goal is indispensable in encouraging discourse. Imagine being in a classroom where the teacher makes this comment: "Wow! Brittni, that was a terrific, insightful response! Who's next?" Not many middle school students have the confidence to follow a response that has been praised so highly by a teacher. If a student's response reveals a misconception and the teacher replies in a negative way, the student may be discouraged from volunteering again. Instead, encourage more discussion and move on to the next comment. Often, students disagree with one another, discover their own errors, and correct their

> Students feel comfortable sharing and discussing ideas

thinking. Allowing students to listen to fellow classmates is a far more positive way to deal with misconceptions than announcing to the class that an answer is incorrect. If several students remain confused, I might say, "I'm hearing that we do not agree on this issue. Your comments and ideas have given me an idea for an activity that will help you clarify your thinking." I then plan to revisit the concept with another activity as soon as possible.

Try not to repeat students' answers. If students are to listen to one another and value one another's input, I cannot repeat or try to improve on what they say. If students realize that I will repeat or clarify what another student says, they no longer have a reason to listen. I must be patient and let students clarify their own thinking and encourage them to speak to their classmates, not just to me. All students can speak louder—I have heard them in the halls! Yet I must be careful not to embarrass someone with a quiet voice. Because students know that I never accept just one response, they think nothing of my asking another student to paraphrase the soft-spoken comments of a classmate.

"Is this the right answer?" Students frequently ask this question. My usual response to this question might be that "I'm not sure. Can you explain your thinking to me?" As soon as I tell a student that the answer is correct, thinking stops. If students explain their thinking clearly, I ask a "What if?" question to encourage them to extend their thinking.

Participation is not optional! I remind my students of this expectation regularly. Whether working in small groups or discussing a problem with the whole class, each student is expected to contribute his or her fair share. Because reminding students of this expectation is not enough, I also regularly apply several of the following techniques:

1. Use the think-pair-share strategy. Whole-group discussions are usually improved by using this technique. When I pose a new problem; present a new project, task, or activity; or simply ask a question, all students must think and work independently first. In the past, letting students begin working together on a task always allowed a few students to sit back while others took over. Requiring students to work alone first reduces this problem by placing the responsibility for learning on each student. This independent work time may vary from a few minutes to the entire class period, depending on the task.

    After students have had adequate time to work independently, they are paired with partners or join small groups. In these groups, each student is required to report his or her findings or summarize his or her solution process. When teams have had the chance to share their thoughts in small groups, we come together as a class to share our findings. I do not call for volunteers but simply ask one student to report on a significant point discussed in the group. I might say, "Tanya, will you share with the class one important discovery your group made?" or "James, please summarize for us what Adam shared with you." Students generally feel much more confident in stating ideas when the responsibility for the response is being shared with a partner or group. Using the think-pair-share strategy helps me send the message that participation is not optional.

    A modified version of this strategy also works in whole-group discussions. If I do not get the responses that I expect, either in quantity or quality, I give students a chance to discuss the question in small groups. On the basis of the difficulty of the question, they may have as little as fifteen seconds or as long as several minutes to discuss the question with their partners. This strategy has helped improve discussions more than any others that I have adopted.

Let students clarify their own thinking

2. If students or groups cannot answer a question or contribute to the discussion in a positive way, they must ask a question of the class. I explain that it is all right to be confused, but students are responsible for asking questions that might help them understand.

3. Always require students to ask a question when they need help. When a student says, "I don't get it," he or she may really be saying, "Show me an easy way to do this so I don't have to think." Initially, getting students to ask a question is a big improvement over "I don't get it." Students soon realize that my standards require them to think about the problem in enough depth to ask a question.

4. Require several responses to the same question. Never accept only one response to a question. Always ask for other comments, additions, clarifications, solutions, or methods. This request is difficult for students at first because they have been conditioned to believe that only one answer is correct and that only one correct way is possible to solve a problem. I explain that for them to become better thinkers, they need to investigate the many possible ways of thinking about a problem. Even if two students use the same method to solve a problem, they rarely explain their thinking in exactly the same way. Multiple explanations help other students understand and clarify their thinking. One goal is to create a student-centered classroom in which students are responsible for the conversation. To accomplish this goal, I try not to comment after each response. I simply pause and wait for the next student to offer comments. If the pause alone does not generate further discussion, I may ask, "Next?" or "What do you think about _____'s idea?"

5. No one in a group is finished until everyone in the group can explain and defend the solution. This rule forces students to work together, communicate, and be responsible for the learning of everyone in the group. The learning of any one person is of little value unless it can be communicated to others, and those who would rather work on their own often need encouragement to develop valuable communication skills.

6. Use hand signals often. Using hand signals—thumbs up or thumbs down (a horizontal thumb means "I'm not sure")—accomplishes two things. First, by requiring all students to respond with hand signals, I ensure that all students are on task. Second, by observing the responses, I can find out how many students are having difficulty or do not understand. Watching students' faces as they think about how to respond is very revealing.

7. Never carry a pencil. If I carry a pencil with me or pick up a student's pencil, I am tempted to do the work for the student. Instead, I must take time to ask thought-provoking questions that will lead to understanding.

8. Avoid answering my own questions. Answering my own questions only confuses students because it requires them to guess which questions I really want them to think about, and I want them to think about all my questions. I also avoid rhetorical questions.

9. Ask questions of the whole group. As soon as I direct a question to an individual, I suggest to the rest of the students that they are no longer required to think.

10. Limit the use of group responses. Group responses lower the level of concern and allow some students to hide and not think about my questions.

11. Do not allow students to blurt out answers. A student's blurted out answer is a signal to the rest of the class to stop thinking. Students who develop this habit must realize that they are cheating other students of the right to think about the question.

Summary

Like most teachers, I entered the teaching profession because I care about children. It is only natural for me to want them to be successful, but by merely telling them answers, doing things for them, or showing them shortcuts, I relieve students of their responsibilities and cheat them of the opportunity to make sense of the mathematics that they are learning. To help students engage in real learning, I must ask good questions, allow students to struggle, and place the responsibility for learning directly on their shoulders. I am convinced that children learn in more ways than I know how to teach. By listening to them, I not only give them the opportunity to develop deep understanding but also am able to develop true insights into what they know and how they think.

Making extensive changes in curriculum and instruction is a challenging process. Much can be learned about how children think and learn, from recent publications about learning styles, multiple intelligences, and brain research. Also, several reform curriculum projects funded by the National Science Foundation are now available from publishers. The Connected Mathematics Project, Mathematics in Context, and Math Scape, to name a few, artfully address issues of content and pedagogy.

BIBLIOGRAPHY

Burns, Marilyn. *Mathematics: For Middle School*. New Rochelle, N.Y.: Cuisenaire Co. of America, 1989.

Johnson, David R. *Every Minute Counts*. Palo Alto, Calif.: Dale Seymour Publications, 1982.

National Council of Teachers of Mathematics (NCTM). *Professional Standards for Teaching Mathematics*. Reston, Va.: NCTM, 1991.

Redefining Success in Mathematics Teaching and Learning

Margaret Schwan Smith

A MAJOR goal of current reform efforts is to help students learn mathematics with understanding. "Good" mathematical tasks are an important starting point for developing mathematical understanding, but selecting and setting up good tasks does not guarantee a high level of student engagement (Smith and Stein 1998). Using such tasks can, and often does, present challenges for teachers and students.

Teachers, accustomed to establishing rules and procedures for students to follow, may have difficulty with a more nebulous set of responsibilities aimed at supporting students as they construct mathematical knowledge. Students may become frustrated with tasks that they cannot immediately solve and pressure the teacher to show them "how to do it." As a result, "good" tasks are often carried out in ways that remove the opportunities for problem solving and sense making (Doyle 1988; Stein, Grover, and Henningsen 1996) and reduce students' chances to engage in meaningful learning of mathematics (Stein and Lane 1996).

This situation occurs because teachers think that frustration and lack of immediate success are indicators that they have somehow failed their students Teachers have no way to measure their own success when teaching is no longer defined as providing explanations and procedures or evaluating students' learning by the correctness of their solutions. According to J. P. Smith (1996), more traditional "teaching and telling" gives teachers a sense of efficacy—a perception that they have had a positive impact on students' learning—that is undermined by current efforts to reform mathematics instruction. Smith suggests that we need to establish "new moorings" for efficacy that are closely related to reform-oriented teaching to ensure that teachers will sustain their commitment to new ways of teaching.

What would constitute these "new moorings" to which Smith refers, and how will teachers come to establish these new indicators of success? Elaine Henderson (name changed), a teacher who participated in the QUASAR project, is an interesting case for exploring this question. QUASAR, which stands for Quantitative Understanding: Amplifying Student Achievement and Reasoning, was a national project funded by the Ford Foundation to improve mathematics instruction for students attending middle schools in economically disadvantaged communities (Silver and Stein 1996; Silver, Smith, and Nelson 1995). A study of Henderson's teaching practice during her first year of implementing this innovative mathematics curriculum revealed that supporting students' engagement with "good" mathematical tasks required her to redefine what it meant for both herself and her students to be successful in mathematics class (Smith [2000]). Over time, Henderson began to establish for herself new indicators that she was, in fact, having a positive impact on students' learning. I encourage you to reflect on your own experiences as you consider Henderson's practice at the beginning and end of her first year of "doing reform" in light of her changing views of success.

Before Reform

Before her involvement in the QUASAR project, Henderson's goal in teaching mathematics was to ensure that her students would successfully learn the algorithms that

REFLECTION

What challenges have you faced in incorporating good mathematical tasks in the classroom?

• • • • • • •

What do you take as evidence of your success as a teacher of mathematics? What does it mean for students to be successful in your mathematics class?

Mathematics Teaching in the Middle School 5 (February 2000): 378–86

they needed. She explained the procedures to be learned, demonstrated a small number of sample problems, monitored students' completion of a few problems, and had students work individually on a larger set of similar problems, using the strategies that she had taught.

Henderson had always considered herself to be a successful mathematics teacher—her students did well on the district standardized tests, teachers in subsequent grades who had her students always remarked that they were well prepared, and parents often requested that their children be placed in her classroom. Henderson saw herself as someone who related well to students and was able to motivate them to learn. She was always on the lookout for new ways to help her students experience success.

Henderson made the decision to participate in the QUASAR project in spring 1990 because she was intrigued by its approach to teaching mathematics, which emphasized thinking and reasoning and encouraged collaboration among students. She believed that such an approach could be beneficial to students, and she was eager to try something new. Henderson had always liked problem solving and found the tasks in the new curriculum to be more interesting than ones that she had typically used in the past. She also liked the fact that the project would give her the opportunity to work closely with her colleagues in mathematics and with teacher educators at a local university who would support the teachers' implementation efforts.

Making a Change

Henderson spent several weeks working through the new curriculum with her school and university colleagues and found that she learned much about mathematics in the process. She was eager to give her students the same opportunity for learning. Early in the school year, however, Henderson began to notice that students were struggling with the tasks in the new curriculum. If they could not solve a problem immediately, which was often the case, they would just say, "I don't know" and give up. Henderson was concerned that her students were not actively participating in class. She believed that if students were not involved, they would not learn and that if they did not feel successful, they would lose their motivation to stay involved. As Henderson commented, "[I will] guarantee them success by asking them to do things they couldn't fail to do right. I can't ignore that success breeds success. Too many are starting out with what I'm sure they perceive to be failure."

REFLECTION

How might Henderson's alterations of tasks affect the level of challenge in the task, the kind of thinking required of students, or the mathematics that they might learn?

To ensure success, Henderson began altering problems from the curriculum. At times, she put in an extra step or took out something that she thought was too hard; she rewrote problem instructions to be clearer and created easier problems to lead up to more challenging ones. In addition, during classroom instruction, Henderson often "broke a task down" into a set of subtasks, each of which the students could complete successfully. Henderson believed that the more correct answers a student gets, the greater the learning he or she experiences. She tended to ask questions that were, as she explained, "not designed for deep thinking, just success."

A lesson on exploring patterns that took place in late October furnishes an example of how Henderson started supporting students' involvement with the new tasks. The curriculum materials suggested providing students with the first three trains in a pattern sequence, then asking students to build the fourth train. Next, the students were asked to build a larger train in the sequence, such as the tenth or fifteenth, without building all the trains in between, then discuss why they believed that the tenth or fifteenth train looked as it did. She presented the pattern shown in **figure 1** and gave student partners eleven minutes to write down what they noticed about each of the trains, which was a deviation from what the curriculum materials suggested. Henderson then asked students to share their observations orally and at times sent a student to the overhead projector to demonstrate an observation using overhead pattern tiles.

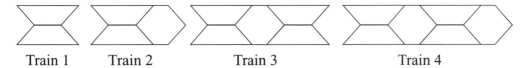

Train 1   Train 2   Train 3   Train 4

Fig. 1. Train pattern

Some of the observations made by students were nonmathematical (e.g., "[the first train] looks like a squished pop can," "[the third train] looks like a belt buckle"). Others showed a basic understanding of a mathematical concept that pertained to only one train (e.g., "[the first train] has four equal sides," "[the second train has] two trapezoids and a hexagon"). A few attempted to look across trains (e.g., "they will have two [sides] that are equal and four [sides] that are equal," "trapezoids are double the number [of the train, if you replace the hexagons with trapezoids]"). Through teacher questioning, two generalizations emerged during the lesson: (1) for the even-number trains, the number of trapezoids was equal to the train number and (2) for the even-number trains, the number of hexagons was half the train number. Consider the following excerpt from the class in which Henderson tries to make sure that all her students "see" this pattern:

Ms. H. Let's do a couple more. Listen to what he's saying, and see if you can do it also. Charles, in train 12, how many red trapezoids will there be?

Charles. Twelve.

Ms. H. And then how many hexagons will there be?

Charles. Six.

Ms. H. Can you do it for any even number I give you? Tony, What's he doing?

Tony. On the hexagons, he's doing the same number as the problem.

Ms. H. On the what? Trapezoids or hexagons?

Tony. Trapezoids.

Ms. H. The number of trapezoids is the same number as the trains.

William. And the hexagons, he's taking it in half.

Ms. H. Take it in half for the hexagons, and it's the same for the other. OK, let's everybody try one. I can pick any even number: train 50. How many trapezoids will there be? Everybody!

SS. Fifty.

Ms. H. How many hexagons?

SS. Twenty-five.

Ms. H. Train 20, how many trapezoids?

SS. Twenty.

Ms. H. How many hexagons?

SS. Ten.

Ms. H. Train 100, how many trapezoids?

SS. One hundred.

Ms. H. How many hexagons?

SS. Fifty.

Ms. H. All right, look up there. On train 50, will it end with two trapezoids or will it end with a hexagon?

SS. Hexagon.

Ms. H. How do you know that? What up there tells you that? Michele, I haven't heard from you in a while.

Michele. The odds are the trapezoid, and evens are the hexagon.

Ms. H. So you're saying to me that any number I tell you, you can tell me how it ends? Eleven.

Michele. It's odd. It would be trapezoid.

When the class was over, Henderson commented, "The lesson was all I could have asked of the kids . . . it was very exciting." She thought that the students had remained engaged throughout the period, the choral responses gave all the students an opportunity to feel good about themselves, and everyone seemed to be able to predict future trains using the generalizations.

Reflecting on the Lesson a Few Weeks Later

At regular intervals over the course of the year, the classes of Henderson and her colleagues were videotaped. The videotapes and related observations were part of the ongoing efforts of the QUASAR project to document classroom practices. The video-tapes were also used by teachers to reflect on their practices. At a staff-development session, Henderson volunteered to show the clip of the videotaped segment from the October lesson featuring the pattern-block sequence shown in **figure 1** and share their thoughts about the lesson. She commented that she had discovered that her students were not very good at observations, but she thought that they were definitely verbalizing more when she "broke it down" for them, focusing their attention on each part of a sequence rather than on the entire series.

One of the university teacher educators who was working with Henderson and her colleagues asked Henderson whether the students had progressed to the point at which they could make observations without her "breaking it down." The educator asked Henderson how long teachers need to break tasks down for students and whether Henderson thought that some of the observations would come out naturally if the students were given time and opportunity to develop them. Henderson responded that she hoped she did not always have to structure things but that her students were still at the "comfort stage" and needed this support. Once students experienced success, Henderson believed that they would try harder and would no longer need to have their learning experiences structured in this way.

The exchange with the educator appeared to have a powerful impact on Henderson. In her journal later that day, she reflected, "I need to make sure I'm not structuring too much. It is easy to be too leading and feel OK about it because the kids seem happy. After all, many kids are happy with 'shut up and add.'" Over time, Henderson continued to question her approach, wondering whether structuring the learning opportunities to ensure success would help students become competent and confident problem solvers—a central goal of the new curriculum—or whether the structuring would remove important challenging aspects from the students' mathematical experiences.

In an attempt to provide more problem-solving opportunities, during a quiz a few days later, Henderson asked students to divide a square into four congruent parts in five different ways. After students had found two solutions, they claimed that the problem could not be solved in other ways and started to give up. Henderson encouraged them to try different things to find the answer and explained to them "that they weren't really doing problem solving until they reached the point where they thought they couldn't do it." As she noted, "I'm used to the idea that their confusion means that I haven't intro-duced the lesson properly or have given kids something too hard. Sometimes that is true, but sometimes it is necessary to go through panic before we find solutions." Later that month, Henderson watched the entire videotape of the October pattern train lesson again and wrote the following entry in her journal:

> Students had ample opportunity to successfully predict visual pattern-block trains in
> this lesson, but it was set up too much for success and not enough for the frustration that

Do you agree
with Henderson's
assessment of how
the lesson went?
Why or why not?
What questions
might you ask of
Henderson to find
out more about
whether the lesson
was successful?

goes with problem solving. Unfortunately, the lesson contained too much whole-group teacher questioning and students' explaining and not enough time for students to stretch and discover independently/collaboratively. I made the lesson safe for the kids—no fail—which was my goal at the time. I now think I need to let them go through the frustration that goes with problem solving. The lesson probably wouldn't have looked as smooth, but I think it would have stretched the kids more. I am at a different point in my thinking than I was at the time of the lessons.

Redefining Success

Throughout the year, Henderson continued to struggle with her feelings of failure, concerns about her students' need for success, and her commitment to giving her students genuine problem-solving opportunities. For Henderson, redefining success required that she create new expectations for students of what it means to know and do mathematics and new expectations for her role in supporting students' learning. She came to believe that daily success should be measured by the extent to which students meet the expectations that she set for them and the extent to which she supported students in meeting those expectations. Henderson was effectively establishing "new moorings" for her own teaching of mathematics.

A series of lessons that Henderson conducted the next spring offers some insight into how these new moorings played out in her classroom. When the students were working in small groups or pairs on challenging problems, Henderson encouraged them to use diagrams and sketches as tools for solving problems, insisted that they be able to explain how they figured out problems, and encouraged them to consider alternative strategies. Many of Henderson's questions focused on trying to understand what students were doing as they solved the problems (e.g., "How did you get that?" "Why do you think that?" "What's happening here?") and on encouraging students to communicate (e.g., "Can you explain it to him?" "Why don't you ask your partner what she was doing first?"). When students presented their solutions at the overhead projector, the class was responsible for understanding the solutions and asking questions that would illuminate any errors in the approach. The teacher's periodic statement, "I assume that everyone understands if you have no questions," was a way of reminding students that the responsibility for understanding was theirs. Students assumed that they had given correct answers and had appropriately justified their solutions if no questions arose from their peers or the teacher, and this assumption was, in fact, true.

Although some problems were particularly difficult for students, rather than guide them step-by-step through the frustration, Henderson would ask one student to begin a problem at the overhead projector but not complete it. Henderson monitored the progress of the class but left her students to make sense of the presenter's explanation and determine how it would further their work on the problem.

Table 1 summarizes Henderson's new expectations for her students, her approach to supporting students in meeting the new expectations, and what she took as classroom-based evidence that students were successful. Classroom-based indicators of success, such as those found in **table 1**, present early evidence that changes in practice have occurred, which is a necessary first step before changes in student learning outcomes can be expected. In Henderson's situation, data collected on student performance showed that her students had grown over the year, as measured by a standardized test and an innovative performance assessment. By the end of the school year, Henderson thought that students were meeting the expectations that she had set for them and that she had done a good job in building a scaffolding for her students to support high-level thinking and reasoning.

REFLECTION

How would you say that Henderson is now defining success in mathematics teaching and learning? What appears to have changed about her definition of successful mathematics teaching and learning since the October lesson?

Table 1

Key Elements in Henderson's Efforts to Redefine Success for Herself and Her Students

REFLECTION

What can you conclude from the spring lessons about Henderson's new expectations for her students? How was she supporting students to enable them to meet her new expectations?

New Expectations for Students	Teacher Actions Consistent with Expectations	Classroom-Based Indicators of Success
Most "real" tasks take time to solve; frustration may occur; perseverance in the face of initial difficulty is important.	Use "good" tasks; explicitly encourage students to persevere; find ways to support students without removing all the challenges in a task.	Students engaged in the tasks and did not give up too easily. The teacher supported students when they "got stuck" but did so in a way that kept the task at a high level.
Correct solutions are important, but so is being able to explain how you thought about and solved the task.	Ask students to explain how they solved a task. Make sure that the quality of the explanations is valued equally as part of the final solution.	Students were able to explain how they solved a task.
Students have a responsibility and an obligation to make sense of mathematics by asking questions when they do not understand and by being able to explain and justify their solutions and solution paths when they do understand.	Give students the responsibilty for asking questions when they do not understand, and have students determine the validity and appropriateness of strategies and solutions.	With encouragement, students questioned their peers and provided mathematical justifications for their reasoning.
Diagrams, sketches, and hands-on materials are important tools for students to use in making sense of tasks.	Give students access to tools that will support their thinking processes.	Students were able to use tools to solve tasks that they could not solve without them.
Communicating with others about your thinking during a task makes it possible for others to help you make progress on the task.	Ask students to explain their thinking, and ask questions that are based on students' reasoning, as opposed to how the teacher is thinking about the task.	Students explained their thinking about a task to their peers and the teacher. The teacher asked probing questions based on the student's thinking.

Sharing Your Reflections

The purpose of this article is to raise questions about what it means to be successful as a teacher and as a student in reform-inspired mathematics classrooms. For Henderson, this questioning meant a year-long process of reflection on her practice and a redefinition of the factors that contribute to success. It is important to note that Henderson was not alone on her journey. She had the invaluable support of her teacher and university colleagues. I encourage you to discuss with your colleagues the issues raised in this article and to share […] your perspective on what it means to be successful.

REFERENCES

Doyle, Walter. "Work in Mathematics Classes: The Context of Students' Thinking during Instruction." *Educational Psychologist* 23 (February 1988): 167–80.

Silver, Edward, and Mary Kay Stein. "The QUASAR Project: The 'Revolution of the Possible' in Mathematics Instructional Reform in Urban Middle Schools." *Urban Education* 30 (January 1996): 476–521.

Silver, Edward, Margaret S. Smith, and Barbara Nelson. "The QUASAR Project: Equity Concerns Meeting Mathematics Education Reform in the Middle School." In *New Directions for Equity in Mathematics Education,* edited by Walter Secada, Elizabeth Fennema, and Lisa Adajian, pp. 9–56. New York: Cambridge University Press, 1995.

Smith, John P. "Efficacy and Teaching Mathematics by Telling: A Challenge to Reform." *Journal for Research in Mathematics Education* 27 (July 1996): 387–402.

Smith Margaret S. "Balancing Old and New: An Experienced Middle School Teacher's Learning in the Context of Mathematics Instructional Reform." *Elementary School Journal* [100 (March 2000): 351–75].

Smith, Margaret S., and Mary Kay Stein. "Selecting and Creating Mathematical Tasks: From Research to Practice." *Mathematics Teaching in the Middle School* 3 (February 1998): 344–50.

Stein, Mary Kay, and Suzanne Lane. "Instructional Tasks and the Development of Student Capacity to Think and Reason: An Analysis of the Relationship between Teaching and Learning in a Reform Mathematics Project." *Educational Research and Evaluation* 2 (October 1996): 50–80.

Stein, Mary Kay, Barbara W. Grover, and Marjorie Henningsen. "Building Student Capacity for Mathematical Thinking and Reasoning: An Analysis of Mathematical Tasks Used in Reform Classrooms." *American Educational Research Journal* 33 (October 1996): 455–88.

This manuscript is based on the author's dissertation.... The author wishes to acknowledge the helpful comments of Marjorie Henningsen on a previous draft of this article.

Signposts for Teaching Mathematics through Problem Solving

James Hiebert

ALL students need a deep, rich understanding of the mathematics they study in school. It is not enough for students to know how to calculate the area of a rectangle, how to add fractions, or how to find percents, or even to memorize procedures for all these things and execute them with blinding speed. It is not enough, because knowing how to execute procedures does not ensure that students understand what they are doing. And unless students understand what they are doing, these procedures will not be very useful. Understanding is the key to remembering what is learned and being able to use it flexibly.

If, as the authors in this book emphasize, understanding mathematics is of utmost importance, why is the book about problem solving? In simplest terms, problem solving leads to understanding (Davis 1992). Students develop, extend, and enrich their understandings by solving problems. Some readers might find this approach odd and inefficient. Problem solving takes time. And if the problems really are problems, some students might not even solve them completely. Why not just teach students the concepts you want them to understand? The answer to this question is not simple. If it were simple, this book would not have been written. Understanding is best supported through a delicate balance among engaging students in solving challenging problems, examining increasingly better solution methods, and providing information for students at just the right time (Brownell and Sims 1946; Dewey 1933; Hiebert et al. 1997). Because the traditional teaching approach often has tipped the balance toward telling students too quickly how to solve problems, educators must correct the balance by thinking again about how to allow students to do more of the mathematical work. Students' understanding depends on it.

> Allow mathematics to be problematic for students

Signposts for Classrooms That Promote Students' Understandings

On the basis of observations of experts and the convergence of research evidence, one can identify several signposts to guide classroom teachers in the direction of giving students opportunities to develop deep mathematical understandings. Just as signposts along the road can highlight for travelers important information for reaching their destination, so signposts for the classroom can highlight for teachers essential features for helping students achieve the intended learning goals. If the learning goals are deep mathematical understandings, then the signposts all point to problem solving as the core activity. They stake out a new kind of balance among allowing students to struggle with challenging problems, helping them examine increasingly better solution methods, and providing appropriate information at the right times. The following paragraphs discuss three signposts; other chapters in this book point out additional signposts.

Signpost 1: Allow Mathematics to Be Problematic for Students

The idea that mathematics should be problematic for students is the most radical of the signposts. It is radical because it is very different from how most of us have thought about mathematics and students. Teachers have been encouraged to make mathematics less problematic for students. Parents assume that teachers should make mathematics less of a struggle; a good teacher helps students learn in a smooth, effortless way.

Teaching Mathematics through Problem Solving, Lester & Charles, eds. (Reston, Va.: NCTM, 2003, chapter 4)

Allowing mathematics to be problematic for students requires a very different mindset about what mathematics is, how students learn mathematics with understanding, and what role the teacher can play. Allowing mathematics to be problematic for students means posing problems that are just within students' reach, allowing them to struggle to find solutions, and then examining the methods they have used. Allowing mathematics to be problematic requires believing that all students need to struggle with challenging problems to learn mathematics and understand it deeply. Allowing mathematics to be problematic does not mean making mathematics unnecessarily difficult, but it does mean allowing students to wrestle with what is mathematically challenging.

For many mathematics teachers, this way of thinking is new. They have learned that they are supposed to step in and remove the struggle, and the challenge, for students. The extent to which U.S. teachers hold this belief is revealed in a study of classroom mathematics teaching as part of the Third International Mathematics and Science Study (TIMSS). Margaret Smith (2000) looked at a subsample of eighth-grade lessons from Germany, Japan, and the United States and examined the kinds of problems teachers presented to students and the way in which they helped students solve the problems. She found, in her sample, that about one-third of the problems presented in U.S. classrooms offered students the opportunity to explore relationships and develop deeper understandings. This percent was not dramatically different from the ones found in other countries. However, after presenting the problems, U.S. teachers almost always stepped in to show students how to solve them; the mathematics they left for students to think about and do was rather trivial. Teachers in the other two countries allowed students more opportunities to wrestle with the challenging aspects of the problems.

As will be pointed out in other chapters (see, e.g., those by Van de Walle and by Russell et al.), allowing mathematics to be problematic (or "problem-based," to use the term as Van de Walle prefers) for students does not require importing numerous special problems for students. Rather, it requires allowing the problems that are taught every day to be *problems*. For example, suppose third graders have just learned to solve problems like 324 − 156 = ? and then encounter 402 − 258 = ? Ordinarily, this problem is treated as a new one requiring new procedures, and teachers step in and show students how to solve problems with zeros in the subtrahend. Allowing mathematics to be problematic means allowing even these simple arithmetic problems to be real problems for students (Carpenter 1985; Hiebert et al. 1997).

This first signpost clearly changes everything. It affects the way in which all mathematics is treated, even the routine computation contained in the preceding examples. It points classroom practice in a new direction. It means that solving problems is the heart of doing mathematics, not a supplement to one's ongoing program. Allowing students to learn by solving problems leads to the next signpost: focusing classroom activity on the methods used to solve problems.

Focus on the methods used to solve problems

Signpost 2: Focus on the Methods Used to Solve Problems

John Dewey (1933) pointed out that the best way to gain deeper understandings of a subject is to search for better methods to solve problems. From a student's point of view, this quest requires the opportunity to share one's own method, to hear others present alternative methods, and then to examine the advantages and disadvantages of these different methods.

Suppose that a third-grade class is solving the problem posed previously: 402 − 258 = ? If the class has not solved a problem with a zero in the subtrahend before, they will likely produce several different solution methods. If they have base-ten materials, some students might break up one of the hundreds into 10 tens, then break up one of the tens into 10 ones, and then subtract 258, leaving 144. Other students might use a (flawed) written procedure and "borrow" 1 from the 4 (hundred); then change the 2 (ones) to 12, leaving the 0 unchanged; and then subtract to get 154. Still other students might add up from 258 to 402, counting the amount they had to add on and getting 144. Other methods will probably be suggested as well.

Learning opportunities for these students begin as they search for a method to solve the problem. Learning opportunities continue as they formulate a way to explain their methods to their classmates and justify their validity. Learning opportunities intensify as students listen to the methods of others and examine them, considering their relative advantages and deciding whether the alternatives provide better choices that they can adopt for subsequent problems.

The process of analyzing the adequacy of methods and searching for better ones drives the intellectual, and social, life of the class. Students should be permitted to choose their own method to solve problems but should commit themselves to searching for better ones. Discussions in class should revolve around sharing, analyzing, and improving methods. The focus always should be on the merits of the method, not the status of the presenter. Whether presented by the teacher or a student, correct methods should win popularity because of their mathematical advantages—they are efficient, or easy to understand, or easy to adjust to solve new problems, and so on.

Why place so much emphasis on examining and improving the methods used to solve problems? The payoffs are substantial. A first benefit is that examining methods encourages students to construct mathematical relationships, and constructing relationships is at the heart of understanding (Brownell 1947). Returning to the three-digit subtraction problem, examining the different methods that are likely to be presented provides a perfect opportunity to look again at how numbers can be decomposed and recomposed using hundreds, tens, and ones, and how subtraction can be conceptualized as taking away, finding a difference, or adding on. The problem presents many relationships to construct, relationships that extend students' understanding well beyond this particular problem.

A second benefit of focusing on methods is that students can learn from analyzing a range of methods, from flawed to primitive to sophisticated. By learning why some methods do not work, students can gain special insights that deepen their understanding and prevent them from making similar mistakes in the future. For example, examining the incorrect written procedure mentioned previously of "borrowing" 1 from the hundreds and making it 10 ones (changing 2 ones to 12 ones) presents another opportunity to think about how numbers can be decomposed and recomposed. Subtracting 5 from 0 to get 5 presents a chance to discuss the meaning of 0 as well as the "take away" meaning for subtraction. In short, focusing on methods creates an environment in which mistakes become sites for learning. This aspect is important because when mathematics is allowed to be problematic for students, making mistakes becomes a natural part of learning.

Because students learn from analyzing correct and incorrect methods, the class benefits from hearing a variety of methods. Variety might be more likely when the class is made up of a diverse population of students than when it is homogeneously grouped. Students with different backgrounds and different achievement levels are likely to think of the problem in different ways and produce different methods of solution. The interesting implication is that individual differences in a classroom become a resource that can enhance instruction rather than a difficulty that hinders good instruction.

A third benefit of focusing on methods is that the spotlight shifts from people to ideas. By focusing on methods and the ideas they contain, teachers can show students that all methods and ideas are sites for learning. Every student's contribution can help the class think about ways to improve the correct methods and avoid incorrect ones.

When students see that the goal is to help the class, as a group, search for better methods and construct new mathematical relationships, the attention shifts from evaluating their response as correct or not toward examining what others can learn from the response. This focus is important because it provides a way for teachers to build a classroom culture in which all students feel welcome and appreciated. If mathematics is allowed to be problematic for students and they are encouraged to explain and justify their methods and examine the methods of others, students might become self-conscious and withdraw from the discussions (Lampert, Rittenhouse, and Crumbaugh 1996). Teachers should create an environment in which students enjoy participating as respected members of the group.

Signpost 3: Tell the Right Things at the Right Times

Tell the right things at the right times

A third signpost that can guide classrooms toward providing opportunities for students to develop deep understandings is best phrased as a question: What mathematical information should teachers present, and when should they present it? For teachers using a traditional approach, this signpost is curious. They wonder, Doesn't good teaching mean presenting mathematical information clearly? Why even ask such questions as "What should be presented?" or "When should it be presented?" Shouldn't all mathematical information be presented as it comes along in the curriculum?

The fact that telling the right thing at the right time even becomes an issue shows how radical the first signpost is. Allowing the mathematics to be problematic for students changes everything. In this environment, teachers must think carefully about what information should be shared and when it should be shared. Presenting too much information too soon removes the problematic nature of problems. Presenting too little information can leave students floundering.

John Dewey faced the same situation in the 1920s. He had recommended that students be allowed to problematize their school subjects. Some educators interpreted this statement to mean that students should not be told anything. Dewey (1933) tried to correct this misinterpretation by saying that although teachers should not present "ready-made intellectual pabulum to be accepted and swallowed" and later regurgitated (p. 257), in many cases teachers can and should provide information for students.

Here are a few rules of thumb for telling students the right things at the right time. First, teachers should show students the words and written symbols that commonly are used to represent quantities, operations, and relationships (e.g., written notation for fractions, decimals, and percents; formats for writing equations; words such as *quotient* and *equivalent*). These are social conventions, and students cannot be expected to discover them. When should teachers present them? The best time is when students need them—when the ideas have been developed and students need a way to record the ideas and communicate with others about them. Rather than burden students with memorizing these conventions, the teacher should present them as beneficial aids.

A second rule of thumb is that teachers can present alternative methods of solutions that have not been suggested by students. Over time, teachers can develop a good sense for which solution methods for particular problems help students understand the main ideas and relationships that are contained in a problem. If students do not come up with these methods, teachers should feel free to present them.

Suppose that third graders are finding the area of a rectangle, say, 4 inches by 6 inches. After working on the problem for a few minutes, some students suggest using their square-inch pieces to cover the rectangle and counting the squares by ones; other students suggest counting the squares by fours; and one student suggests multiplying 4 × 6 (the class recently had been solving simple multiplication problems). After discussing these methods, the teacher suggests a slightly different one: counting the squares by sixes. This additional method allows the teacher to then lead a discussion comparing counting by fours and counting by sixes, which, in turn, introduces the concept of commutativity, including how 4×6 and 6×4 capture the act of counting and the ways in which they are the same and different.

Teachers should feel free to reveal alternative methods during the discussion of problems that students have already solved. The trick is for the teacher to present the method as one that students should examine, just as they examine other methods, rather than a method that is automatically preferred just because the teacher presented it. The teacher should ensure that students use a method because they understand why it is correct and because they find it useful.

A third rule of thumb for presenting information is that teachers should highlight the mathematical ideas embedded in students' methods. Students can invent and present methods for solving problems without being aware of all the ideas on which the methods depend. Consider again the third graders finding the area of a 4-by-6 rectangle. Many ideas and relationships are embedded in the various solution methods, and the teacher can make these ideas explicit. By restating and clarifying students' methods and pointing to the important ideas in them, teachers not only show respect for students' thinking but help guide students' attention to the important mathematics, thereby guiding the mathematical direction of the class.

How Different Is This Approach from Traditional Practice?

The kind of classroom practice described in this chapter, and in this entire volume, represents a fundamental change from business as usual. It will not be achieved by making superficial changes. It will be achieved only by rethinking our beliefs about two issues that lie at the core of mathematics teaching: the learning goals we set for students and how students can best achieve these goals.

> **This kind of classroom practice represents a fundamental change from business as usual**

This chapter is built on the premise that the traditional learning goals for students must be expanded and reshaped to include a deep understanding of mathematics (National Research Council 2001). Understanding has long been advertised as a goal for students. Many teachers obviously would like their students to understand the mathematics they study, but when asked to specify the goal for a particular lesson, most U. S. teachers in the TIMSS Video Study talked about skill proficiency; few mentioned understanding (Hiebert and Stigler 2000).

Valuing deep understanding of mathematics as an important goal for students is the first step. The second step is to help students achieve this goal. One way to support students' efforts to understand is to allow mathematics to be problematic for them. Why has this point received major emphasis in this chapter? It is based on a theory about how students construct understanding that is very different from the beliefs that many people hold about how students learn. In particular, the theory places great importance on the role of *struggle* in developing understanding. As mentioned previously, many teachers believe that their job is to remove struggles from students' learning experiences. But struggling, in a healthy sense and on the right kinds of problems for an appropriate amount of time, prepares students to make sense of relevant information, to piece together ideas in new ways, and to see the benefits of better methods of solution.

An important aspect of this theory is that deep understanding develops over time. Although quick insights can occur—sometimes while working on a single problem—significant and lasting understanding of mathematics develops gradually and accumulates over time as students solve increasingly challenging problems.

The changes called for in this book are fundamental. Teachers who take the recommendations seriously will face many obstacles as they revise their instruction to help students develop understanding through solving problems (Ball 1993; Lampert 2001;

Schifter and Fosnot 1993). But the changes need not occur overnight and all at once. Change of this importance often happens in small steps, small steps that build on one another. Small successes can yield dramatic improvement if they are saved and shared and accumulate over time. Steady, gradual improvement often is more lasting than overnight reform.

Some encouraging signs are appearing. Many teachers are embracing new learning goals for students, creating for themselves new images of practice, and displaying rich models of what is possible in classrooms. If the mathematics education community can create a system in which teachers can record, accumulate, and share these images and models with others, classroom practice throughout the country might gradually become more aligned with the visions portrayed in this book.

> Small successes can yield dramatic improvement if they are saved and shared and accumulate over time

REFERENCES

Ball, Deborah L. "With an Eye on the Mathematical Horizon: Dilemmas of Teaching Elementary School Mathematics." *Elementary School Journal* 93 (March 1993): 373–97.

Brownell, William. A. "The Place of Meaning in the Teaching of Arithmetic." *Elementary School Journal* 47 (January 1947): 256–65.

Brownell, William A., and Verner M. Sims. "The Nature of Understanding." In *Forty-fifth Yearbook of the National Society for the Study of Education, Part I: The Measurement of Understanding,* edited by Nelson B. Henry, pp. 27–43. Chicago: University of Chicago, 1946.

Carpenter, Thomas P. "Learning to Add and Subtract: An Exercise in Problem Solving." In *Teaching and Learning Mathematical Problem Solving: Multiple Research Perspectives,* edited by Edward A. Silver, pp. 17–40. Hillsdale, N.J.: Lawrence Erlbaum Associates, 1985.

Davis, Robert B. "Understanding 'Understanding.' " *Journal of Mathematical Behavior* 11 (1992): 225–41.

Dewey, John. *How We Think: A Restatement of the Relation of Reflective Thinking to the Educative Process.* Boston: D.C. Heath & Co., 1933.

Hiebert, James. "Relationships between Research and the NCTM Standards." *Journal for Research in Mathematics Education* 30 (January 1999): 3–19.

Hiebert, James, Thomas P. Carpenter, Elizabeth Fennema, Karen Fuson, Diana Wearne, Hanlie Murray, Alwyn Olivier, and Piet Human. *Making Sense: Teaching and Learning Mathematics with Understanding.* Portsmouth, N.H.: Heinemann, 1997.

Hiebert, James, and James W. Stigler. "A Proposal for Improving Classroom Teaching: Lessons from the TIMSS Video Study." *Elementary School Journal* 101 (September 2000): 3–20.

Lampert, Magdalene. *Teaching Problems and the Problems of Teaching.* New Haven, Conn.: Yale University Press, 2001.

Lampert, Magdalene, Peggy Rittenhouse, and Carol Crumbaugh. "Agreeing to Disagree: Developing Sociable Mathematical Discourse." In *Handbook of Education and Human Development: New Models of Learning, Teaching, and Schooling,* edited by David R. Olson and Nancy Torrance, pp. 731–64. Cambridge, Mass.: Blackwell Publishers, 1996.

National Research Council. *Adding It Up: Helping Children Learn Mathematics.* Edited by Jeremy Kilpatrick, Jane Swafford, and Bradford Findell. Washington, D.C.: National Academy Press, 2001.

Schifter, Deborah, and Catherine T. Fosnot. *Reconstructing Mathematics Education: Stories of Teachers Meeting the Challenge of Reform.* New York: Teachers College Press, 1993.

Smith, Margaret S. "A Comparison of the Types of Mathematics Tasks and How They Were Completed during Eighth-Grade Mathematics Instruction in Germany, Japan, and the United States." Unpublished doctoral dissertation, University of Delaware, 2000.

Van de Walle, John A. *Elementary and Middle School School Mathematics: Teaching Developmentally.* 4th ed. New York: Longman, 2001.

COMMENTARY

In the words of Emily Style of the National SEED (Seeking Educational Equity and Diversity) Project in Inclusive Curriculum, we need a curriculum that provides both windows to see the diverse experiences of others and mirrors to reflect our own reality (Style 1988). The windows-and-mirrors metaphor is crucial in all professional development as well as in classroom teaching. Regardless of the content or topic of the professional development effort (e.g., questioning, geometry, or inquiry), it should include discussions of how to support a range of learners through appealing to their experience and stretching their knowledge of others. Part of this support comes from developing instructional strategies that give all learners access to the content, and another part lies in creating an environment in which students care for and value one another. The articles in this section target these approaches.

REPRODUCED BY PERMISSION OF THE PUBLISHER FROM DAVID SIPRESS, *IT'S A TEACHER'S LIFE* (NEW YORK: DUTTON SIGNET, A DIVISION OF PENGUIN GROUP USA INC., 1993)

Synopses of Articles

"Helping English-Language Learners Develop Computational Fluency," by Bresser (2003), offers a set of ten general strategies that will benefit all students, particularly students with varying linguistic abilities. Using communication as a focus in instruction, the author shares actual samples of teacher-student interactions that support all students' full participation in the classroom activities.

Perkins and Flores, in their 2002 article "Mathematical Notation and Procedures of Recent Immigrant Students," point out the myth of mathematics as a universal language. By sharing differences in notation, measurement, symbols, and algorithms, they allow teachers to experience varied knowledge bases that they may encounter with immigrant students while helping identify ways to validate, honor, and incorporate students' diverse experiences into instruction.

"Building Responsibility for Learning in Students with Special Needs," by Karp and Howell (2004), describes how to move responsibility for learning from the teacher to the student. The approach is appropriate for all learners but targeted to students with special needs. By identifying six common barriers for students and using classroom examples, the authors show how instructional accommodations can help students become more independent learners.

In "Differentiating the Curriculum for Elementary Gifted Mathematics Students," Wilkins, Wilkins, and Oliver (2006) present a framework for offering mathematically gifted students challenging activities that lend depth to topics being explored by the entire class. The authors share nine types of independent activities from such categories as mathematics across the curriculum, logical thinking and problem solving, and data projects.

Helping English-Language Learners Develop Computational Fluency

Rusty Bresser

STUDENTS who are computationally fluent can solve problems accurately, efficiently, and with flexibility. These students draw on a repertoire of strategies when solving problems, and their choice of strategies often depends on the type of problem they are solving and the numbers involved. Computational fluency is rooted in an understanding of arithmetic operations, the base-ten number system, and number relationships. Communicating mathematical ideas is fundamental to developing computational fluency. When students share their solution strategies with others, they learn that there are many ways to solve problems and that some strategies are more efficient than others.

Although we can define computational fluency and explain how it can be nurtured, the challenge is to ensure that all students attain fluency. How can we help all students, especially English-language learners, develop computational fluency if they have experienced mathematics as quiet, solitary practice of standard procedures? How do we make communication the focus of mathematics class so that mathematical conversations are productive and accessible to everyone? What sensitivity, awareness, and skills do teachers need when working with students from diverse backgrounds with differing experiences and skills who may be learning English as a second language?

Communication: An Instructional Feature That Can Promote Fluency

Research in mathematics education identifies specific instructional features that promote conceptual understanding of mathematics and are associated with higher levels of performance. One such feature is communication. Many researchers of mathematics learning have found that students benefit from communicating their mathematical ideas (Cobb et. al 1997; Heibert and Wearne 1993; Khisty 1995; Lampert 1990; Wood 1999). When teachers ask effective questions, they prompt students to articulate their various solution strategies, which can create a cross-pollination of ideas. Students become flexible problem solvers, and through shared dialogue they begin to build computational fluency. But what happens when students learning English as a second language are expected to communicate mathematical ideas in English?

Communication in mathematics class has the potential to facilitate understanding and develop computational fluency, but the practice of discussing ideas in English may place children who are learning English as a second language at a distinct disadvantage. For example, English-language learners can become confused during a discussion if the mathematics vocabulary has different meanings in everyday usage, as with co*lumn, table,* and *rational.* They also may be confused if the same mathematical operation can be signaled with a variety of mathematics terms, such as *add, and, plus, sum,* and *combine.* A word such as left—as in "How many are left?"—can be confusing when the directional meaning of the word is most commonly used in everyday English. The words sum and whole also can cause confusion because they have nonmathematical homonyms. Furthermore, a symbolic statement such as $9 - 4 = 5$ can be expressed verbally in several different ways, such as "Nine take away four is five" or "Four from nine leaves five." Unless teachers thoughtfully construct conversations that are intended to promote

Teaching Children Mathematics 9 (February 2003): 294–99

an understanding of mathematics, English-language learners are less likely to benefit from mathematical discussions.

If students talk about their mathematical ideas in order to develop their computational fluency, teachers must make sure that communication does not result in inequity for English-language learners. The Equity Principle in *Principles and Standards for School Mathematics* (NCTM 2000) states that all students, regardless of their personal characteristics, backgrounds, or physical challenges, must have opportunities to study and learn mathematics. NCTM recognizes that equity requires an accommodation of differences. Teachers must pay special attention to classroom discussions so that students who are not native speakers of English can equally participate. Secada (1996) suggests that mathematics teachers use what bilingual educators have learned to help English-language learners communicate their mathematical ideas in English, thereby helping them develop computational fluency.

Strategies for Helping English Learners

Computational fluency requires the kind of flexibility in thinking that enables a student to choose an efficient strategy for a particular problem to yield a correct answer. In order for students to become flexible problem solvers, they must be aware that numerous strategies for finding an answer may exist. By participating in mathematical discussions, students can become aware of a variety of strategies. If these discussions involve English-language learners, teachers must provide extra support.

I based the strategies in **figure 1** on my own teaching experiences and the ideas of bilingual educators (Diaz-Rico and Weed 1995; Garrison 1997; Khisty 1995; Secada 1996). These suggestions have two goals: to help English learners communicate their mathematical ideas in English and to support their thinking of different strategies for solving problems.

Classroom Discussions That Develop Computational Fluency

While spending time with a class of second and third graders during their mathematics class last year, I was reminded of how difficult it can be for students to communicate their mathematical ideas and to think of different strategies for solving problems. Trying out my discussion strategies helped me assist the students, all of whom were English-language learners with varying levels of English language competency. Some

1. **Ask questions and use prompts.**
    *   What do you think the answer will be? Why do you think that?
    *   What did you do first? Second? What do you need to do next?
    *   You figured it out by . . .
    *   You said that you counted. How did you count?
    *   What is the problem about? Do you think you will add or subtract?
    *   Is there another way to solve that problem? How about starting in the tens place instead of the ones?
    *   Is that number close to a friendly number you know?
    *   Can you break apart that number in some way?

2. **Practice wait time.** After asking a question, wait for a while before calling on a volunteer. This gives English-language learners time to process questions and formulate responses.

3. **Modify teacher talk.** Speak slowly and use clear articulation. Reduce the amount of teacher talk, use a variety of words for the same idea, exaggerate intonation, and place more stress on important new concepts or questions. Model or gesture when possible to supplement the verbal discussion.

4. **Recast mathematical ideas and terms.** Mathematics has many linguistic features that can be problematic for English-language learners. Use synonyms for mathematical words, such as *subtract, take away,* and *minus.* At the same time, be aware that using too many terms simultaneously can confuse the English-language learner.

5. **Pose problems that have familiar contexts.** When a problem is embedded in a familiar context, English-language learners have an easier time understanding the problem's structure and discussing how to solve it.

6. **Connect symbols with words.** When strategies for solving problems are described, write the number sentences and point to the symbols (such as +, ×, =), stressing the words in English.

7. **Reduce the stress level in the room.** Create a low-stress environment that encourages expression of ideas, where mathematical mistakes are seen as opportunities for learning and linguistic mistakes such as incorrect grammar do not inhibit the recognition of good mathematical thinking.

8. **Use "Think-Pair-Shares."** In this activity, students think about an idea, share the idea with a partner, then share the idea with the class.

9. **Use "English experts."** A student explains a strategy in her native language to a more capable English speaker, then the "English expert" translates the strategy for the teacher.

10. **Encourage students to "retell."** This is when a student is asked to explain a strategy, in English, that someone else in the group might have used.

Fig.1. Ten strategies for helping English-language learners

students were experiencing instruction in English for the first time, while others were fairly skilled but not quite fluent. Spanish was the native language of all the students.

The following discussion occurred while four students were playing "The Game of Pig" (Burns 2000). As I watched them play, I used several different strategies to help students carry on a productive mathematical conversation. In this game, students roll two number cubes and add the numbers to find the sum. The goal is to be the first to reach one hundred. If a player rolls a one, he loses all the points earned in a round. If he rolls double ones, he loses all his points and must return to zero. Games of this type engage and motivate English-language learners and provide a familiar context for discussing arithmetic strategies. In the following, Noel acts as the "English expert" and Lenny retells Iziquiel's "doubles minus one" strategy.

Noel. Seventeen plus 10 equal 27.

Teacher. How did you get 27?

Noel. I add 10 plus 10; that makes 20, and 20 plus 7 is equal to 27.

Teacher. Noel, will you tell Anton in Spanish how you solved the problem?

Noel. Diez mas diez igual a veinte. Veinte mas siete igual a veinte y siete. *[Ten plus 10 equals 20. Twenty plus 7 equals 27.]*

Teacher. Anton, do you understand?

Anton. Yes. [rolls the number cube] Six plus 3 equals 9.

Teacher. How did you get 9? [Anton does not respond.] Noel, will you ask Anton to tell you how he got the answer to 6 plus 3, then tell me in English?

Noel. ¿Como supiste que seis mas tres es nueve? [How did you know that 6 plus 3 is 9?]

Anton. Empeze a contra del seis. [I counted on from 6.]

Noel. Anton counted on from 6.

Iziquiel. [rolls a 5 and then a 4] Five plus 5 is 10, and 10 minus 1 is 9.

Teacher. Iziquiel, will you explain your strategy to us?

Iziquiel. I know 5 and 5 is 10. But 5 plus 4 is like, 4 is 1 less than 5, so 5 plus 4 is 9.

Teacher. Lenny, please retell how Iziquiel solved the problem in English.

Lenny. [pause] He did 5 and 5 is 10, then he takes away 1 to make it 9.

Having a group discussion during the game allowed several strategies for combining numbers to surface. Iziquiel's "doubles minus one" strategy reflects a certain level of computational fluency: He uses his knowledge of number relationships to solve the problem. Having "English experts" provide translations gives beginning English speakers like Anton access to new ideas. And asking students like Lenny, with more highly developed English language skills, to "retell" someone else's strategy helps those participating in the conversation understand more efficient methods.

The next discussion took place after I posed the following problem: "Maria had 49 playing cards. Antonio gave her some cards. Now she has 84 cards. How many cards did Antonio give her?"

I asked the students to solve the problem mentally; whenever possible, I also provide models and manipulatives. Unlike the small-group conversation during "The Game of Pig," the discussion that follows included the entire class. Although facilitating a discussion with a large group can be challenging, it exposes students who are less skilled in English, like Anton and Jesús, to a broader range of ideas.

As the students shared their strategies, I recorded their ideas on the chalkboard. I used the "Think-Pair-Share" activity to give students some initial practice with describing their strategy to a friend before beginning a whole-group discussion.

Teacher. Think about a way to solve the problem I just gave you. Then pair up with a partner and share your strategy. [After giving the class some time to think and share their thoughts, I called on Marla.]

Marla. I did 84 take away 49. I couldn't do 4 take away 9 so I borrowed a 10 from the 8 and gave it to the 4 to make 14. Then I did 14 take away 9 is 5 and 7 take away 4 is 3, so it's 35 cards Antonio gave Maria.

Teacher. Who solved it a different way?

Jesús. I counted up from 49 to 84 and I got the same answer as Marla.

Teacher. Who solved it in a different way?

Christina. I did 49 plus 13, but I didn't get 84 for an answer, so I tried 49 plus 30 and I got 79. That didn't work, so I tried different numbers till I got up to 84. I finally got 49 plus 35.

Teacher. So Christina used trial and error. She tried different numbers until she got the right answer. Another way?

Anton. I count back by ones.

Teacher. What number did you start with?

Anton. Eighty-four.

Teacher. What number did you stop at?

Anton. Forty-nine.

Teacher. And how many numbers did you count back?

Anton. I got lost.

Daniela. I started with 49 and I added 5 to get 54. Then I did 54 plus 10 and got 64. Then I did 64 plus 20 to get to 84. I added 5 plus 10 plus 20 and got 35.

Teacher. Any other ways? [no response] OK. I want you to look at the numbers in the problem. Are the numbers close to any friendly numbers you know of? If so, can this help you think of another way to solve the problem?

Daviel. You could make the numbers 84 and 50. Fifty is close to 49. So 84 take away 50 is 34. But I have to add one on 34 'cause I added one to 49. That makes the answer 35.

Daniela. I know! You could do 85 minus 50. If you add one to 84 you can add one to 49 to make it easy. The answer is 35.

In this discussion, several of the students' strategies were inefficient or prone to error. Although Marla arrived at the correct answer by using the standard algorithm, children who use this method often make mistakes because keeping track of the regrouping is difficult, especially when solving problems mentally. Jesús and Anton's strategy of counting by ones also was inefficient and prone to error, especially because 84 is much greater than 49. Although creative, Christina's trial-and-error strategy also was inefficient. Daniela's use of making leaps of five, ten, and twenty to solve the problem was the most efficient strategy and revealed her working knowledge of base-ten number concepts.

After it seemed that the students had exhausted all their ideas, I tried to help them consider more efficient strategies by prompting them to use friendlier numbers that are easier to work with. This hint led Daviel to use a strategy that Carpenter et al. (1999)

call "compensating." First, he added 1 to 49 to make the friendlier 50. Then he subtracted 50 from 84 and got 34. Finally, he added 1 to 34 to compensate for the 1 that he initially added to 49. Daniela used a similar strategy that Fosnot and Dolk (2001) describe as "constant difference": She added the same amount—1—to both numbers, then subtracted. Both Daviel's and Daniela's strategies reflect computational fluency: They are efficient and appropriate for the problem type and the numbers involved. Furthermore, both methods often yield correct answers when students understand the ideas. If Daviel and Daniela had not come up with their ideas, I might have modeled these methods for the class and posed other problems for which they are appropriate.

Teachers can facilitate mathematical discussions such as the one above through activities like "Think-Pair-Shares," by asking questions that prompt student reflection such as "What number did you start with? How many numbers did you count back?" and by giving children access to one another's ideas. Although these teaching strategies are likely to be effective with all students, they are especially important for English learners, who many need extra support when engaging in oral language activities in English.

Making Communication a Focus During Mathematics Class

Garrison (1997) points out that many educators share the misconception that because it uses symbols, mathematics is not associated with any language or culture and is ideal for facilitating the transition of recent immigrant students into English instruction. If teachers view mathematics as language-free, they likely will breathe a sigh of relief when mathematics time comes and think, "Finally! A subject where my second-language learners do not have to struggle with English!" Consequently, students often end up working on "word-free" worksheets, practicing only standard procedures for solving problems, and experiencing mathematics as something that is done quietly in isolation.

The students I worked with last year had indeed experienced mathematics as the drill and practice of standard algorithms, which they practiced alone by doing worksheets. When I initially asked them to solve problems in ways other than by using standard procedures, many of these English-language learners reverted to inefficient methods such as counting. They did not have a repertoire of strategies from which to choose because they had not been exposed to the idea that many ways to solve problems exist. They also were not accustomed to discussing their mathematical ideas in English.

As soon as communication became the focus of mathematics class, however, students began to make progress in their mathematical thinking. This process did not come easily; it took time and did not happen merely because students were allowed to talk. My role as the teacher was to create a safe environment for expressing ideas, model mathematical talk, provide mathematics games for students to work on in small groups that encourage conversations, and moderate discussions to make sure that the talk was productive and focused on the mathematics. Over time, the class became less resistant to my expectation that an explanation must accompany an answer to a mathematics problem.

When facilitating productive talk during mathematics class, teachers can help ensure that emergent English speakers fully participate by structuring discussions in ways that provide access to students with varying linguistic expertise. By using prompts, asking questions, and encouraging mathematics conversations, teachers accomplish two valuable goals: English-language development and computational fluency.

REFERENCES

Burns, Marilyn. *About Teaching Mathematics*. Sausalito, Calif.: Math Solutions Publications, 2000.

Carpenter, Thomas P., Elizabeth Fennema, Megan Loef Franke, Linda Levi, and Susan B. Empson. *Children's Mathematics: Cognitively Guided Instruction*. Portsmouth, N.H.: Heineman, 1999.

Cobb, Paul, Ada Boutfi, Kay McClain, and Joy Whitenack. "Reflective Discourse and Collective Reflection." *Journal for Research in Mathematics Education* 28 (1997): 258–77.

Diaz-Rico, Lynne T., and Kathryn Z. Weed. *The Crosscultural, Language, and Academic Development Handbook*. Needham Heights, Mass.: Simon & Schuster, 1995.

Fosnot, Catherine Twomey, and Maarten Dolk. *Young Mathematicians at Work*. Portsmouth, N.H.: Heinemann, 2001.

Garrison, Leslie. "Making the NCTM's Standards Work for Emergent English Speakers." *Teaching Children Mathematics* 4 (November 1997): 132–38.

Hiebert, James, and Diana Wearne. "Instructional Tasks, Classroom Discourse, and Students' Learning in Second-Grade Arithmetic." *American Educational Research Journal* 30 (Summer 1993): 393–425.

Khisty, L. L. "Making Inequality: Issues of Language and Meanings in Mathematics Teaching with Hispanic Students." In *New Directions for Equity in Mathematics Education,* edited by Walter Secada, Elizabeth Fennema, and Linda Byrd Adajian. New York: Cambridge University Press, 1995.

Lampert, Magdalene. "When the Problem Is Not the Question and the Solution Is Not the Answer: Mathematical Knowing and Teaching." *American Educational Research Journal* 27 (Spring 1990): 29–63.

National Council of Teachers of Mathematics (NCTM). *Principles and Standards for School Mathematics*. Reston, Va.: NCTM, 2000.

Secada, Walter. "Urban Students Acquiring English and Learning Mathematics in the Context of Reform." *Urban Education* 30 (January 1996): 422–48.

Wood, Terry. "Creating a Context for Argument in Mathematics Class." *Journal for Research in Mathematics Education* 30 (1999): 171–91.

Mathematical Notations and Procedures of Recent Immigrant Students

Isabel Perkins
Alfinio Flores

MATHEMATICS is often referred to as a universal language. Compared with the differences in language and culture faced by students who are recent immigrants, the differences in mathematical notation and procedures seem to be minor. Nevertheless, immigrant students confront noticeable differences between the way that mathematical ideas are represented in their countries of origin and the manner that they are represented in the United States. If not addressed, the differences in notation and procedures can add to the difficulties that immigrants face during their first years in a new country.

Exposing teachers to these differences expands their repertoires and gives them an appreciation of their students' previous experiences and struggles. For example, the statement $59 : 8 = 7 + 3 : 8$ might cause a teacher in the United States to pause until he or she realizes that the colon can also denote division; the statement is another way to express the conversion of 59/8 to 7 3/8. In the same way, an immigrant student might hesitate when faced with unfamiliar notation. Confusion may also arise when parents who were schooled in other countries try to help their children by using procedures that are different from those taught in the United States (Ron 1998).

Teachers are often unaware that an immigrant student is confused or has a question, because the traditions of many of these students discourage questioning the teacher; students may need several weeks before they feel confident enough to ask questions. The teacher can help to include immigrant students by introducing alternative algorithms with the preface "In other countries, you might see this problem done in the following way."

As teachers encounter algorithms taught in other countries, they realize that the algorithms that they have learned are just some of the possible ways to compute answers. This realization can help teachers become more accepting when students deviate from the procedures or algorithms taught in class and use their own procedures.

This article describes some differences in representations of mathematical concepts and procedures that recent immigrants from Latin America face in schools in the United States. We do not attempt to describe all the differences that immigrants face in school mathematics. Rather, our purpose is to help students and teachers be aware of those differences and use them to the advantage of the students. Of course, immigrants from other parts of the world will also bring their own notations and procedures, and the teacher can include those conventions, too.

Most of the differences reported here were collected over a period of three years by the first author from her students, who included recent immigrants. The second author, schooled in Mexico, provided additional examples. We first discuss notational differences, then examine algorithmic differences; we finish with some considerations for success of recent immigrants in their new school systems.

Mathematics Teaching in the Middle School 7 (February 2002): 346–51

Notational Differences

Numbers

The same symbols for numbers are used in Latin America and the United States, but some differences exist in the way that numbers are written, in the names that are read for numbers, in the use of the decimal point, and in the separation of digits in large numbers.

In writing numerals, immigrant students may put a crosshatch in the number 7 to distinguish it from 1. Most people in the United States recognize this character, although some say that the 7 looks like a handwritten F. The problem for immigrants is that they cannot always distinguish a 1 or a 7 written by someone from the United States. See **table 1** for a summary of differences in handwritten numerals.

We can also find differences in the ways in which numbers are read in Latin America and the United States. Both systems are identical through the millions; they differ, however, at the point that textbooks in the United States identify as *billions*. Both a student in the United States and one from Latin America will read the number 782,621,751 as "782 million, 621 thousand, 751." Students from Latin America, and some other countries, such as Great Britain, however, will not designate *billions* until a number has at least thirteen digits. The number 23,500,000,000,000 is read by a student in the United States as "23 trillion, 500 billion" and by a Latin American as "23 billion, 500 thousand million" (Serralde et al. 1993). A student schooled in the United States will read 10,782,621,751 as "10 billion, 782 million, 621 thousand, 751." In some students' countries of origin, the number is read as "10 mil 782 *millones*, 621 mil, 751"; or it is read as "10 thousand 782 million, 621 thousand, 751."

In the United States, people separate numbers into groups of three digits by commas. In some countries of Latin America, the point is used to separate such groups. Consequently, the number 10,752,101 is equivalent to 10.752.101 (Secada 1983). Other textbooks (Serralde et al. 1993) leave a space between groups of three digits and write 10 753 101. In Mexico, a third convention is used (Secretaría de Educación Pública 1993): Millions are separated by an apostrophe, and multiples of thousands are separated by commas, as in 10'752,101. The semicolon is also used in Mexico to separate millions from thousands, as in, for example, 1;958,201.

In Mexico, negative numbers can be expressed in two ways: with a preceding negative sign (–2) or with a bar over the number ($\overline{2}$). Some students who are accustomed to this last notation have problems with the bar for repeating decimal fractions used in the United States. For example, the repeating decimal 0.3333333 . . . is designated as 0.$\overline{3}$. Some Mexican textbooks indicate a repeating decimal with an arc, for example, 0.$\overset{\frown}{3}$ (Beristáin and Campos 1993).

Measurement

Immigrants, most of whom are accustomed to the metric system, in which all measures are interrelated, find the units of measure used in the United States confusing. Divisions in the metric system are based on powers of ten, but the English system does not have consistent subdivisions; for example, 12 inches is equal to 1 foot and 3 feet is equal to 1 yard. Immigrant students from Latin America may also find that textbooks in the United States use linear metric measures (such as the decameter and the hectometer) that are seldom used in their countries of origin.

Symbols

A dot or point may be used in decimal notation and to represent multiplication in the United States. Differences in placement of the point may create several problems for

TABLE 1

Differences in Handwritten Numerals

United States	Latin America	Comments
7	7	Sometimes the numeral 7 is drawn from the bottom up.
8	8	The numeral 8 is often drawn from the bottom up.
4	y	The numeral 4 is sometimes drawn from the bottom up. Students may confuse 4s and 9s.
9	g	The numeral 9 may resemble a lowercase g, particularly when written by Cuban students.

immigrant students. In designating decimal fractions, the point rests on the line and between the numbers. The number 2.54, for example, refers to 2 whole parts and 54 hundredths. Some countries in Latin America use the comma to separate the fractional parts from the whole. The decimal number 2.54 would be written as 2,54 (Secada 1983). In the United States, the point is also used to indicate multiplication. In that use, it is placed in the middle of the line and between the given numbers; 2•54 means 2 times 54. A bolder or larger raised point, such as in 2•54, is also used in Mexico to indicate multiplication (Almaguer et al. 1994). In other countries, the dot used to multiply is placed in the lower position and between the given numbers; the notation 2.54 indicates the product of 2 and 54, not a decimal fraction. Because of these differences, students may interpret the notation 3.789 in a variety of ways. Students may assume that the notation refers to the number three thousand, seven hundred, eighty-nine, deleting the point in the belief that it is merely a "spacer" between thousands. Others may assume that the notation calls for multipling 3 and 789 or that it is the decimal expression of 3 whole parts and 789 thousandths of another.

Four different symbols are used to denote division in Mexican textbooks, namely, $\div$, /, :, and $\overline{)}$. All four symbols are used in the United States, but some confusion may arise with the use of the colon. In the United States, the colon is seen primarily in ratios and proportions, but in Latin America, it is also used to designate division. The division of the fractions

$$\frac{3}{4} \text{ and } \frac{3}{5}$$

might be written as

$$\frac{3}{4} : \frac{3}{5} ;$$

the division of 16 by 2 can be written as 16 : 2 = 8. Similarly, the equation 4 : 5 = 8 : x has the solution $x = 10$ (Serralde et al. 1993).

Another notation difference is that used in angles. Textbooks in the United States might note angles in one of several ways:

$$\angle \alpha \qquad \angle ABC \qquad \angle 1$$

Some Mexican textbooks write the same angles as follows:

$$\overset{\wedge}{\alpha} \qquad \overset{\frown}{ABC} \qquad \overset{\wedge}{1}$$

In addition, some immigrant students may try to write the angle symbol as the teacher directs, but they make the symbol in the manner that they were previously taught and place it on the side. The result often looks like this:

$$<\alpha \qquad <ABC \qquad <1$$

Point out notational differences before the lesson begins

Summary of Notational Differences

The confusion is apparent when students begin to work with inequalities. The symbols look the same to them as those for angles do. Students may wonder why they are doing operations on angles, whereas the teacher may ask why the students are writing the angles as if they were inequalities.

Teachers need to point out notational differences before the beginning of any lesson. Doing so not only validates the experiences of the students but also reminds them that understanding differing systems enriches their knowledge and establishes the idea that mathematics can be examined and represented in different ways.

Algorithmic Differences

Immigrant students often learn different algorithms for operations. Frequently, teachers know only the algorithms that they were taught and are unaware of alternative methods. In some countries, intermediate steps in a procedure are computed mentally. Many recent immigrants take pride in the ability to compute quickly and accurately in their heads. As a result, an immigrant student may merely write the answer to a problem and omit the intermediate steps in his or her written work. Teachers who are unfamiliar with the student's background might wrongly assume that the student has copied another's work.

Subtraction

One subtraction algorithm taught in Mexico is based on the fact that the difference will not change if the same number is added to both subtrahend and minuend. For example, if 10 is added to both numbers, as ten units in the minuend and as a unit in the tens place in the subtrahend, the difference remains the same (see **fig. 1**). This method transforms the problem $42 - 19$ into $(40 + 12) - (20 + 9)$, which is solved by two easy subtraction steps, $12 - 9$ and $40 - 20$. This algorithm eliminates the need to "borrow" from the column to the left when the number at the top of a column is smaller than the number below. Rather, the student adds ten units to one column in the number on top and, in the next step, adds one unit to the column to the left of the number below to compensate (see **table 2**). The reaction of many teachers in the United States is that the student does subtraction backward.

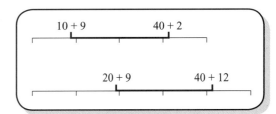

Fig. 1. Constant difference

Division

Division is done in a similar manner in the United States and Mexico, but more steps are written in long division, the version taught in the United States. In the Mexican division algorithm, numbers are multiplied and subtracted mentally, and these steps are not written down. Often, immigrant students take pride in being able to do intermediate steps mentally. Some of them consider writing all the

Table 2
Subtraction Algorithm Based on Missing Addend

Written Form	Thought Process
542 − 269 —— 3	9 from 12? 3. (Notice that ten units were added mentally to convert the 2 into a 12.) Write the 3, and add 1 (ten) mentally to the 6 (tens). (Notice that we add 1 ten to the lower number in a different column.)
542 − 269 —— 73	7 (tens) from 14 (tens)? 7 (tens). Write down the 7 (tens), and mentally add 1 (hundred) to the 2 (hundreds) in the next column.
542 − 269 —— 273	3 (hundreds) from 5 (hundreds)? 2 (hundreds). Write down the 2 (hundreds) to complete the problem.

steps in long division to be an exercise for younger children: *"Nomás para los niños en la primaria, maestra"* (Only for the children in the elementary school). The long-division algorithm for $126 \div 3$ is illustrated in the left column of **figure 2**. In the shorter version, the subtractions are done mentally. The steps are illustrated in the middle column of **figure 2**. The third column of **figure 2** shows another way of writing the steps that students learn in such countries as Honduras and Cuba.

Long Division	Short Division	Another Form of Short Division
$\begin{array}{r} 42 \\ 3\overline{)126} \\ -12 \\ \hline 06 \\ -6 \\ \hline 0 \end{array}$	$\begin{array}{r} 42 \\ 3\overline{)126} \\ 06 \\ 0 \end{array}$	$\begin{array}{r} 126\,\rfloor\,3 \\ 06 \quad 42 \\ 0 \end{array}$

Fig. 2. Three ways to divide 126 by 3

Parentheses and distributive property

Students in the United States are taught to do all work inside the parentheses before any other operations. The expression $2(3 + 5 - 2)$ is evaluated by doing $3 + 5 - 2$ first, then multiplying the resulting 6 by 2. In Mexican textbooks, the expression is evaluated by using the distributive property, $2(3 + 5 - 2) = 2 \cdot 3 + 2 \cdot 5 - 2 \cdot 2 = 6 + 10 - 4 = 12$. The two procedures are mathematically equivalent, but to a learner, they can be confusing if the connection between the two is not made explicit.

Prime factorization

Textbooks in the United States generally use a factor tree to find prime factors systematically. Mexican textbooks use a vertical line to accomplish the same process (see **fig. 3**). The first prime is 2; 140 divided by 2 is 70; 70 divided by 2 is 35; 35 is not divisible by 2 nor by the next prime, 3; therefore, the next consecutive prime, 5, is tried. The final prime factor is 7. All prime factors appear on the column to the right. In contrast, in a factor tree, prime factors appear only at the end of each branch. Often, students have trouble keeping track of the final factors because they are spread all over the tree.

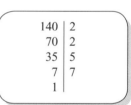

Fig. 3. Factorization in Mexico

Common denominators

To change fractions to obtain common denominators, Mexican textbooks show both denominators decomposed into primes. The lowest common denominator is found by multiplying all the common prime factors and the prime factors that appear in at least one of the two denominators. **Figure 4** shows the steps for finding the prime factors of 12 and 18 to obtain the least common multiple of the two numbers, which is $2 \times 2 \times 3 \times 3$, or 36.

12	18	2	(prime common factor of 12 and 18)
6		2	(prime factor of 12)
3	9	3	(prime common factor of 12 and 18)
1	3	3	(prime factor of 18)
	1		

Fig. 4. Prime factors of 12 and 18

Division of fractions

The most common algorithm in the United States to divide fractions is to invert the second fraction, then multiply. In Mexico, a common algorithm for a fraction division problem, such as

$$\frac{1}{2} \div \frac{3}{4},$$

Growing Professionally

is to cross-multiply, as shown in **figure 5**. Teachers can help students see how the two algorithms are equivalent. In this method, the numerator of the first fraction is multiplied by 4 and its denominator is multiplied by 3. This approach is equivalent to multiplying the first fraction by 4/3, the inverse of the second fraction.

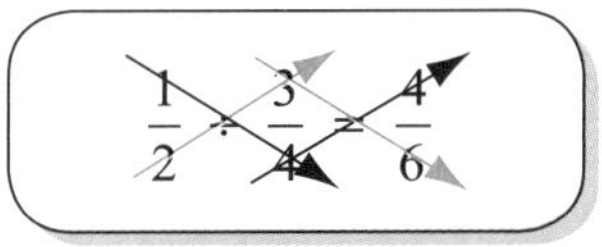

Fig. 5. Mexican algorithm for division of fractions

Algebraic equations

The equation $x + 35 = 75$ is solved in the United States by subtracting 35 from both sides to arrive at $x = 40$. Some Mexican students write the original problem the same way, but their thought processes are different. They ask themselves, "What number and 35 add up to 75?" The answer, of course, is that 40 and 35 add up to 75. Often, these students may write only the answer, and some teachers think that they have cheated. Students' written work for the procedure may also look different, as in the following:

$$x + 35 = 75$$
$$40 + 35 = 75$$
$$75 = 75$$

Students may not write $x = 40$ explicitly. These students may be able to do one-step equations, but they may balk at two-step equations because their "internal" algorithm does not work with those equations.

Multiplication of binomials

Mexican textbooks use arrows to show multiplication of binomials and polynomials (Serralde et al. 1993) (see **fig. 6**). To multiply two binomials, the diagram indicates that the first term in the first set of parentheses is multiplied by each of the terms in the second set of parentheses; then the second term in the first set of parentheses is multiplied by each of the terms in the second set of parentheses. This process results in $9x^2 + 6x + 18x + 12$. Notice that the emphasis in this method is in systematically multiplying successive terms of the first algebraic expression by all the terms of the second expression. The algorithm can be readily extended when the algebraic expressions have more than two terms. In contrast, a shift of emphasis is necessary to generalize the popular FOIL method (first, outer, inner, last) used in the United States.

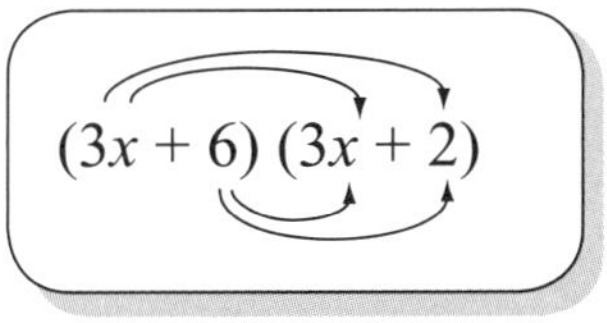

Fig. 6. Order for multiplication of terms

Summary of Algorithmic Differences

These algorithms and notational differences are not all the mathematical distinctions that immigrant students face in the United States. The differences outlined here, however, establish the idea that students from other countries may use different methods and may become confused when even basic ideas are presented through different representations.

Considerations for Success

We have identified mathematical situations that immigrant students from Latin America may find difficult. What can the classroom teacher do to help such students be successful in a school in the United States? The following paragraphs offer some recommendations.

First, validate students' previous experiences both linguistically and mathematically. Emphasize the richness and diversity of students' knowledge and their experiences with other systems of learning and expression. Most students have had schooling in their countries of origin, and those experiences differ from the school experience in a new country. Refer to those experiences in a positive light, and eliminate the notion that the methods used in schools in the United States are inherently better. Take time to explain the differences that students encounter so as to create a comfortable environment in which students know that they can express themselves without fear. When students know that different approaches will be accepted, they are more relaxed and confident.

Validate students' experiences and establish rapport

Second, find common beginning points for students to start their experiences in the United States. Two successful beginning points are graphing calculators and writing. Most immigrant students did not use calculators in their native countries. In mainstream classrooms, these students can be teamed with others to give immigrant students an opportunity to learn about graphing calculators and to hone English skills. In a sheltered English, ESL, or bilingual classroom, an introduction to graphing calculators becomes direct instruction, as few of the students will have had any experience with this tool. Using written explanations is also a good beginning point; students are not used to writing explanations in mathematics. All students benefit because they develop the ability to analyze and express mathematics in a written format, and immigrant students increase their English skills. Classroom experience shows that students balk in the beginning, but they quickly develop the ability to listen critically to teacher explanations and raise questions when they feel explanations are unclear.

Of course, in addition to differences in representations, differences can also be found in the sequence of presentation of mathematical topics in other countries. The teacher can provide enrichment and challenging activities for a topic that may be known to immigrant students but is new to those from the United States.

Finally, establish a sense of rapport in which both students and teachers are learners. For example, the teacher's mastery of Spanish may not be as good as that of the students. Students can help improve the teacher's Spanish as the teacher helps them master English. Allow class time for students to practice the variety of language used in mathe-matics, in both formal and informal settings and in both English and Spanish. Allow students to share algorithms learned in different countries to enrich the class as a whole.

The teacher plays a pivotal role in the success of immigrant students. The essential element is the teacher's decision to actively guide that success.

REFERENCES

Almaguer, Guadalupe, José Manuel Bazaldúa, Francisco Cantú, and Leticia Rodríguez. *Matemáticas 2.* México, D.F.: Limusa, 1994.

Beristáin, Eloísa, and Yolanda Campos. *Matemátias y realidad 1.* México, D.F.: Ediciones Pedagógicas, 1993.

Ron, Pilar. "My Family Taught Me This Way." In *The Teaching and Learning of Algorithms in School Mathematics,* 1998 Yearbook of the National Council of Teachers of Mathematics (NCTM), edited by Lorna J. Morrow and Margaret J. Kenney, pp. 115–19. Reston, Va.: NCTM, 1998.

Secada, Walter G. *The Educational Background of Limited English Proficient Students: Implications for the Arithmetic Classroom.* Arlington Heights, Ill.: Bilingual Education Service Center, 1983. (ERIC Document Reproduction no. ED 237 318)

Secretaría de Educación Pública. *Educación básica secundaria: Plan y programas de estudio.* México, D.F.: Secretaría de Educación Pública, 1993.

Serralde, Eulalio, Jorge Zúñiga, Héctor Zúñiga, and Enrique Zúñiga. *Matemáticas Uno.* México, D.F.: Ediciones Pedagógicas, 1993.

Building Responsibility for Learning in Students with Special Needs

Karen Karp
Philip Howell

M S. ALEXANDER smiled excitedly at the beginning of her mathematics instruction. "I have a very interesting activity for you," she announced to her students. She could tell from the twenty-seven expectant fourth-grade faces in front of her that she had piqued their curiosity. On the desk beside her sat a new activity that she had brought back from the recent National Council of Teachers of Mathematics (NCTM) regional conference. The teachers participating in the session had actually solved the problem presented in the activity. After experiencing the lesson herself, she knew that it was designed to foster interaction and investigative problem solving, which was perfect for her students.

Ms. Alexander prepared a necklace for each student with a loop of string tied to a large index card. Each card had a representation of a two- or three-dimensional geometric shape. She placed a necklace on each child, with the card hanging against the child's back rather than his or her chest. This way, students could view other people's cards but did not know what was on their own.

She announced, "When I say, 'Begin,' I want you to move around the room, asking about ten of your classmates a question about your geometric shape that can be answered with either a 'yes' or 'no' response." To highlight what she expected, the teacher asked the students to give a sample question that would require only a "yes" or "no" answer. After several sample questions she continued, "After you ask a question and get the answer, jot it down on your paper, move on, and ask another person a question. Based on these questions, try to determine what shape is drawn on your necklace. Ready? Begin."

Ms. Alexander moved around the classroom to observe her class in action and to provide guidance as needed. To her dismay, however, she did not witness the kinds of behaviors she had anticipated. Jonathan began tearing around the room, bumping into other students. Ms. Alexander quickly stepped in and redirected him, but a moment later he was tugging on another student's necklace, almost choking her as he pretended to bring the card closer to his eyes to read. Ms. Alexander sent Jonathan to his seat, asking him to write three questions that he could ask other students before he moved around the room, so that the rest of the class could better focus on the activity.

Meanwhile, Eliza began telling students the shapes that were on their cards. Ms. Alexander quickly soothed the upset students, gave them new necklaces, and temporarily removed Eliza from the activity so she could speak with her.

Although no other students were as disruptive, Ms. Alexander recognized that several others were getting nowhere on the task. Lena was wildly guessing a shape to each student she approached; her strategy was to look at another student's necklace and ask if her card showed the same shape. Ms. Alexander found that Jerome was not modifying his questions as he moved from person to person; he did not seem to know how the answers he was receiving could give him clues about the shape on his back. During the NCTM conference session, Ms. Alexander had been so sure that this activity would work well, but now she was having her doubts.

Teaching Children Mathematics 11 (October 2004): 118–26

What Went Wrong?

Although well intentioned, Ms. Alexander had not considered how to ensure success for her students with special needs. She had fallen into believing one of two common myths about teaching students with learning disabilities. The first myth is that students with special needs are vastly different from the regular school population and must be spoon-fed information or they will not be able to learn it. The second myth is that students with special needs are just like all other children in the class and "good teaching" is good teaching for all students. Both of these myths limit the success that students with learning disabilities can attain. Teachers who accept the first myth foster students who are passive learners (Poplin 1988). These students rely on an authority figure to tell them how to approach each new problem and seek the help of others to evaluate their answers to problems. They often are learning to be helpless (Seligman and Altenor 1980) and lack the confidence and analytical skills necessary for independent learning (Pressley and Harris 1990). Teachers such as Ms. Alexander who accept the second myth—that the answer lies in good teaching—fail to understand why students with special needs are referred for special status in the first place, namely, that they require different learning conditions and methods than do the majority of their peers (Kauffman 1999; Levine 1993; Thurlow 2000; Ysseldyke et al. 2001). These teachers set high expectations but do not equip their students to reach those expectations.

Although Ms. Alexander presented a meaningful problem-solving activity in which students could construct their own knowledge through inquiry and interaction with peers, she did not fully consider the following three questions:

- What organizational, behavioral, and cognitive skills are necessary for students with special needs to derive meaning from this activity?
- Which students have important weaknesses in any of these skills?
- How can I provide additional support in these areas of weakness so that students with special needs can focus on the conceptual task in the activity?

Simply put, Ms. Alexander did not consider the need to individualize her instruction.

> Equity demands that reasonable and appropriate accommodations be made as needed

Individualizing Instruction

Individualization of content taught and methods used with students with special needs is one of the basic tenets of special education. *Principles and Standards for School Mathematics* (NCTM 2000) states, "Equity does not mean that every student should receive identical instruction; instead, it demands that reasonable and appropriate accommodations be made as needed to promote access and attainment for all students" (p. 12). Calls for individualization such as this lead some teachers to abandon the philosophy of "one approach fits all" (myth number 2), only to adopt the opposite extreme (myth number 1) as their new philosophy. But if the goal is to prepare students to be autonomous learners of mathematics, teachers must find new ways to support each student while still encouraging independent learning. Students who are teacher-dependent will need help as they move to self-reliance in tackling novel mathematical situations and learning new concepts.

Teachers must consider the following four components of individualization:

- Remove specific barriers.
- Structure the environment.
- Incorporate more time and practice.
- Provide clarity.

Remove specific barriers

All students have a unique profile of relative strengths and weaknesses, including how they process different types of information. This profile is not painted in the broad strokes of subject areas, such as "strong in reading, moderate in mathematics," but in the finer detail of underlying skills (see **fig. 1**), including—

- memory (Mastropieri and Scruggs 1998; Thornton, Langrall, and Jones 1997; Wilson and Swanson 2001);
- self-regulation (Lyon and Krasnegor 1996; Swanson 1996);
- visual processing (Badian 1999; Ginsburg 1997; Rourke and Conway 1997; Thornton, Langrall, and Jones 1997);
- language processing (Cawley et al. 1998; Ginsburg 1997);
- related academic skills (Deshler, Ellis, and Lenz 1996); and
- motor skills (Miller and Mercer 1997; Rourke and Conway 1997).

Students with learning disabilities usually experience a dramatic deficit in one or more of these areas. These deficits create a roadblock between the student and the learning of skills and concepts. A teacher cannot be effective in teaching until barriers to students' learning are removed.

Barriers can be removed in several ways. Ultimately, the goal of teaching is to strengthen areas of weakness so that they no longer impede student learning. Remedial techniques are often geared to such goals. In the meantime, some types of deficits must be accommodated so that they do not impede learning in other areas. For example, Sean is a fifth grader who has particular difficulty with written expression. His teacher, Mr. Gage, noted deficits in Sean's sentence structure, along with near-phobic responses when asked to do writing activities. When asked to communicate his mathematical thinking processes through an open-ended writing prompt, Sean typically produced a near-wordless response, as **figure 2** shows.

In accordance with the NCTM Communication Standard, Mr. Gage knows that he needs to help Sean develop his ability to "communicate [his] mathematical thinking coherently and clearly to peers, teachers, and others" (NCTM 2000, p. 60). Sean's weak written-communication skills present a barrier to his full understanding of mathematical ideas. Mr. Gage must assist in removing this barrier so that Sean can focus on learning to communicate his ideas more effectively.

The day after an open-ended mathematics activity about comparing fractions, Mr. Gage met with his students one-on-one for a brief conference. When he met with Sean, however, he did not focus on Sean's written response; with a tape recorder at their side, he interviewed Sean to elicit an oral explanation of his understanding. The transcripts from this recording (see **fig. 3**) demonstrate two important facts: (1) Sean did not understand his purpose or responsibility in communicating his ideas and relied on Mr. Gage's adept questioning skills; and (2) Sean actually understood the problem and related concepts better than some other students in the class. The day after the interview, Mr. Gage conferenced with Sean once again. This time, he had the typed transcript of their interview, which he treated as if it were Sean's written response. He also had colored Post-It flags, which he had used in other writing activities in the classroom. Mr. Gage helped Sean evaluate the interview by using a flag to mark the parts of the interview in which Sean demonstrated his understanding. They discussed what was so impressive about Sean's insights. Then Sean identified with different colored flags the parts of his responses that gave little or no indication of his understanding. "The next time I ask the class members to explain their answers in writing, I will interview you again," Mr. Gage explained. "Then I want you to focus on your main job, which is to demonstrate your understanding without my asking so many specific questions." By

Memory: visual memory, verbal/auditory memory, working memory

Self-regulation: excitement/relaxation, attention, inhibition of impulses

Visual processing: visual memory, visual discrimination, visual/spatial organization, visual-motor coordination

Language processing: expressive language, vocabulary development, receptive language, auditory processing

Related academic skills: reading, writing, study skills

Motor skills: writing legibly, aligning columns, working with small manipulatives, using one-to-one-correspondence, writing numerals

Fig. 1. Potential barriers for students with special needs

The Submarine Sandwich

Hank got a submarine sandwich for dinner. Hank is going to share the sandwich with his brother, Bill. This is what they say.

Hank: "Would you rather have $\frac{1}{4}$ or $\frac{1}{5}$ of the sandwich?"

Bill: "I'm really hungry, so I want $\frac{1}{5}$ because that's more than $\frac{1}{4}$."

Hank: "Wait! That doesn't make sense."

Bill: "Yes, it does! Anybody knows that 5 is more than 4, so $\frac{1}{5}$ of a sandwich has to be more than $\frac{1}{4}$ of it."

Tell who is correct and explain why.

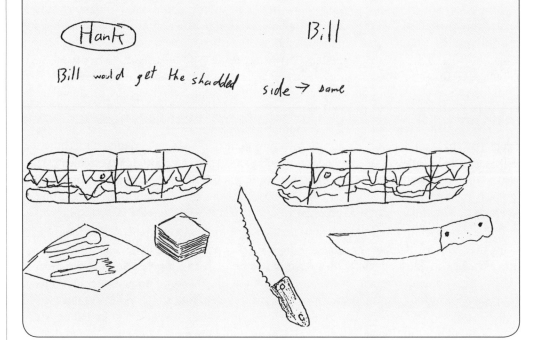

Fig. 2. Sean's near-wordless response

removing Sean's barrier in writing effectively, Mr. Gage helped Sean develop his communication skills and become more responsible for his own learning.

Structure the environment

To many children with learning disabilities, school is a place of competing stimuli. Written symbols are confusing; lessons seem abstract and hard to follow; directions are difficult to remember; and their own desks, backpacks, and notebooks have no organization. Their mathematics lessons, which some teachers consider "creative, stimulating, and interactive," are simply overwhelming and difficult to follow. In order to learn, such students need structure, eliminating the disorder.

To individualize instruction, then, a teacher must determine the type of structure the child needs. The teacher must consider several types of structure: information, environment, and behavior (Lerner 2003).

Often, the teacher will need to structure the information so that the students understand it. For such students, the wording of directions or the steps in presenting new concepts are important.

For many students with learning disabilities, the structure of the environment determines success or failure. These students are often easily distracted by the variety of sights and sounds in the room, so the teacher should choose the area of the classroom that presents the fewest distractions and keep visual displays purposeful rather than distractingly entertaining.

Students who are impulsive or who become easily overexcited need an environment that helps them structure their behavior. Transitions between activities must have clear directions and limited opportunities to get off-task. For these students, periods of time without clear purpose and expectations are invitations to behavioral difficulties. For example, step-by-step guidance is necessary to transition successfully from a discussion about comparing fractions to one about the use of fraction manipulatives. The teacher could say, "One person at each table should get the fraction pieces for his or her group. When you receive the fraction pieces, arrange them across the top of your desk according to size, starting with the whole at the left and moving in decreasing order to the eighths at the right. We will begin in two minutes." Limiting the number of students moving, the time to accomplish the work, and the number of requested tasks reduces the breaks in action that invite some students with special needs to move in less productive directions. A structured mathematics environment will still have noisy activities with animated discussions and student-led activities. The teacher, however, can plan to keep the environment structured so that the experience is meaningful and organized for students with special needs.

Incorporate more time and practice

While realizing that students with learning disabilities require more repetition in order to master concepts and skills (Carnine 1997; Miller 1996), teachers often grow uncomfortable with the notion of mindless drilling of facts and skills. The term "drill and kill" has become a popular expression of disapproval among educators (see Kohn 1998), and for good reason. Endless drilling without the initial conceptual understanding not only frustrates students but also leads to strong negative feelings about the subject area. Drill also tries to use memory as a significant learning strategy when memory is often a weak area for these students.

The key to successful practice is neither the amount of time spent on the skill in one sitting nor the use of time-pressured tests. Successful practice depends on repeated interactions with mathematics content, in small doses, throughout the day and week as the opportunity arises. Students with memory-related difficulties must continue to practice a new skill beyond the point of just achieving correct responses. The skill should be repeated periodically after some time passes to help lock information into long-term memory.

Mr. Gage. Tell me about how you figured this out.

Sean. (re-reads the problem) Bill is stupid.

Mr. Gage. Why is Bill stupid?

Sean. Because he thought 1/5 was bigger.

Mr. Gage. Why was that stupid?

Sean. Because 1/4 is bigger than 1/5.

Mr. Gage. How do you know that?

Sean. I just know it.

Mr. Gage. If I gave you the same kind of problem with different numbers, you would know it?

Sean. Yes.

Mr. Gage. OK, I am going to share a submarine sandwich with my brother. I ask my brother if he would rather have one-seventh or one-third. If he is hungry, what would be the right answer for him to say?

Sean. One-third.

Mr. Gage. How do you know that?

Sean. Because the smaller number on the bottom is actually bigger. (To explain, Sean discusses his drawing: two sandwiches, cut into fourths and fifths . . .)

Mr. Gage. Can you think of a problem like this one, but with different numbers, that would be harder than this problem?

Sean. (thinks) Two-fifths and one-fourth.

Mr. Gage. Why would that be harder?

Sean. Because fifths are smaller than fourths, but there are two of those pieces now, so it would be more.

Mr. Gage. What would you do if you had that problem?

Sean. I'd draw it.

Mr. Gage. That's how you would solve it?

Sean. Mmm-hmm.

Mr. Gage. OK, show me what you would draw.

Sean. (Draws two rectangles, carefully measuring them out to be equivalent, and shades in 2/5 and 1/4. Shows that 2/5 is more than 1/4.)

Fig. 3. Transcripts from interview with Sean

To achieve a deep learning, students with special needs require extended time per topic for adequate practice and application. This is particularly problematic when considering the number of teachers and textbooks that implement the spiral approach to instruction. The spiral approach does not often meet the needs of students with learning disabilities because topics are often covered too quickly and too much time lapses between the repeated coverage each year (Miller and Mercer 1997). Over time, fluency-building practice with concepts helps students have the facility they need to solve problems and answer mathematical questions (Johnson and Layng 1994).

Provide clarity

Undeniably, clarity is necessary for the solid learning of concepts and skills by all students. This is a particularly significant issue for teachers of students with learning disabilities. As demonstrated in reading instruction, methods that rely heavily on constructivist approaches are sometimes not as effective for the learning-disabled population as are approaches that focus on more explicit instruction (Torgesen 1998). The desire to provide clarity can lead us to overcompensating for students who are struggling, however, and never challenging them to take risks and grapple with the unknown.

Depending on the mathematics content and the student, a mathematics teacher may use direct modeling of a new task, guide the student's thought processes through the use of open-ended questions, or provide insight when necessary after a period of student-led inquiry. No one approach fits all students. The goal is for the teacher to ensure clarity of understanding, giving students multiple opportunities for practice and application.

The need for clarity in instruction can be seen with Paquin, a fourth grader in Ms. Boone's mathematics class. Ms. Boone gave her students a story problem related to a school's upcoming "Spring Fling," for which students were selling raffle tickets. The problem was a released item from the Ohio Proficiency test. Ms. Boone saw the problem as a way to reinforce evaluation and editing skills, which were troubling several of her students. Paquin's response was rather perplexing to her; a strange scribbled image on his sheet showed that he did not know how to approach the problem (see **fig. 4**). Upon reflection, however, Paquin's response made sense. Paquin often seemed overwhelmed by a new task, especially one that contained multiple steps and had no clear beginning. This characteristic is quite common among students with learning disabilities or other learning differences, such as ADHD, for which difficulty in organizing tasks is a diagnostic criterion (American Psychiatric Association 2000).

Ms. Boone began to plan a follow-up activity. How could she help students discover where to begin without taking responsibility for the thinking away from them? The next day she presented the Spring Fling problem again. This time, however, she began with a clarifying activity. "Look at the mathematics that was a part of solving this problem," she said. "What kinds of mistakes do people sometimes make when they do this kind of work?" She wrote the students' responses on a large sheet of paper (see **fig. 5**). When the students finished brainstorming, she left their list of ideas in full view and asked them to edit the problems. This time, all her students were more successful in correcting multiple errors hidden in the problem. Paquin's revised response (see **fig. 6**), although not completely successful, showed that he could now approach the task without being overwhelmed.

Revising the Necklace Activity

In light of these four fundamental components of individualization, let us look back at Ms. Alexander's geometry necklace activity. Not to be defeated, this courageous teacher repeated the activity again several days later, after considering particular students' strengths and weaknesses. She considered the following issues: How could she

Tom and Amanda are selling raffle tickets for the Spring Fling. The chart below tells how many tickets they have sold, so far.

	Monday	Tuesday	Wednesday	Thursday	Friday
Tom	19	4	10		
Amanda	16	18	12		

Who has sold more tickets? How many more did he/she sell? To find out, Tom worked out the following problem:

$$
\begin{array}{r} Tom\ \ \overset{1}{1}9 \\ \overset{\times}{1}4 \\ +\ 10 \\ \hline 43 \end{array}
\qquad
\begin{array}{r} Amanda\ \ 16 \\ 18 \\ +12 \\ \hline 35 \end{array}
\qquad
\begin{array}{r} 43 \\ -35 \\ \hline 1\ 2 \end{array}
$$

Tom figured out that he has sold 12 more tickets than Amanda. Tell whether Tom solved the problem correctly. If he did not, identify each of his mistakes. Use words, numbers, or pictures to explain your answer. Show your work on the next page.

Fig. 4. Paquin's scribbled response

- Copying things incorrectly from the chart
- Reading the chart the wrong way (vertically instead of horizontally)
- Reading or writing the numbers wrong (25 instead of 52)
- Adding and not regrouping correctly
- Subtracting up instead of down
- Not regrouping

Fig. 5. List of student mistakes

modify this activity in order for her students to better focus and gain from the learning experience? What barriers could she remove?

Ms. Alexander revisited what she observed during the initial experience with the lesson and decided on several changes. She noticed that Jerome previously had trouble making cause-and-effect connections—that is, seeing the relationships between ideas. In fact, he frequently had difficulty solving open-ended problems. Some support at the beginning of the task would help give Jerome the clarity to work on his own. When she repeated the activity, Ms. Alexander decided to start the lesson by asking all the students for specific examples of a good geometry-based question and a not-so-good question. Then they discussed how each leads to either helpful information or a dead end. She listed a few helpful questions on an overhead, then wrote "yes" or "no" under each question. Students talked about what shapes this set of answers eliminated and why. Ms. Alexander then changed the answers to each question, and the students talked about how the change altered the next question. She began the necklace activity. This clarification was very productive for a student such as Jerome, who often seemed not to know where to begin when confronted with an open-ended task. Although it provided the clarity necessary for him to approach the problem, it left the actual problem-solving task in his own hands.

Lena seemed to face several barriers. First, during the original activity, the purpose of the activity was unclear to her. She did not understand that the goal of the activity was to do detective work, asking important questions and gathering clues to solve a mystery instead of just randomly guessing a shape. Identifying her barriers and structuring the environment helped provide this clarity, but Ms. Alexander believed that Lena needed additional support to begin the activity. She helped Lena decide on her first question to ask a student, then asked Lena to explain what clues an answer to the question gave her. From her response, Ms. Alexander saw that Lena needed explicit

Fig. 6. Paquin's revised response

help in narrowing the task to questions that emphasized particular characteristics of the shapes. Ms. Alexander continued to serve as Lena's "sounding board" throughout the activity, having Lena voice her thoughts clearly in order to decide on the next steps. Ms. Alexander was careful not to tell Lena what to do at any stage, but required her to communicate clearly. She asked Lena open-ended questions when Lena seemed to be at a roadblock. An additional roadblock existed, as well: In several previous activities, Lena demonstrated some difficulty in visualizing shapes and representations. Upon reflection, Ms. Alexander realized that such a deficit could explain Lena's random guessing. For this activity, then, Ms. Alexander provided a sheet of paper that showed all possible two- and three-dimensional shapes that could be on the necklaces. Lena carried this paper around so that as she asked questions, she could cross off shapes that did not fit the answers she received to her questions.

Jonathan and Eliza both needed structure and a limiting of options. They were too easily overstimulated by loosely structured activities, but in different ways. Jonathan, who bumped into others, had difficulty controlling his physical excitement level, whereas Eliza, who told students their secret shape, was verbally impulsive. They needed clearer boundaries and specific expectations. Ms. Alexander devised more explicit directions, getting them to think ahead and remain calm. She thought that the clarification also helped them both by showing them how to approach the task and narrow possibilities. She gave Jonathan a sheet of illustrations, just as she did for Lena, but for a different purpose. Lena had difficulty picturing the shapes in her head, whereas Jonathan needed a hands-on tool to organize his approach and to keep his hands busy. Finally, to help Eliza, Ms. Alexander pulled her aside and privately asked her to describe what types of talking were appropriate for the activity. This served as one more clarifying moment to help ward off impulsive actions.

Ms. Alexander also decided to change the shape representations on each card for the second trial. In place of a single illustration, she drew several examples and orientations of the shape on each necklace. This change provided clarity for all students and practice and repeated interaction with the geometric concepts. Ms. Alexander saw this as a way to prevent overgeneralizations and expand thinking during the task.

The teacher learned to provide the four fundamental components of individualization that her students needed without reducing the learning expectations placed on them. The important issue was to remove specific barriers between the student and the learning task while still challenging each student to take risks and to have responsibility for his or her own learning. At times, this support can be very teacher-directed, involving modeling and immediate feedback. At other times, it can involve more subtle interventions from the teacher. As a teacher becomes more in tune with her students and their needs, she will become more successful in planning an activity adequately the first time and in learning from her successes.

Ms. Alexander found that confusion could result from not thinking about how to structure the environment, how to anticipate barriers that students might encounter (such as visualization difficulties or impulsivity), and how to ensure clarity so that students truly understood the task.

Knowing how to successfully teach students with special needs is essential

Conclusion

At a time when the No Child Left Behind Act has led to a nationwide movement toward more rigorous mathematics coursework to achieve a high school diploma, knowing how to successfully teach students with special needs is essential. Given the diversity of the ability levels in most inclusive classrooms, a continuum of responsibility for learning, from high teacher responsibility to high student responsibility, is useful in thinking about instruction. Selecting the pedagogical strategies that require the most student responsibility for learning is the constant goal of the teacher. Often, though, other characteristics of the students with special needs prevent the consistent use of more student-centered approaches. Therefore, a number of important elements must be considered as strategies are integrated in the development of instruction. Teachers must make decisions about the characteristics of the learner, the task, and the setting. As teachers assess students through intensive observation, they also can take into account the need for identifying and removing barriers, structuring the environment, incorporating more time and practice, and providing clarity so that they can adjust their approach for all students, particularly those with special needs.

REFERENCES

American Psychiatric Association. *Diagnostic and Statistical Manual of Mental Disorders*. 4th ed. Washington, D.C.: American Psychiatric Association, 2000.

Badian, Nathalie. "Persistent Arithmetic, Reading or Arithmetic and Reading Disability." *Annuals of Dyslexia* 49 (1999): 45–70.

Carnine, Douglas. "Instructional Design in Mathematics for Students with Learning Disabilities." *Journal of Learning Disabilities* 30 (2) (1997): 134–41.

Cawley, John, Rene Parmar, Wenfan Yan, and James Miller. "Arithmetic Computation Performance of Students with Learning Disabilities: Implications for Curriculum." *Learning Disabilities Research and Practice* 13 (2) (1998): 68–74.

Deshler, Donald, Edwin S. Ellis, and B. Keith Lenz. *Teaching Adolescents with Learning Disabilities: Strategies and Methods*. Denver, Co.: Love Publishing, 1996.

Ginsburg, Herbert. "Mathematics Learning Disabilities: A View from Developmental Psychology." *Journal of Learning Disabilities* 30 (1) (1997): 20–33.

Johnson, Kent R., and Terrence V. Layng. "The Morningside Model of Generative Instruction." In *Behavior Analysis in Education: Focus on Measurably Superior Instruction*, edited by Ralph Gardner, Diane Sainato, John Cooper, Timothy Heron, William Heward, John Eshleman, and Teresa Grossi, pp. 173–97. Monterey, Calif.: Brooks/Cole, 1994.

Kauffman, James M. "How We Prevent the Prevention of Emotional and Behavioral Disorders." *Exceptional Children* 65 (4) (1999): 448–68.

Kohn, Alfie. *What to Look for in a Classroom*. San Francisco: Jossey-Bass, 1998.

Lerner, Janet. *Learning Disabilities: Theories, Diagnosis, and Teaching Strategies*. Boston: Houghton Mifflin, 2003.

Levine, Mel. *All Kinds of Minds*. Cambridge, Mass.: Educator's Publishing Service, 1993.

Lyon, G. Reid, and Norman Krasnegor, eds. *Attention, Memory, and Executive Function*. Baltimore: Paul H. Brookes, 1996.

Mastropieri, Margo, and Thomas Scruggs. "Constructing More Meaningful Relationships in the Classroom: Mnemonic Research into Practice." *Learning Disabilities Research & Practice* 13 (3) (1998): 138–45.

Miller, Susan P. "Perspectives on Mathematics Instruction." In *Teaching Adolescents with Learning Disabilities*, edited by Donald Deshler, Edwin Ellis, and B. Keith Lenz, pp. 313–68. Denver, Co.: Love Publishing, 1996.

Miller, Susan P., and Cecil D. Mercer. "Educational Aspects of Mathematics Disabilities." *Journal of Learning Disabilities* 30 (1) (1997): 47–56.

National Council of Teachers of Mathematics (NCTM). *Principles and Standards for School Mathematics*. Reston, Va.: NCTM, 2000.

Poplin, Mary S. "The Reductionistic Fallacy in Learning Disabilities: Replicating the Past by Reducing the Present." *Journal of Learning Disabilities* 21 (1988): 389–400.

Pressley, Michael, and Karen R. Harris. "What We Really Know about Strategy Instruction." *Educational Leadership* 48 (1) (1990): 31–34.

Rourke, Byron, and James Conway. "Disabilities of Arithmetic and Mathematical Reasoning: Perspectives from Neurology and Neuropsychology." *Journal of Learning Disabilities* 30 (1) (1997): 34–46.

Seligman, Martin E., and Aidan Altenor. "Coping Behavior: Learned Helplessness, Psychological Change, and Learned Inactivity." *Behaviour Research and Therapy* 18 (1980): 459–512.

Swanson, H. Lee. "Informational Processing: An Introduction." In *Cognitive Approaches to Learning Disabilities,* edited by D. Kim Reid, Wayne Hresko, and H. Lee Swanson, pp. 251–86. Austin, Texas: Pro-Ed, 1996.

Thorton, Carol, Cynthia Langrall, and Graham Jones. "Mathematics Instruction for Elementary Students with Learning Disabilities." *Journal of Learning Disabilities* 30 (2) (1997): 142–50.

Thurlow, Martha. "Standards-Based Reform and Students with Disabilities: Reflections on a Decade of Change." *Focus on Exceptional Children* 33 (3) (2000): 1–16.

Torgesen, Joseph K. "Assessment and Instruction for Phonemic Awareness and Word Recognition Skills." In *Language and Reading Disabilities,* edited by Hugh W. Catts and Alan G. Kamhi, pp. 128–53. Needham Heights, Mass.: Allyn & Bacon, 1998.

Wilson, Kathleen, and H. Lee Swanson. "Are Mathematics Disabilities Due to a Domain-General or a Domain-Specific Working Memory Deficit?" *Journal of Learning Disabilities* 34 (3) (2001): 237–48.

Ysseldyke, Jim, Martha Thurlow, John Bielinski, Allison House, Mark Moody, and John Haigh. "The Relationship between Instructional and Assessment Accommodations in an Inclusive State Accountability System." *Journal of Learning Disabilities* 34 (3) (2001): 212–20.

Recommendations for using this article include the following:

Ask teachers to create a Mathematics Investigation Center (MIC) matrix that correlates to their mathematics program and their grade level.

—Mari Muri.

Use the unit to prompt the task of differentiating a unit for a teacher's classroom of diverse learners.

—Tanna Nicely.

Use in a study group as a tool for reflection on how such accommodations might be accomplished or enhanced in classrooms.

—Heidi Shepard

Differentiating the Curriculum for Elementary Gifted Mathematics Students

Michelle Muller Wilkins
Jesse L. M. Wilkins
Tamra Oliver

Carlos, a first grader, comes home from school one day looking upset. When Mom asks what is wrong, Carlos says that he got in trouble for working ahead in his mathematics workbook and that his teacher took it away from him. He had been asked not to work ahead because it would mean that he would not have anything to do at mathematics time when those pages were assigned. Carlos was disobedient, but Mom's heart goes out to him. He was just trying to fill the mathematics lesson time with something he enjoyed. It is not his fault that he already knows how to do the mathematics that will be taught several weeks in the future and that he can finish a 45-minute lesson's worth of curriculum in 10 minutes. Carlos is just a normal gifted mathematics student, but he is a real challenge for his classroom teacher.

"Imagine a classroom, a school, or a school district where all students have access to high-quality, engaging mathematics instruction. There are ambitious expectations for all, with accommodation for those who need it."
—Principles and Standards for School Mathematics *(NCTM 2000, p. 3)*

GIFTED mathematics students need accommodation in the mathematics curriculum, but frequently they are not appropriately challenged. Often, gifted mathematics students are asked to work independently or help others with their work (Winebrenner 2001). Sometimes, these students are given extra worksheets to complete or more difficult problems of the same type (Galbraith 1998). Sometimes, as in the true story of Carlos in the opening vignette, gifted students are just asked to sit quietly in their seats for long periods while others in the class are finishing assigned work. Although there are clear benefits to helping others (Cohen 1982) and some learning may be gleaned by doing extra work, gifted students are properly challenged when they are asked to move beyond computation into higher-order mathematical thinking processes such as applying computational skills to everyday problems, problem solving, problem posing, and creating new mathematics (Sheffield 2003).

This article presents a framework for creating a Mathematics Investigation Center (MIC) for gifted mathematics students. The goal of the framework is to make it easier for elementary teachers to provide challenging activities for students working above their grade level in mathematics without having to plan a separate lesson every day. The concept is to provide enrichment activities that generally fit the theme of the unit that the whole class is working on, instead of developing enrichment activities that are tailored to individual lessons. The activities focus on processes of mathematics rather than computation skills in an attempt to provide depth rather than breadth. Because gifted education is widely underfunded (DeLacy 2004) and many schools do not have aides or specialists who can regularly work with gifted children in the classroom, the students

Teaching Children
Mathematics 13
(August 2006): 6–13

are expected to work semi-independently on MIC activities, and the teacher's role is to facilitate rather than instruct.

The Framework

Classroom setup

The Mathematics Investigation Center (MIC) for a given unit consisted of nine activities kept in separate folders in a box with any necessary manipulatives. Initially, there were nine activities because the Winebrenner format that we used to set up the center called for a 3 × 3 menu in tic-tac-toe style (see **fig. 1**). The activities remained available to the students for the duration of the unit. Students worked at the MIC during mathematics time once or twice a week at the teacher's discretion. The center was used in conjunction with informal curriculum compacting (Smutny 2001; Strip 2000). Students were asked to complete many of their regular assignments but were not asked to work at the MIC outside of the designated mathematics period so that gifted students would not feel that they were being asked to do extra work.

The MIC was available to any students whom the teacher felt would benefit from more challenging mathematics activities, not just students identified as gifted. One or two of the activities in each unit were appropriate for the whole class and provided enrichment for everyone at some level. However, the intent of the MIC was to provide activities for the gifted students, and we soon discovered that most of the activities we designed were too difficult for the majority of the students without intensive support and assistance.

Student menu guide

For each unit, a menu of center choices was created and laid out in a 3 × 3 grid using Winebrenner's format (2001). **Figure 1** presents a sample menu for a third-grade measurement unit. The students were usually given the freedom to select any activity in the center that they wished. At other times, the teacher asked this mathematics group to work on a certain activity. If a teacher wanted to guide the students' choices, he or she could

Integrating Mathematics and Science	Problem Solving	Writing about Mathematics
Measuring What You Observe	*A Horse Has Got to Eat*	*A Citizen Writes to the President*
Science is all about being able to communicate what you observe. Measure water using different measuring tools and think about what is the best tool for the job!	Build a fence for your hungry horse that lets him eat the most grass.	Write a letter to the president telling him which system of measurement (metric or U.S. customary) Americans should use and why!
Mathematics Game	Data Project	Logic Problem
Poison	*Walking a Million Steps*	*Treachery*
Figure out how *not* to be the person who takes the last penny!	Find out how far you would have walked if you walked a million steps!	Help Christopher Columbus solve the mystery of how fast he should go in order to prevent Captain Devious from becoming the first one to sail to San Salvador Island!
Mathematics and Social Studies	Building Project	Literature and Mathematics
Accuracy on the Appalachian Trail	*The Dunking Booth Challenge*	*Crazy Measuring*
Working with an enlarged map, use Cuisenaire rods to measure the part of the Appalachian Trail that runs through Virginia.	Can you measure out exactly 4 cups of water using only a 3-cup jar and a 5-cup jar? Solve this challenge so that you can dunk Ms. Lester!	Read *Counting on Frank* (Clement 1994), then make up your own crazy measuring problem!

Fig. 1. MIC menu for measurement unit

set up the menu strategically and ask students to select activities according to a particular rule (Winebrenner 2001). For example, students could be asked to do "three in a row," "four corners," or "the middle, plus any four outer activities." Occasionally, activities built on one another in such a way that one needed to be completed before another.

Types of activities

All nine types of activities in the MIC were related to a theme that an elementary teacher might be covering in a given unit in her class (such as multiplication, measurement, geometry, and so on). Although the activities themselves changed from unit to unit, the nine types remained the same across all units. Four were intended to stretch mathematics across the curriculum (Integrating Mathematics and Science, Writing about Mathematics, Literature and Mathematics, and Mathematics and Social Studies); these provided enrichment by making connections to other subjects. Four other types of activities were intended to develop logical thinking and problem-solving skills (Mathematics Game, Building Project, Logic Problem, and Problem Solving); these activities provided enrichment by creating opportunities to focus on processes of mathematics. The ninth type of activity was based on the use of data, in which the students were asked to collect, tally, and represent findings from data they collected themselves (Data Project); this provided enrichment by encouraging students to organize mathematical information and look for patterns. **Figure 2** describes each of these types of activities in more detail.

Characteristics of individual activities

The activities were the cornerstone of differentiating the curriculum for gifted mathematics students and focused on problem solving, cross-curricular connections,

1. Integrating Mathematics and Science: This activity is a short science experiment or research project within which a mathematics problem is embedded. It is not just a mathematics problem with science words. The closer it is related to what the class is currently studying, the better.

2. Writing about Mathematics: The idea is to devise an activity in which children have to write about their mathematical thinking processes so their thinking becomes more deliberate. This activity is a writing activity, not just documentation of what happened in a problem.

3. Mathematics and Social Studies: This activity is a mathematics problem or activity that is contextually based in everyday family, cultural, or citizenship activities. The more it is truly an activity a teacher would do in social studies and the more related it is to what the class is currently studying, the better.

4. Literature and Mathematics Connection: This activity consists of having the children read literature and then asking a question about the mathematics content. Children's literature with mathematics content can also be used to elaborate on a mathematics concept or illuminate a problem situation. Many books are available to guide you in choosing literature for this activity. Depending on the difficulty of the book or its mathematics content, this activity can be good for sharing with the whole class, not just gifted students.

Logical Thinking and Problem Solving

5. Mathematics Game: Children have fun playing games with their friends. Lots of good practice in basic skills can be embedded in mathematics games, but problem-solving skills can also be developed while strategizing. This activity often turned out to be one in which the whole class could participate in playing but which the gifted students could be encouraged to analyze and strategize.

6. Logic Problem: This can be any activity that requires logical thinking. Fair-share games, matrix logic puzzles, table logic puzzles, or Venn diagram logic puzzles are all excellent choices for which reproducible activities are easily available.

7. Building Project: This activity is intended to appeal to the kinesthetic learner. The kind of activity could range from paper folding to block building to collage making. This is an excellent opportunity to design an activity that uses manipulatives.

8. Problem Solving: Although almost every activity in this framework is a problem to solve, teachers can use this square to specifically target different problem-solving strategies over the course of the school year by creating problems that are most efficiently solved using a particular strategy.

Data Project

9. Data Project: Collecting and tabulating data provide rich opportunities for children to use their mathematical skills and learn something interesting. Organizing data to make it easier to see patterns and trends is an important skill in problem solving. One potential for this kind of problem is long-term project work in which children might collect data over a period of time.

Fig. 2. Nine activity types

and processes of mathematics. Using the same mathematical theme that the rest of the class was studying, the activities provided depth for the gifted students by shifting from a computation level to a problem-solving level. Forging connections to other content areas of the curriculum also provided depth to the students' mathematics experiences. Processes of mathematics became the focus of many of the activities, so we chose to ground our activities in an investigative approach (Baroody 1998) in which the process was the critical element.

Principles and Standards (NCTM 2000) outlines five broad types of processes: problem solving, reasoning and proof, communication, connections, and representation. Individually, each activity met one or more of the Process Standards. Taken as a whole unit, the nine activities gave students the opportunity to develop all these processes.

Selecting, modifying, and creating activities

Everyday Mathematics (by the University of Chicago School Mathematics Project) is the adopted curriculum for our county; however, not enough challenging activities were available in this curriculum to meet the gifted students' needs. Therefore, we selected individual activities from a variety of sources, including Everyday Mathematics curriculum suggestions, puzzle books, gifted education resources, literature and mathematics lists, Web sites, and mathematics education texts. We modified some activities because we wanted to make them more discovery-oriented. We created others from scratch because we wanted them to coordinate closely with another lesson (such as social studies or science).

Our criteria for selecting, modifying, and creating activities were based on suggestions made by Sheffield (2003) and Baroody (1998). Tomlinson (1999) also spells out guidelines for differentiation that match the goals we had when creating activities. See **figure 3** for details.

The specific activities for each unit were selected, modified, or created to be "worthwhile mathematical tasks" (NCTM 2000, pp. 18–19) and were intended to be seeds for a good mathematical investigation experience. According to *Principles and Standards,* this type of task bundles mathematical thinking, concepts, and skills; sparks students' curiosity; and encourages speculation and investigation. Baroody (1998) notes that a worthwhile task can be approached in different ways and that it leads naturally to mathematical discussion. We deliberately tried to avoid activities that were structured like recipes or that walked the students through unfamiliar territory step by step. Instead, we chose inquiry-based activities.

By design, most activities had more than one solution path, so there was no answer key. We required that solutions to the activities be written so that the teacher could understand what the solution was and how the student figured it out. Having the students answer questions along the way demonstrated their thought processes as they worked toward a solution. Sheffield (2003) states that students explore problems in depth when they go beyond a solution to consider generalizations, comparisons, and relationships to other mathematical situations. Questions cmbcddcd in the activity push students in this direction. **Figure 4**, A Horse Has Got to Eat, lists important components of a problem intended for third-grade gifted mathematics students.

1. The activity is investigative and will require some initiative and discovery on the student's part. Recipe-like activities in which the students are walked through the steps of one solution are to be specifically avoided.

2. The activity can be approached in different ways. There may be more than one pathway to the solution or more than one solution.

3. The activity is complex and will require a variety of mathematical skills to solve. The activity may or may not be solvable in one mathematics period, allowing students to learn to cope with frustration and develop perseverance.

4. The activity is structured so that gifted students of a variety of abilities can begin the problem at their own level. However, the activity is *not* designed to be simple enough that the lowest-achieving student can succeed without assistance.

5. The activity will provide practice or fresh insight into the skills being presented in the regular mathematics unit.

6. The activity is engaging for elementary school students.

7. The activity is structured so that it can be worked on individually or in small groups, thus providing opportunities to discuss mathematical ideas.

8. The activity is structured to encourage reflection and communication about mathematical ideas.

9. For each unit, attention will be given to different learning styles. For example, some of the nine activities will be geared toward kinesthetic learners, others to visual learners, and so on.

10. For each unit, attention will be given to Bloom's taxonomy. For example, some of the nine activities will be designed to promote analysis, synthesis, or evaluation. Comprehension is a prerequisite, not a goal, of all activities.

Fig. 3. Ten critieria for selecting, modifying, and creating activities

Unit: Measurement[a]
Materials: Graph paper, string[b]
Introduction: You love horses,[c] and your parents have just bought you one! Your neighbor has agreed to let you build a fence in her field to let your horse graze. You have only one piece of electric fence wire 100 feet long, and you have as many stakes as you need to hold the fence wire. Because the fence is electric, assume that you need just one strand of fence all the way around your horse. Build a fence around your horse with this wire.[d]

1. Draw a picture of your fence on a piece of graph paper.[e] Make sure your picture shows that your wire is 100 feet long.
2. What shape is the fence you made?[f]
3. How can you use the string to represent the 100 feet of wire?
4. Your horse is very hungry! Experiment with different shapes to solve this problem[g] so that when you build the fence around your horse, the horse is able to eat the most grass.[h]
5. What shapes are your fences?[i]
6. Draw pictures of your fences that show how you know that your horse will eat the most grass in the fenced area you drew.[j]
7. Is this problem similar to another kind of problem you have worked on before? Explain the similarity.[k]

a. This is the theme of the unit the teacher was teaching in mathematics.
b. This activity will appeal to a kinesthetic learner and can be done alone or in pairs.
c. The fantasy of owning a horse may motivate some elementary school students to dive into the problem.
d. There will be more than one solution to this problem, allowing students to approach it in their own way.
e. The use of graph paper (and later, string) addresses NCTM's Standards of Representation and Communication and helps students think about scale.
f. A direct question helps the students write down their thoughts for the teacher. This particular question about shapes highlights the way in which a complex problem touches on several areas of mathematics at once (in this case, geometry, measurement).
g. A direct suggestion for how to approach the problem may be necessary if you want to see evidence of trial-and-error problem solving in the write-up. If you do not, this language can be omitted, making the problem more open.
h. This is a more difficult problem to solve. It will offer opportunities for abstract thinking about the properties of shape and area and give the students practice in representing measurements if they try out several solutions.
i. You may choose to accept only solutions that are circles, or you may choose to accept the "best" shape of the ones the student tried. Some students may even begin to articulate a conceptualization of limits and infinity.
j. These are more opportunities to communicate and represent mathematical ideas.
k. This probe encourages the students to connect their solution to other things they have done in mathematics or other domains. By comparing and contrasting, the students move beyond comprehension to an analysis level of Bloom's taxonomy. Implicitly, the question says that the learning is not over when an answer to the problem is found.

Fig. 4. Problem solving: A Horse Has Got to Eat

Assessment

Students' work on the activities at the Mathematics Investigation Center could be assessed by using multiple criteria including time on task, finding a workable solution to the problem posed in an activity, and communicating thought processes orally or in writing to the teacher. Tomlinson (1999) points out that teachers use assessment procedures to assign grades but also to determine whether the work is appropriately challenging for the student. Depending on the assessment goal, a teacher could choose one or more criteria for evaluating students' work.

Evaluative criteria may also depend on how the teacher presents the MIC work to the students. If students expect to work at the MIC because that is what their mathematics group does, then more specific feedback to the students about how well they are accomplishing the tasks may be warranted. However, if the students perceive work at the MIC as a privilege or as extra work, then they may find ways to sabotage themselves at the center in order to return to work where they get the easy A. In this case, Winebrenner (2001) suggests that students be given an A for all work they have tested out of and be given a daily effort grade reflecting their diligence at the MIC, instead of having the more challenging work influence their grade-point average.

Responding to the Needs of Gifted Children Enrichment

Gifted students can benefit from opportunities to grapple with worthwhile tasks because gifted students think differently from other students (Strip 2000). On the bell curve of IQ scores, gifted students are as different from their age-mates as they are from

Recommendations
for using this
article include the
following:

- Ask teachers
  to create a
  Mathematics
  Investigation
  Center (MIC)
  matrix that
  correlates to their
  mathematics
  program and their
  grade level.
  —*Mari Muri.*

- Use the unit to
  prompt the task
  of differentiating
  a unit for
  a teacher's
  classroom of
  diverse learners.
  —*Tanna Nicely.*

- Use in a study
  group as a tool
  for reflection
  on how such
  accommodations
  might be
  accomplished
  or enhanced in
  classrooms.
  — *Heidi Shepard.*

students whose IQs fall below 85 (Winebrenner 2001). Unlike many of their age-mates, for example, gifted students have an ability to work with abstract or complex concepts; to be enriched, they need activities that move beyond a comprehension level. Gifted students tend to make more rapid progress through new material and have an outstanding memory for information; to be enriched, they need to be able to pace themselves, and they do not need much review or practice. Gifted students may approach tasks in unique ways, sometimes because they see a new way to accomplish the task or a connection to another process that gives them insight; to be enriched, they need opportunities for creativity and independence. In short, gifted students do not need more work of the same type that is offered in many traditional curricula; instead, they need different types of activities (Galbraith 2001).

Semi-independence

Students who are gifted in mathematics need both teacher direction and opportunities for independence. These students are sometimes perceived as not needing help with mathematics because they can accomplish the mainstream curriculum without much assistance. When we challenge them to apply concepts in new ways or grapple with problems that take them in new directions, however, we cannot expect them to learn entirely on their own. Gifted students also need opportunities to develop personality attributes such as creativity, curiosity, insight, perseverance, and imagination (Piirto 1998). These attributes may be best developed when students are given opportunities to struggle with complex problems on their own. A teacher in a facilitator role leaves students alone when appropriate and offers instruction or assistance when appropriate.

An important area in which gifted students may need help is motivation. Although many gifted students are capable of sustained attention and focus on a task when it is of interest to them, the intrinsic motivation may not exist when we challenge gifted students to do more difficult work than they are accustomed to doing. Teacher-facilitators can assist students with staying on task and applying themselves fully to the problems. Also, many gifted students struggle with perfectionism when they are asked to work on problems for which they cannot see an immediate, clear-cut solution (Smutny 2001; Strip 2000; Winebrenner 2001). A teacher-facilitator can be supportive to gifted students by helping them learn self-discipline, take risks, and develop tolerance for ambiguity.

Gifted students need some instruction and guidance. Yet teachers are asked to work with a range of ability levels in their classrooms and cannot always provide instruction to all groups at once. Using a center approach, in which the teacher acts as a facilitator rather than a direct instructor and in which students work in pairs or small groups at times, allows gifted mathematics students to work independently with support. We call this semi-independence.

In the field

The Mathematics Investigation Center started small but is growing. This framework was conceived by a parent looking for ways to help her child, organized by a gifted education specialist committed to facilitating any efforts to help gifted children in her school district, implemented by a pioneering third-grade teacher looking for ways to serve higher-level mathematics students in her classroom, and advised by a university mathematics educator. Last year this small collaboration was picked up by third-grade teachers across the school district who refined and extended the Mathematics Investigation Center to cover the whole year's worth of curriculum. They were joined by fifth-grade teachers across the district trying to accomplish the same thing.

The MIC framework is generic enough that each teacher can implement the center in accordance with his or her classroom management style. The menu and activity descriptions seem to help start the process of planning for differentiation in mathematics. For example, this year the first- and fourth-grade teachers in the district have picked up the framework and are planning to use a version of it in their classrooms. Initially, there were nine activities; however, the fourth-grade teachers in our county have chosen to use a seven-activity "pie chart" menu with six activities around a circle and one activity in the center. The nine types of activities have been reserved to rotate into the seven slots, and a tenth type of activity, Math in the Media, has been proposed.

Conclusion

"I don't want things to be simple!" states third-grader Kiana after the adult volunteer suggests she try a less complicated shape. She is creating a highly irregular polygon "fence" with beautiful symmetry for the problem A Horse Has Got to Eat. Figuring out the area and perimeter of the shape she has drawn will be tough, but she is enjoying her own creativity and would rather struggle a bit before looking for an easier path to the solution. She has already come up with several shortcuts to replace counting all the squares to determine the area of her polygon. The adult volunteer looking at her drawing asks how long the sides of the shape are where they cut diagonally across a graph-paper square and suggests that Kiana test her assumption that the diagonals are the same length as the sides. Using a measuring tape, Kiana discovers that the diagonals of the graph-paper squares are longer than the sides of the squares, making the perimeter less straightforward to determine than a rectangle. At times, she is frustrated. In fact, she eventually decides to start over with a simpler polygon but still refuses to use a rectangle. Working through this problem and the snags she encounters, Kiana deepens her understanding of area and perimeter and also strengthens her problem-solving skills and her perseverance. Because she already comprehends how area and perimeter are defined and because she has an ease with the arithmetic skills she needs to compute the solutions, this is a challenging problem for her (and other gifted mathematics students) that is neither overwhelming nor too simple.

Differentiating the mathematics curriculum for gifted students is an important task worth doing. Kiana's real-life experience with the A Horse Has Got to Eat problem demonstrates how a complex problem can provide opportunities for gifted children to deepen their mathematical understandings of concepts they already comprehend well enough to pass a test. Students who are gifted in mathematics are short-changed when they are not given appropriate levels of mathematics work; they may misbehave in class or begin to lose interest in a topic that they used to enjoy. Worst of all, they may begin to perceive mathematics as a subject in which they do not have to struggle to succeed, a subject that they finish as quickly as possible. Without appropriate opportunities, students with great potential in mathematics, such as Carlos and Kiana, may never develop the skills, motivation, and perseverance they need to reach their potential.

Differentiating the mathematics curriculum may be a challenge to teachers, but it is possible. The pressure on elementary teachers to bring as many students up to grade level as needed to pass nationally or state-mandated standardized tests is strong. The No Child Left Behind act may have weakened the pressure to differentiate the curriculum for gifted students. A teacher can use a Mathematics Investigation Center with a few well-chosen mathematical investigation activities to enrich the theme of the regular curriculum for weeks at a time instead of having to plan a second or third set of lessons every day. Our hope is that the framework we have outlined simplifies the task of differentiation, making it more manageable for elementary school teachers.

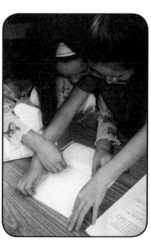

REFERENCES

Baroody, Arthur J., with Ron T. Coslick. *Fostering Children's Mathematical Power: An Investigative Approach to K–8 Mathematics Instruction*. Mahwah, NJ: Lawrence Erlbaum Associates, 1998.

Clement, Rod. *Counting on Frank*. Boston: Houghton Mifflin School Division, 1994.

Cohen, Peter A. "Educational Outcomes of Tutoring: A Meta-analysis of Findings." *American Educational Research Journal* 19, no. 2 (1982): 237–48.

DeLacy, Margaret. "The 'No Child' Law's Biggest Victim? An Answer That May Surprise." *Education Week* 23, no. 41 (2004): 40.

Galbraith, Judy. *The Gifted Kids' Survival Guide for Ages 10 and Under*. Minneapolis, MN: Free Spirit Publishing, 1998.

National Council for Teachers of Mathematics (NCTM). *Principles and Standards for School Mathematics*. Reston, VA: NCTM, 2000.

Piirto, Jane. *Understanding Those Who Create*. 2nd ed. Scottsdale, AZ: Great Potential Press, 1998.

Sheffield, Linda J. *Extending the Challenge in Mathematics: Developing Mathematical Promise in K–8 Students*. Thousand Oaks, CA: Corwin Press, 2003.

Smutny, Joan F. Stand Up for Your Gifted Child: *How to Make the Most of Your Kids' Strengths at School and at Home*. Minneapolis, MN: Free Spirit Publishing, 2001.

Strip, Carol A., with Gretchen Hirsch. *Helping Gifted Children Soar: A Practical Guide for Parents and Teachers*. Scottsdale, AZ: Gifted Psychology Press, 2000.

Tomlinson, Carol A. *The Differentiated Classroom: Responding to the Needs of All Learners*. Alexandria, VA: Association for Supervision and Curriculum Development, 1999.

Winebrenner, Susan. *Teaching Gifted Kids in the Regular Classroom: Strategies and Techniques Every Teacher Can Use to Meet the Academic Needs of the Gifted and Talented*. Minneapolis, MN: Free Spirit Publishing, 2001.

The authors would like to acknowledge the contribution of their first cooperating teacher, Karen Lester, without whom the Mathematics Investigation Center would have remained just an idea.

COMMENTARY

"I'm going to prove that Math comes in handy later in life."

Carefully selected tasks can motivate and inspire students and capture their interest by establishing meaning and significance. "Without engagement (and persistence) there is little likelihood that students will learn that which it is intended they learn" (Schlechty 2001, p. 64). Students become engaged when they are dealing with important content and do not want to stop what they are doing. In addition, students and teachers should see and use mathematics in the world in which we live. "Numeracy is the ability to cope confidently with the mathematical demands of adult life" (Cockcroft 1986). The articles in this section elaborate on this notion of what makes a high-quality task.

Synopses of Articles

The classic article "Selecting and Creating Mathematical Tasks: From Research to Practice," by Smith and Stein (1998), provides a set of tasks A–H and asks the reader to sort the tasks into one of four levels of demands, two of which are at a lower level cognitive demand and two of which are at a higher level. Discussion focuses on the kinds of thinking generated by engaging in such tasks. The tasks in this article can be used in professional development, and the article can be used as a follow-up. For more sorting tasks across grades K–12, see the *Professional Development Guidebook for Perspectives on the Teaching of Mathematics,* the 2004 NCTM Yearbook supplement, edited by George W. Bright and Rheta N. Rubenstein.

Kabiri and Smith (2003), in "Turning Traditional Textbook Problems into Open-Ended Problems," provide rich examples of the types of questions offered in a traditional "closed" form and how each can be rewritten in an open form. This article can prompt such work within the context of a workshop and is particularly useful when teachers are using textbooks that do not supply many meaningful tasks.

Reys and Bay-Williams (2003) contrast a traditional textbook lesson with a lesson in an NSF-sponsored middle school curriculum in their article titled "The Role of Textbooks in Implementing the Curriculum and Learning Principle." In the comparison, they discuss the opportunities each task provides for student learning and engagement. For teachers or stakeholders considering what a standards-based curriculum has to offer in comparison with a curriculum from a mainstream textbook company, this article is particularly helpful.

Selecting and Creating Mathematical Tasks: From Research to Practice

Margaret Schwan Smith
Mary Kay Stein

I have used this article (and task-sorting activity) in a graduate class to develop a unit designed to maintain rigor in tasks as students used them to learn. A unit plan was designed and later implemented.

—*Karen Brannon*

W HAT features of a mathematics classroom really make a difference in how students come to view mathematics and what they ultimately learn? Is it whether students are working in small groups? Is it whether students are using manipulatives? Is it the nature of the mathematical tasks that are given to students? Research conducted in the QUASAR project, a five-year study of mathematics education reform in urban middle schools (Silver and Stein 1996), offers some insight into these questions. From 1990 through 1995, data were collected about many aspects of reform teaching, including the use of small groups; the tools that were available for student use, for example, manipulatives and calculators; and the nature of the mathematics tasks. A major finding of this research to date, as described in the article by Stein and Smith in the January 1998 issue of *Mathematics Teaching in the Middle School,* is that the highest learning gains on a mathematics-performance assessment were related to the extent to which tasks were set up and implemented in ways that engaged students in high levels of cognitive thinking and reasoning (Stein and Lane 1996). This finding supports the position that the nature of the tasks to which students are exposed determines what students learn (NCTM 1991), and it also leads to many questions that should be considered by middle school teachers.

In particular, results from Stein and Lane (1996) suggest the importance of starting with high-level, cognitively complex tasks if the ultimate goal is to have students develop the capacity to think, reason, and problem solve. As was noted in our earlier discussion of Ron Castleman (Stein and Smith 1998), selecting and setting up a high-level task well does not guarantee students' engagement at a high level. Starting with a good task does, however, appear to be a necessary condition, since low-level tasks almost never result in high-level engagement. In this article, we focus on the selection and creation of mathematical tasks, drawing on QUASAR's research on mathematical tasks and on our own experiences with teachers and teacher educators.

Knowing a Good Task When You See One

When classifying a mathematical task as "good," that is, as having the potential to engage students in high-level thinking, we first consider the students—their age, grade level, prior knowledge and experiences—and the norms and expectations for work in their classroom. Consider, for example, a task in which students are asked to add five two-digit numbers and explain the process they used. For a fifth- or sixth-grade student who has access to a calculator, the addition algorithm, or both, and for whom "explain the process" means "tell how you did it," the task could be considered routine. If, however, the task is given to a second grader who has just started work with two-digit numbers, who has base-ten pieces available, and for whom "explain the process" means "you need to explain your thinking," the task may indeed be high level. Therefore, when a teacher selects a task for use in a classroom setting, all these factors need to be considered to determine the extent to which a task is likely to afford an appropriate level of challenge for her or his students.

I have used this task-sort multiple times with pre-service teachers in a functions course. The students sort the tasks using their own categories to begin with (at this point in the course, we have already read a variety of articles and chapters on worthwhile mathematical tasks). We typically take several of the tasks, and I ask groups of students to take a task that they thought was a worthwhile mathematical task and make a list of specific mathematical goals for that task. I then ask groups of students to pick a task that they do not consider a worthwhile mathematical task and modify it to make it more worthwhile (again listing mathematical goals for the task). After doing our own modified task-sort, students read the article and we discuss and compare Smith and Stein's sorting categories with those the students identified. We always have good discussions about some of the tasks in the sort, and the activity really helps the students think carefully about what makes a good, worthwhile mathematical task.

—*Pamela Wells.*

Mathematics Teaching in the Middle School 3 (February 1998): 344–50

Can you think of a task you used that was harder or easier for students than you had anticipated? What factors do you think contributed to the level of difficulty of the task for your students?

• • • • • • • •

Consider the eight tasks shown in figure 1. How would your students go about solving these tasks? Using the four categories of cognitive demand, how would you categorize each of the tasks for your students?

• • • • • • • •

Can you think of other factors that might make a task appear to be high level on the surface but that actually require only recall of memorized information or procedures?

A second step we use in classifying tasks as good is to consider the four categories of cognitive demand described in Stein and Smith (1998):

- Memorization
- Procedures without connections to concepts or meaning
- Procedures with connections to concepts and meaning
- Doing mathematics

Using these categories as templates, we ask ourselves what kind of thinking a task will demand of the students. Tasks that ask students to perform a memorized procedure in a routine manner lead to one level of thinking; tasks that ask students to think conceptually lead to a very different set of thinking processes.

In our work with teachers, we have found that they do not always agree with one another—or with us—on how tasks should be categorized. For example, some have categorized task D (shown in **fig. 1**) as a high-level task because it says that students must "explain the process you used" or because it is a word problem. Similarly, some have thought that task F (shown in **fig. 1**) was high level because it used manipulatives and featured a diagram. But we have classified both tasks as low level because each required the use of a procedure as stated (task F) or as implied by the problem (task D). Neither task presented any ambiguity about what needed to be done or how to do it or had any connection to meaning. So even though the problem might look high level, an observer must move beyond its surface features to consider the kind of thinking it requires.

A Tool for Analyzing Cognitive Demands

On the basis of the findings regarding the importance of using cognitively demanding tasks in classroom instruction, we, along with our colleague and collaborator, Marjorie Henningsen, created a task-sort activity and a task-analysis guide for use in professional-development sessions to help teachers with the selection and creation of tasks. The task-sort activity consists of twenty carefully selected instructional tasks that represent the four categories of cognitive demand for middle school students. The eight tasks shown in **figure 1** are a subset of the tasks that are included in the sort.

In addition to differing with respect to cognitive demand, the tasks in this activity also differ with respect to other features that are often associated with reform-oriented instructional tasks (NCTM 1991; Stein, Grover, and Henningsen 1996). For example, some tasks require an explanation or description (e.g., tasks A, C, D, and G); can be solved using manipulatives (e.g., tasks A, E, and F); have real-world contexts (e.g., B, C, and D); involve multiple steps, actions, or judgments (e.g., A, B, C, D, E, and G); and make use of diagrams (e.g., A, E, F, and G). Varying tasks with respect to these features across categories of cognitive demand requires an analysis of the task that goes beyond superficial features to focus on the kind of thinking in which students must engage to complete the tasks.

The task-analysis guide (**fig. 2**) consists of a listing of the characteristics of tasks at each level of cognitive demand. It serves as a judgment template—a kind of scoring rubric—that can be applied to all kinds of mathematical tasks, permitting a rating of the tasks. Also included in the task-analysis guide is an example of a task at each level, as shown in **fig. 3**. Note that each of the four tasks shown in **figure 3** involves fraction multiplication, yet the tasks vary with respect to the demands they place on students.

Using the Tool to Facilitate Discussion

To date, the task-sort activity and the task-analysis guide have been used in a range of settings with preservice and in-service teachers and with teacher educators. In one situation, thirty-three preservice teachers were asked to place each of the twenty

TASK A

Manipulatives/Tools: Counters

For homework Mark's teacher asked him to look at the pattern below and draw the figure that should come next.

Mark does not know how to find the next figure.

A. Draw the next figure for Mark.

B. Write a description for Mark telling him how you knew which figure comes next.

(QUASAR Project—QUASAR Cognitive Assessment Instrument—Release Task)

TASK B

Manipulatives/Tools: None

Part A: After the first two games of the season, the best player on the girls' basketball team had made 12 out of 20 free throws. The best player on the boys' basketball team had made 14 out of 25 free throws. Which player had made the greater percent of free throws?

Part B: The "better" player had to sit out the third game because of an injury. How many baskets, out of an additional 10 free-throw "tries," would the other player need to make to take the lead in terms of greatest percentage of free throws?

(Adapted from *Investigating Mathematics* [New York: Glencoe Macmillan/McGraw-Hill, 1994])

TASK C

Manipulatives/Tools: Calculator

Your school's science club has decided to do a special project on nature photography. They decided to take a few more than 300 outdoor photos in a variety of natural settings and in all different types of weather. They want to choose some of the best photographs and enter the state nature photography contest. The club was thinking of buying a 35 mm camera, but one member suggested that it might be better to buy disposable cameras instead. The regular camera with autofocus and automatic light meter would cost about $40.00, and film would cost $3.98 for 24 exposures and $5.95 for 36 exposures. The disposable cameras could be purchased in packs of three for $20.00, with two of the three taking 24 pictures and the third one taking 27 pictures. Single disposables could be purchased for $8.95. The club officers have to decide which would be the better option and justify their decisions to the club advisor. Do you think that they should purchase the regular camera or the disposable cameras? Write a justification that clearly explains your reasoning.

TASK D

Manipulatives/Tools: None

The cost of a sweater at a department store was $45. At the store's "day and night" sale it was marked 30 percent off the original price. What was the price of the sweater during the sale? Explain the process you used to find the sale price.

TASK E

Manipulatives/Tools: Pattern blocks

1/2 of 1/3 means one of two equal parts of one-third

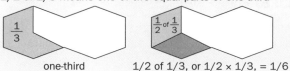

Find 1/3 of 1/4. Use pattern blocks. Draw your answer.

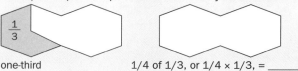

Find 1/4 of 1/3. Use pattern blocks. Draw your answer.

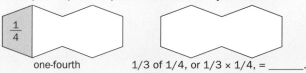

TASK F

Manipulatives/Tools: Square pattern tiles

Using the side of a square pattern tile as a measure, find the perimeter of, or distance around, each train in the pattern-block figure shown.

TASK G

Manipulatives/Tools: Grid paper

The pairs of numbers in (a)–(d) represent the heights of stacks of cubes to be leveled off. On grid paper, sketch the front views of the columns of cubes with these heights before and after they are leveled off. Write a statement under the sketches that explains how your method of leveling off is related to finding the average of the two numbers.

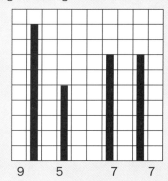

(a) 14 and 8   (b) 16 and 7   (c) 7 and 12   (d) 13 and 15

By taking two blocks off the first stack and giving them to the second stack, I've made the two stacks the same. So the total number of cubes is now distributed into two columns of equal height. And that is what average means.

(Taken from Bennett and Foreman [1989/1991])

TASK H

Manipulatives/Tools: None

Give the fraction and percent for each decimal.

0.20 = _____ = _____.
0.25 = _____ = _____.
0.33 = _____ = _____.
0.50 = _____ = _____.
0.66 = _____ = _____.
0.75 = _____ = _____.

Fig. 1. Sample tasks from the task-sort activity

Levels of Demands

Lower-level demands (memorization):

- Involve either reproducing previously learned facts, rules, formulas, or definitions or committing facts, rules, formulas or definitions to memory
- Cannot be solved using procedures because a procedure does not exist or because the time frame in which the task is being completed is too short to use a procedure
- Are not ambiguous. Such tasks involve the exact reproduction of previously seen material, and what is to be reproduced is clearly and directly stated.
- Have no connection to the concepts or meaning that underlie the facts, rules, formulas, or definitions being learned or reproduced

Lower-level demands (procedures without connections):

- Are algorithmic. Use of the procedure either is specifically called for or is evident from prior instruction, experience, or placement of the task.
- Require limited cognitive demand for successful completion. Little ambiguity exists about what needs to be done and how to do it.
- Have no connection to the concepts or meaning that underlie the procedure being used
- Are focused on producing correct answers instead of on developing mathematical understanding
- Require no explanations or explanations that focus solely on describing the procedure that was used

Higher-level demands (procedures with connections):

- Focus students' attention on the use of procedures for the purpose of developing deeper levels of understanding of mathematical concepts and ideas
- Suggest explicitly or implicitly pathways to follow that are broad general procedures that have close connections to underlying conceptual ideas as opposed to narrow algorithms that are opaque with respect to underlying concepts
- Usually are represented in multiple ways, such as visual diagrams, manipulatives, symbols, and problem situations. Making connections among multiple representations helps develop meaning.
- Require some degree of cognitive effort. Although general procedures may be followed, they cannot be followed mindlessly. Students need to engage with conceptual ideas that underlie the procedures to complete the task successfully and that develop understanding.

Higher-level demands (doing mathematics):

- Require complex and nonalgorithmic thinking—a predictable, well-rehearsed approach or pathway is not explicitly suggested by the task, task instructions, or a worked-out example.
- Require students to explore and understand the nature of mathematical concepts, processes, or relationships
- Demand self-monitoring or self-regulation of one's own cognitive processes
- Require students to access relevant knowledge and experiences and make appropriate use of them in working through the task
- Require students to analyze the task and actively examine task constraints that may limit possible solution strategies and solutions
- Require considerable cognitive effort and may involve some level of anxiety for the student because of the unpredictable nature of the solution process required

These characteristics are derived from the work of Doyle on academic tasks (1988) and Resnick on high-level-thinking skills (1987), the *Professional Standards for Teaching Mathematics* (NCTM 1991), and the examination and categorization of hundreds of tasks used in QUASAR classrooms (Stein, Grover, and Henningsen 1996; Stein, Lane, and Silver 1996).

Fig. 2. Characteristics of mathematical instructional tasks

tasks into one of the four categories of cognitive demand, without the aid of the list of characteristics in **figure 2**. Thus, the teachers were not only sorting but also engaging in discourse about students' levels of thinking as they negotiated definitions for the categories. Once each group had accomplished its assignment, the classifications were tallied in a table. The tally revealed that several tasks had complete or near consensus!

Lower-Level Demands

Memorization

What is the rule for multiplying fractions?

Expected student response:

You multiply the numerator times the numerator and the denominator times the denominator.

or

You multiply the two top numbers and then the two bottom numbers.

Procedures without Connections

Multiply:

$$\frac{2}{3} \times \frac{3}{4}$$

$$\frac{5}{6} \times \frac{7}{8}$$

$$\frac{4}{9} \times \frac{3}{5}$$

Expected student response:

$$\frac{2}{3} \times \frac{3}{4} = \frac{2 \times 3}{3 \times 4} = \frac{6}{12}$$

$$\frac{5}{6} \times \frac{7}{8} = \frac{5 \times 7}{6 \times 8} = \frac{35}{48}$$

$$\frac{4}{9} \times \frac{3}{5} = \frac{4 \times 3}{9 \times 5} = \frac{12}{45}$$

Higher-Level Demands

Procedures with Connections

Find 1/6 of 1/2. Use pattern blocks. Draw your answer and explain your solution.

Expected student response:

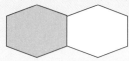 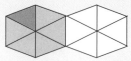

First you take half of the whole, which would be one hexagon. Then you take one-sixth of that half. So I divided the hexagon into six pieces, which would be six triangles. I only needed one-sixth, so that would be one triangle. Then I needed to figure out what part of the two hexagons one triangle was, and it was 1 out of 12. So 1/6 of 1/2 is 1/12.

Doing Mathematics

Create a real-world situation for the following problem:

$$\frac{2}{3} \times \frac{3}{4}.$$

Solve the problem you have created without using the rule, and explain your solution.

One possible student response:

For lunch Mom gave me three-fourths of a pizza that we ordered. I could only finish two-thirds of what she gave me. How much of the whole pizza did I eat?

I drew a rectangle to show the whole pizza. Then I cut it into fourths and shaded three of them to show the part Mom gave me. Since I only ate two-thirds of what she gave me, that would be only two of the shaded sections.

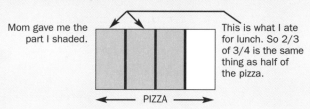

Mom gave me the part I shaded.

This is what I ate for lunch. So 2/3 of 3/4 is the same thing as half of the pizza.

PIZZA

Fig. 3. Examples of tasks at each of the four levels of cognitive demand

Many of these tasks had the hallmarks of a particular category of cognitive demand. For example, task E was categorized by all groups as procedures with connections. The discussion brought out the facts that the task focused on what it means to take a fraction of a fraction, as opposed to using an algorithm, such as "multiply the numerators and multiply the denominators"; and that it could not be completed without effort, that is, students needed to think about what their actions meant as they worked on the problem. From the specifics of the example, we began to extract characteristics of the category more generally. In this example, the tasks categorized as procedures with connections

REFLECTION

How might you use this tool in professional-development sessions to stimulate rich and lively discussions about mathematical tasks and the levels of thinking required to solve them?

• • • • • • • •

What do the classifications "procedures with connections" and "doing mathematics" mean to you? How are they alike? How are they different? In what ways can these classifications be helpful in selecting and creating worthwhile mathematics tasks for use in your own classroom?

focus on meaning, require effort, and involve a procedure. Similar discussions surrounding consensus tasks served to develop descriptors of the other categories of cognitive demand.

For other tasks, such as task A, little agreement occurred. Some teachers classified task A as procedures without connections; some, as procedures with connections; and others, as doing mathematics. The ensuing discussion highlighted the fact that no procedure or pathway was stated or implied for task A, yet the group had included the use of a procedure as a hallmark of tasks that were classified as procedures without connections and procedures with connections. A more focused look at the characteristics of doing mathematics brought out the fact that tasks in this category required students to explore and understand the nature of relationships—a necessary step in extending and describing the pattern in task A. The discussion concluded with the preservice teachers deciding to classify the task as doing mathematics. By using the established descriptions created by the group as a template against which to judge little-consensus tasks, the group had a principled basis for the [decisions] it made.

Once teachers had fairly refined ideas of the characteristics of each category of cognitive demand, it was time to start digging deeper. We began to discuss tasks for which they disagreed on the category, for example, procedures with connections versus doing mathematics, but, for the most part, we agreed on the level of thinking required, for example, high level. We saw an almost even split in terms of the categorization of task G as either procedures with connections or doing mathematics. After reviewing the criteria established by the group for these two categories, the teachers determined that procedures with connections was a better choice, since a procedure was given—leveling off stacks of cubes—and the procedure was connected to the meaning of average. The discussion focused attention on the various forms that procedures can take, such as algorithms and general pathways through the problem, and on an important characteristic of doing mathematics tasks that this particular task did not possess: the need for students to impose their own structure and procedure.

We concluded the session by distributing the task-analysis guide and comparing the teachers' descriptors with those that appeared in the guide. By distributing the guide after the task-sorting activity was completed, we did not constrain the earlier discussion by the characteristics listed in the guide and participants had the opportunity to construct a listing in their own language. The long-term goal of this activity was twofold: to raise awareness of how mathematical tasks differ with respect to the levels of cognitive engagement that they demand from students and to facilitate teachers' development of a deep and sustained appreciation for the principles of task selection and design.

Sharing Your Reflections

In this article we shared our findings concerning the importance of beginning with a task that has the potential to engage students at a high level if your goal is to increase students' ability to think and reason. The point is that the task you select and evaluate should match your goals for student learning. We encourage you to (a) reflect on the extent to which the tasks you use match your goals for student learning, (b) reflect on the extent to which your students have the opportunity to engage in tasks that require thinking and reasoning, (c) use the eight tasks that are listed in **figure 1** in a discussion with your colleagues, and (d) share the results of your experiences through … this journal.

REFERENCES

Bennett, Albert B., and Linda Foreman. *Visual Mathematics Course Guide: Integrated Math Topics and Teaching Strategies for Developing Insights and Concepts,* vol. 1. Salem, Ore.: Math Learning Center, 1989/1991.

Doyle, Walter. "Work in Mathematics Classes: The Context of Students' Thinking during Instruction." *Educational Psychologist* 23 (February 1988): 167–80.

National Council of Teachers of Mathematics (NCTM). *Professional Standards for Teaching Mathematics.* Reston, Va.: NCTM, 1991.

Resnick, Lauren. *Education and Learning to Think.* Washington, D.C.: National Academy Press, 1987.

Silver, Edward A., and Mary K. Stein. "The QUASAR Project: The 'Revolution of the Possible' in Mathematics Instructional Reform in Urban Middle Schools." *Urban Education* 30 (January 1996): 476-521.

Stein, Mary Kay, Barbara W. Grover, and Marjorie Henningsen. "Building Student Capacity for Mathematical Thinking and Reasoning: An Analysis of Mathematical Tasks Used in Reform Classrooms." *American Educational Research Journal* 33 (October 1996): 455–88.

Stein, Mary Kay, and Suzanne Lane. "Instructional Tasks and the Development of Student Capacity to Think and Reason: An Analysis of the Relationship between Teaching and Learning in a Reform Mathematics Project." *Educational Research and Evaluation* 2 (October 1996): 50–80.

Stein, Mary Kay, Suzanne Lane, and Edward Silver. "Classrooms in Which Students Successfully Acquire Mathematical Proficiency: What Are the Critical Features of Teachers' Instructional Practice?" Paper presented at the annual meeting of the American Educational Research Association, New York, April 1996.

Stein, Mary Kay, and Margaret S. Smith. "Mathematical Tasks as a Framework for Reflection." *Mathematics Teaching in the Middle School* 3 (January 1998): 268–75.

The preparation of this paper was supported by a grant from the Ford Foundation (grant no. 890-0572) for the QUASAR Project. Any opinions expressed herein are those of the authors and do not necessarily represent the views of the Ford Foundation. This paper grew out of a research report by Mary Kay Stein, Barbara Grover, and Marjorie Henningsen (1996). The authors wish to acknowledge the helpful comments of Judith Zawojewski on an earlier draft of this article.

REFLECTION

What other issues might be important to raise in a discussion of tasks? What task would you add to the sort to stimulate additional discussion?

Turning Traditional Textbook Problems into Open-Ended Problems

Mary S. Kabiri
Nancy L. Smith

THE Equity Principle contained in *Principles and Standards for School Mathematics* [*Principles and Standards*] states that we should have ". . . high expectations and strong support for all students" (NCTM 2000, p. 12). Often, when teachers plan instruction for their students, they focus on the middle achievers. However, to truly provide equitable instruction, students with special interests or talents in mathematics may need additional resources to challenge and engage them (NCTM 2000, pp. 12–13) while those students who are lacking prerequisite skills need extra support. To address the difficulty of meeting the needs of *all* learners in their class, many teachers have found success by "basing instruction on problems and activities that invite different solution approaches and many levels of solution so that less talented students can participate in the task with more talented individuals and all can experience individual success . . ." (Bley and Thornton 1994, p. 158). When teachers present these types of problems, they not only support the high achievers but also communicate high expectations and provide opportunities for higher-level thinking for all students in the class. This strategy promotes good classroom management in that it provides enrichment or sponge activities for students who finish early and are ready for more challenge while giving slower finishers the time they need as well. Traditional textbook problems do not always lend themselves to multiple solutions or solution strategies. However, many problems can be made more open-ended and accessible to a wide variety of student ability levels with minimum effort. In this article, we will show how this transformation was accomplished in a mixed-ability seventh-grade classroom.

A mathematical problem involves a situation for which the solution is not immediately obvious. A traditional problem gives a set of constraints or conditions, and strategies are applied to obtain a desired result. See this problem, for example:

Insert parentheses to make the sentence true:

$5 + 2 \times 3 - 7/2 = 7$.

An open-ended problem has an additional dimension—more than one answer is possible. For example:

Use four 4s and any of the four fundamental operations or parentheses to write mathematical sentences that make the number 8.

Ocasionally, this definition of *open-ended* is broadened to include problems for which different approaches or strategies lead to the correct single result. It is the *approach* that is open-ended. For example, "Find several different ways to calculate 38 + 15 + 42 + 18 mentally" asks for an open-ended approach. However, if the teacher defines one approach as "best," then the open-endedness is lost (Shimada 1997, p. 1).

One way that teachers can easily include more open-ended problem solving is to take traditional problems from their textbooks and adapt one or more in a lesson so that all students in the class have access to enrichment beyond the regular textbook lesson. Current textbook publishers seem to be doing a much better job of providing open-ended problems in their lessons. However, traditional problems are also represented and may be adapted. To demonstrate this approach, we took five traditional problems from a

The authors wish to thank Debbie Mummert and her students from Lewis and Clark Middle School in Jefferson City, Missouri.

Mathematics Teaching in the Middle School 9 (November 2003): 186–92

2002 textbook, one each from the first five NCTM Content Standards. We adapted them so that they would be more open-ended and asked some seventh graders to solve them. These seventh graders were in a heterogeneous mathematics class from Lewis and Clark Middle School in Jefferson City, Missouri. What follows are the problems, their adaptations, and some student solutions.

Number and Operations

The traditional form of the menu problem shown in **figure 1** may be found in most textbooks. However, it becomes more open-ended when students may select the items to purchase and are directed to find more than one solution. Providing a budget and a menu with choices also gives students a more real-life problem to solve. Not only are students using problem-solving skills, they are also getting valuable practice adding multiple combinations of decimals. They are often eager to select a meal that they would like to order or enjoy the challenge of getting the most for their money.

At first, several students commented that they thought $13.00 was a lot of money. They thought that they could randomly select any combination and it would fit within their budget. However, once they got started, they found that several combinations cost more than $13.00. When Meghan and Lauren saw that a random selection of items was not working, they decided to select the cheapest item from each category and were then successful.

When given the open-ended opportunity, some students will go beyond the teacher's expectations. Jasma and Keona decided to use tree diagrams to solve this problem (see **fig. 2**), having studied this strategy several months earlier. They filled three pages with tree diagrams to work out all the possible combinations, calculated the totals, and highlighted all the combinations that Randy could buy for $13.00. The Problem Solving Standard recommends that through "problem solving in mathematics, students should acquire ways of thinking, habits of persistence and curiosity, and confidence . . ." (NCTM 2000, p. 52). Open-ended problems like this one involving a menu encourage students to develop those characteristics.

MENU					
Side Dishes		**Main Dishes**		**Desserts**	
Taco Salad	$3.50	Chicken and Rice	$6.95	Flan	$3.25
Quesadillas	$2.25	Beef Fajitas	$8.95	Baked Apples	$2.45
Cheese Nachos	$1.50	Chicken Fingers	$4.50	Empanadas	$3.15
Baked Potato	$1.75	Chimichangas	$5.95		
Cole Slaw	$1.00	Beef Burritos	$4.95		

Traditional form: Randy, Becky, CJ, Lauren, and Ty go to eat dinner at their favorite restaurant. Ty orders quesadillas, beef fajitas, and flan. CJ has chicken fingers and does not order a side dish or dessert. How much do these two meals cost? (Houghton Mifflin 2002, p. 54).

Open-ended form: Randy has $13.00 to spend at his favorite restaurant. He wants to order one main dish, two side dishes, and one dessert. He knows he will spend $1.50 on video games while he waits for his order. Find three different meals that Randy could choose. Show your calculations, and explain how you thought about the problem.

Fig. 1. A Number and Operations problem shown in both a traditional and open-ended form

Fig. 2. Jasma and Keona solve the Menu problem by using tree diagrams.

Algebra

The problem in **figure 3** is traditional because there is only one solution and primarily one way to solve it. Using cards or dice to have students generate problems is one way to develop an infinite number of integer sentences. Challenging students to find as many solutions as they can also encourages them to use their creativity and be persistent.

Using integer cards and a die put the problem in a game format and also encouraged the use of the guess-and-check strategy, since the cards could be arranged and re-arranged to form the number sentences. When in an open-ended setting, some students meet the minimum requirements, whereas others will go beyond our expectations for the problem, therefore receiving much more practice. For example, some students worked primarily with two or three integers at a time forming such sentences as $(-5) + (+8) = 3$ or $(+4) + (+5) + (-1) = 8$. However, several students formed lengthy number sentences using all eight integer cards. For example, Aaron and Michael used $(+9) + (-8) - (-5) + (-9) + (+5) + (+4) + (+1) - (-1) = 8$. They said, "We guessed and checked," and they wrote that "subtracting a negative is the same as adding a positive."

The NCTM's Problem Solving Standard also recommends that students "monitor and reflect on the process of mathematical problem solving" (NCTM 2000, p. 52). Because the problem was also open-ended, several approaches were used and

Traditional form: In a game, a player's scores on five successive turns were +8, –11, +7, –7, +6. After which two turns was the player's total score the same? How many points were scored altogether during those five turns? (Houghton Mifflin 2002, p. 220).

Open-ended form: You have a set of integer cards from –9 to 9 in a bag, a six-sided die, and a set of plus and minus (+/–) operation cards. Shake the bag, and draw out 8 integer cards. Roll the die. This is your target number. Use the 8 drawn integer cards and any of the operation cards to make integer sentences that use addition and subtraction and [that] total the number on your die. Record your sentence, return the cards to the pile of 8, and make as many sentences as you can.

Fig. 3. An algebra problem shown in both a traditional and open-ended form

explained by students. Rachael and David also used what they knew about integers to solve the problem but thought about it differently than Aaron and Michael. They said, "We used opposite numbers to get a total of 0 and then added 3 [their target number]" (see **fig. 4**).

Geometry

The traditional geometry problems shown in **figure 5** are also very common in texts. However, the seventh-grade students enjoyed the open-ended version that still required them to recognize and name geometric figures but also asked them to go beyond that concept. Manipulating two congruent triangles cut out of an index card supported the guess-and-check problem-solving strategy. The results from this problem clearly illustrate how students can work on a problem at different levels. It was also a great way for the teacher to assess students' fluency with naming and creating two-dimensional figures. Students submitted a wide range of responses, and each pair was able to make three to six figures.

Fig. 4. Rachael and David solve the Integer problem.

Terry and Ryan formed a parallelogram, triangle, and kite. When describing the figures' properties, they focused on the length of the sides. For example, they made a kite and said that it has two pairs of sides that are the same length but different from the others. Dusk and his partner went a step further. In addition to finding various polygons like the first pair of students did, they also found and named more specialized figures, such as identifying a triangle as being isosceles rather than just a triangle. When looking at the properties of the figures, these students described the types of angles in the figures

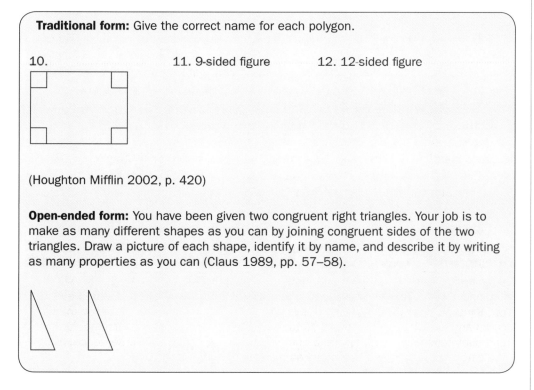

Traditional form: Give the correct name for each polygon.

10.   11. 9-sided figure   12. 12-sided figure

(Houghton Mifflin 2002, p. 420)

Open-ended form: You have been given two congruent right triangles. Your job is to make as many different shapes as you can by joining congruent sides of the two triangles. Draw a picture of each shape, identify it by name, and describe it by writing as many properties as you can (Claus 1989, pp. 57–58).

Fig. 5. A geometry problem shown in both traditional and open-ended form

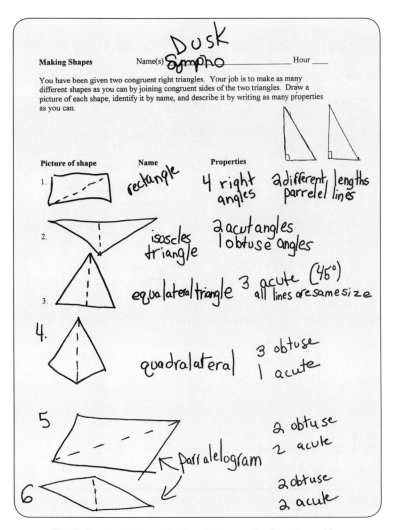

Making Shapes   Name(s) Dusk Sympho   Hour ___

You have been given two congruent right triangles. Your job is to make as many different shapes as you can by joining congruent sides of the two triangles. Draw a picture of each shape, identify it by name, and describe it by writing as many properties as you can.

Picture of shape	Name	Properties
1.	rectangle	4 right angles, 2 different lengths parrelel lines
2.	isoscles triangle	2 acut angles, 1 obtuse angles
3.	equalateral triangle	3 acute (45°) all lines are same size
4.	quadralateral	3 obtuse 1 acute
5.	parralelogram	2 obtuse 2 acute
6.		2 obtuse 2 acute

Fig. 6. Dusk and his partner's solutions to the Triangle problem

as being right, acute, or obtuse (see **fig. 6**). Brandon and Justin's descriptions focused on the degree of the angles. For example, they formed a parallelogram, measured it with a protractor, and wrote that it had two 45-degree angles and two 135-degree angles. Each group was able to focus on different properties in their solutions. By having the students complete this open-ended activity, the teacher learned much more about the students' understanding of two-dimensional figures.

Measurement

The traditional form of the problem in **figure 7** requires only applying the volume formula and no critical thinking. To make it more open-ended, we asked students to look for more than one possibility and to make a judgment about which option will be most appropriate.

Principles and Standards recommends that students "apply and adapt a variety of appropriate strategies to solve problems" (NCTM 2000, p. 53). This problem elicited a variety of strategies and solutions. Although it would have been possible for students to design a dimension using a decimal or a fraction, no one did. They all used whole-number combinations.

Aaron and Michael drew pictures of the options, then asked the teacher for a ruler. Then they measured the dimensions of one classroom wall to get a life-sized, visual picture before they made their selection. They had chosen an aquarium 8 feet long, 1 foot wide, and 3 feet tall and wanted to be sure "an 8 foot length would make sense." They said they selected the 8 × 1 × 3 "so all the teachers in the lounge won't crowd around and they can walk around it."

Emily and Craig listed possible whole-number dimensions in a table. They said, "We recommend the 4 × 2 × 3 because it is like a box and the water will be deeper than the other aquariums." Derek and Liz demonstrated persistence by making an organized list of thirty-six possibilities. For example, using 3, 4, and 2 could produce an aquarium that is 3 feet long, 4 feet wide, and 2 feet tall; 4 feet long, 2 feet wide, and 3 feet tall; or 2 feet long, 3 feet wide, and 4 feet tall (see **fig. 8**).

Brandon and Justin also listed the six whole-number combinations, but they observed that although they found six, they really found three times as many because the numbers could be reordered. Although they did not find all thirty-six possibilities, they were aware that the six number combinations could be rearranged. They also recommended the 4 × 2 × 3 size because "It would give the fish and creatures inside more freedom to move easily."

Data Analysis and Probability

Many probability problems or discussions found in textbooks start like **figure 9**'s illustration of a simple die or sum of two dice. Students are asked to determine the proba-

> **Traditional form:** A rectangular aquarium is 12 in. wide by 14 in. long by 12 in. high. What is the volume of water needed to fill the aquarium? (Houghton Mifflin 2002, p. 475).
>
> **Open-ended form:** You have been asked to design an aquarium in the shape of a rectangular prism for the school visitor's lounge. Because of the type of fish being purchased, the pet store recommends that the aquarium should hold 24 cubic feet of water. Find as many different dimensions for the aquarium as possible. Then decide which aquarium you would recommend for the lounge and explain why you made that choice.

Fig. 7. A measurement problem shown in both a traditional and open-ended form

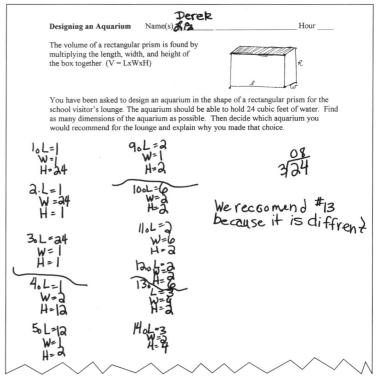

Fig. 8. Derek and Liz's list for the Aquarium problem

bility of various outcomes. However, once that procedure is understood, very little creative thought is needed. Simple probability problems of this type may become more open-ended by asking students to look for ways to make a game or experiment fair or unfair. This revision allows students to examine more than one probability concept.

The students found the open-ended form in **figure 9** to be one of the most challenging problems posed. In fact, one group, Rachael and David, was not able to solve it as written but was able to create a fair and unfair game using only one die. They indicated that the game would be fair if player A earned a point for a toss of 1, 2, or 3, and if player B earned a point for a toss of 4, 5, or 6. They also knew that each player had a 3/6 chance of rolling his or her number. To make the game unfair, they assigned player A the numbers 1–5 and player B the number 6.

Brandon and Justin worked for a long time on this problem. They first eliminated all sums of 7, the middle sums that are diagonal in the chart. To be a fair game, player A gets a point if the sum is 6 or below and player B gets a point if the sum is 8 or above. If a 7 is rolled, no one gets a point. They correctly interpreted this situation as being a conditional probability problem by recognizing that after the diagonal of 7s had been eliminated, each player had a 15/30 chance of getting a point. To make the game unfair, they simply allowed player A to get a point for a sum of 7 or above and for player B to get a point for a sum of 6 or below. Player A had a 21/36 chance of getting a point, and player B had a 15/36 chance of getting a point on each roll (see **fig. 10**).

The beauty of open-ended problems like this one is that all students are able to participate with the mathematical concepts at their own level. Each group of students was able to demonstrate an understanding of simple probability and fairness, and those who were able took it to a higher level.

Conclusions

Providing open-ended problems helps teachers meet the needs of diverse learners, since all students will benefit. The seventh-grade students in this classroom were of mixed ability. Those who displayed the most persistence or the most elegant solutions were not necessarily the best students. When teachers plan a lesson, asking a few questions will help the process.

Traditional form:

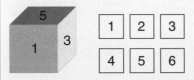

The outcomes for tossing this number cube are equally likely. You have the same chance of tossing 1, 2, 3, 4, 5, or 6. This table shows the probabilities of the events of tossing each number when tossing the number cube shown above.

Event	Probability	Event	Probability	Event	Probability
Toss a 1.	P(1) = 1/6	Toss a 2.	P(2) = 1/6	Toss a 3.	P(3) = 1/6
Toss a 4.	P(4) = 1/6	Toss a 5.	P(5) = 1/6	Toss a 6.	P(6) = 1/6

(Houghton Mifflin 2002, p. 516)

Open-ended form: Suppose that you are rolling one green die and one red die and you compute the sum on each roll. This chart shows all possible sums.

Green Die

Red Die	1	2	3	4	5	6
1	2	3	4	5	6	7
2	3	4	5	6	7	8
3	4	5	6	7	8	9
4	5	6	7	8	9	10
5	6	7	8	9	10	11
6	7	8	9	10	11	12

Your task is to design two games for player A and player B. The first game should be a fair game, and the second game should be an unfair game in favor of player A. State the rules of your games, and explain why you think each game is fair or unfair.

Fig. 9. A Data Analysis and Probability activity shown in both a traditional and open-ended form

- Do I have students who are working below this concept? Who are working above?
- How can I encourage more critical thinking?
- How can I open the tasks to more approaches or solutions to engage more students?

Considering these questions will help us all move toward the NCTM's vision of equity, ". . . high expectations and strong support for all students" (p. 12).

Fig. 10. Brandon and Justin try their fair game.

REFERENCES

Bley, Nancy S., and Carol A. Thornton. "Accommodating Special Needs." In *Windows of Opportunity: Mathematics for Students with Special Needs,* edited by Carol A. Thornton and Nancy S. Bley, pp. 158–59. Reston, Va.: National Council of Teachers of Mathematics, 1994.

Claus, Alison. "Fifth-Graders Investigate Geometry." In *New Directions for Elementary School Mathematics,* 1989 Yearbook of the National Council of Teachers of Mathematics (NCTM), edited by Paul R. Trafton and Albert P. Shulte, pp. 57–58. Reston, Va.: NCTM, 1989.

Houghton Mifflin. *Mathematics, Grade 6.* Boston, Mass.: Houghton Mifflin, 2002.

National Council of Teachers of Mathematics (NCTM). *Principles and Standards for School Mathematics.* Reston, Va.: NCTM, 2000.

Shimada, Chigeru. "The Significance of an Open-Ended Approach." In *The Open-Ended Approach: A New Proposal for Teaching Mathematics,* edited by Jerry P. Becker and Chigeru Shimada, p. 1. Reston, Va.: National Council of Teachers of Mathematics, 1997.

The Role of Textbooks in Implementing the Curriculum Principle and the Learning Principle

Barbara J. Reys
Jennifer M. Bay-Williams

THE Principles outlined in *Principles and Standards for School Mathematics* [*Principles and Standards*] (NCTM 2000) summarize the fundamental elements necessary for establishing and maintaining a high-quality mathematics education for all students. One of these six principles focuses on curriculum:

> A curriculum is more than a collection of activities: it must be coherent, focused on important mathematics, and well articulated across the grades.

The topic of curriculum as used here refers to the set of learning expectations (curriculum standards) that society and the educational community agree should be the focus of school mathematics instruction. The standards establish *what* should be taught and *when* various mathematical content and processes should receive emphasis in the school program. Teachers and textbook publishers use these curriculum standards (sometimes called a curriculum framework) to build lessons and materials to enact the intended curriculum. A good curriculum framework provides a road map to help teachers and textbook publishers develop methods and materials to support students' learning.

As teachers navigate and enact the curriculum, student learning must be the compass by which decisions are made. The Learning Principle states:

> Students must learn mathematics with understanding, actively building new knowledge from experience and prior knowledge.

The experiences that teachers provide for students play a critical role in what and how students learn. Oftentimes, curriculum materials (textbooks) have a great influence on these experiences. In this article, we share the important tenets of the Curriculum Principle and the Learning Principle and use one lesson from two different textbooks to illustrate the role of curriculum materials in curriculum and learning.

The Curriculum Principle and the Learning Principle

Much attention has been focused recently on curriculum standards, prompted by such international comparisons as the Third International Mathematics and Science Study (TIMSS) (Stigler 1997) and by the No Child Left Behind federal legislation. The latter requires states and school districts to articulate curriculum standards for mathematics learning and to regularly (annually, in grades 3–8) assess the extent to which students are learning the mathematics outlined in the standards.

The Curriculum Principle provides guidance regarding criteria for high-quality curriculum standards:

1. *A mathematics curriculum should be coherent.* Curriculum coherence refers to a well-organized set of learning expectations that integrates important mathematical ideas so that students understand how mathematical ideas build on, or connect

Mathematics Teaching in the Middle School 9 (October 2003): 120–24

with, other ideas. A coherent curriculum sequences topics so that students use foundational knowledge to enable development of new understandings and skills.

2. *A mathematics curriculum should focus on important mathematics*. The mathematics that students are expected to learn is determined by many factors, including the needs of society, science, and workforce demands. In addition, mathematics is a subject that can and should be examined for its own intellectual merit.

3. *A mathematics curriculum should be well articulated across the grades*. Teachers in the elementary grades should know what to teach so that middle-grades teachers can capitalize on this knowledge and build on it. Middle-grades teachers count on certain prior knowledge and skills, and they work to extend knowledge and prepare students for continued study in high school. A well-articulated curriculum framework helps avoid a duplication of effort and redundancy.

A major portion of *Principles and Standards* is devoted to articulating a mathematics curriculum framework across pre-K–12. The grade-band mathematics content expectations outlined in the Appendix of *Principles and Standards* reflect the experience and knowledge of teachers, mathematicians, and researchers. Although not a grade-by-grade specification of learning expectations, the curriculum framework serves as a foundation for work at the state and local levels as teachers and administrators develop clearly specified learning expectations for each grade level.

The Learning Principle is closely linked to the Curriculum Principle, describing the importance of understanding how students learn when making decisions about curriculum and teaching. The Learning Principle focuses on the need to integrate factual knowledge, procedural proficiency, and conceptual understanding. The two key tenets of the Learning Principle are that "Learning mathematics with understanding is essential" and "Students can learn mathematics with understanding." The Learning Principle advocates an environment in which much student interaction occurs and where students propose mathematical ideas, make conjectures, and evaluate their own ideas and those of other students.

The Principles in Practice

Curriculum materials (textbooks) play an important role in both what gets taught (the curriculum) and how mathematics is presented (instruction). Teachers often teach only the content that is found in their textbook, even if they use outside resources to supplement it. In addition to what content is taught at each grade level, the way in which the content is presented must be considered. The Curriculum Principle and the Learning Principle advocate for curriculum that integrates important mathematical ideas and focuses on conceptual development, as well as procedural proficiency and factual knowledge. Curriculum materials may present important mathematical ideas, but they may not provide the opportunity for students to see the connections with other mathematics that they have learned. To be truly effective, a curriculum framework requires well-aligned curriculum materials that will guide the day-to-day decisions of teachers and help them focus on the important mathematical learning goals in significant ways. Additionally, *Principles and Standards* emphasizes the need for an active learning environment for students—one that engages and challenges students intellectually and physically in exploring, representing, understanding, and developing mathematical proficiency. The textbook should provide instructional ideas and activities that are consistent with this active learning model.

Not all textbooks are alike and not all are aligned with grade-level content expectations. To illustrate the role that textbooks play in implementing both the Curriculum Principle and [the] Learning Principle, compare the two "textbook lessons" shown in **figures 1 and 2**. Although both figures focus on the same topic (volume of cylinders and cones) described in *Principles and Standards*, each outlines very different ways of

Volume of Cylinders and Cones

Part A. The city water tower is in the shape of a cylinder. The tower is 120 ft. high and has a diameter of 30 ft. Find the volume of the tower.

To find the volume (V) of a cylinder that is 120 ft. high and has a diameter of 30 ft., multiply the area of its base by the height of the cylinder.

$V = \pi r^2 \cdot h$
$V = 3.14 \cdot 15 \cdot 15 \cdot 120$
$V = 84{,}780$

The volume of the water tower is 84,780 ft.[3].

Part B. The volume of any cone is equal to 1/3 the volume of a cylinder that has the same base area and height. To find the volume of a cone that is 120 ft. high and has a radius of 15 ft., use the formula:

Volume of a cone =
1/3 Base $\cdot$ height, or $V = 1/3\ Bh$.

$V = 1/3\ \pi r^2 \cdot h$
$V = 1/3 \cdot 3.14 \cdot 15 \cdot 15 \cdot 120$
$V = 28{,}260$

The volume of the cone is 28,260 ft.[3].

Try it. Find the volume of each solid.

1. A cylinder with height 21.5 cm and radius 10 cm.

2. A cone with height 24 mm and radius 10 mm.

3. A cylinder: Base = 116.8 cm^2, height = 15 cm.

4. A large, conical mound of sand has a diameter of 45 ft. and a height of 20 ft. Find the volume.

5. One of the tallest cylindrical columns known is 90 ft. high and has a diameter of 6.5 ft. Find the volume of the column.

Fig. 1. A traditional textbook lesson

studying this topic. The textbook lesson in **figure 1** begins with a question about the volume of a cylindrical water tower and describes a formula for finding the volume of any cylinder. Next, the reader (students and teacher) are told that the volume of a cone with the same height and base area as the cylinder is 1/3 the volume of the cylinder. The symbolic formula for determining the volume of a cone is provided, along with additional problems for students to practice calculating volume of cylinders and cones. The essential elements of the topic are presented, but the level of intellectual engagement and challenge for students is low. In contrast, the textbook lesson in **figure 2** does not open with a completed example. In this lesson, students are asked to collect some information based on building, measuring, and comparing two related solids (cylinder and cone) and, after studying the data collected, to make a conjecture about a relationship between the volume of a cylinder and cone. The teacher then facilitates a discussion in which students summarize the relationship between volumes of cylinders and [volumes of] cones and students solve additional problems.

The textbook lessons shown here both focus on the same important mathematical topic. However, the nature of the instruction related to each topic is vastly different. Unlike the lesson in **figure 1**, the lesson in **figure 2** allows students to connect mathematical ideas. Rather than learn the volume of a cone as a separate, isolated topic, students have the opportunity to explore the relationship between the volume of a cylinder and [that of a] cone. The lesson in **figure 2** is highly interactive, and students must make conjectures about the formula and test their theories. By relating the new content (volume of a cone) to prior knowledge (volume of a cylinder) and allowing students an opportunity to connect the formula to the concrete objects, students will develop a deeper and more permanent understanding of the mathematics. The examples illustrate that a textbook (if used as presented) influences both the coherence of the mathematics (integrating mathematical ideas) and the way in which it is taught (through explicit explanation or through exploration and discovery).

Selecting Curricula That Align with the Principles

Mathematics textbooks have the potential to promote good instructional methods and a well-articulated, coherent, and comprehensive mathematics curriculum. As teachers and school administrators work to articulate a high-quality mathematics curriculum, they must also demand high-quality instructional materials to enact that curriculum. Some important steps in choosing high-quality textbooks include these:

Comparing Cones and Cylinders

In this lesson, you will look for a relationship between the volume of a cone and the volume of a cylinder.

Problem to Consider: Build each of the models described below. Use the models to explore the relationship between the volume of a cone and cylinder with common dimensions. Review your findings with other members of the class, then write a response to the following questions:

What appears to be the relationship between the volume of a cone and the volume of a cylinder with like dimensions? Does this relationship hold for larger or smaller cones and cylinders?

1. Start with a cylinder you find at home or at school (for example, a cup or can). Roll a piece of stiff paper into a cone shape so that the tip touches the bottom of the cylinder, as shown here.

2. Tape the cone shape along the seam and trim it at the base to form a cone with the same height and base as the cylinder.

3. Compare the space inside the cone and cylinder. Make an estimate of the comparison of volume of the cone to the volume of the cylinder. What fractional part (or percentage) of the cylinder is taken up by the cone?

4. Fill the cone to the top with a dry ingredient (for example, rice or beans) and empty the contents into the cylinder. Repeat as many times as needed to fill the cylinder. Compare your results with those of another student. Write a response to the following questions:

What appears to be the relationship between the volume of a cone and the volume of a cylinder with like dimensions? Does this relationship hold for larger or smaller cones and cylinders?

Applications

The local movie theater is looking at various containers for their popcorn sales as shown below.

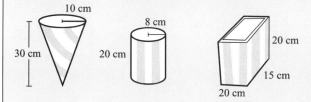

a) Which of the containers will hold the most popcorn?

b) If the theater decides to use all three containers and to charge $2 for the popcorn served in the cone-shaped container, how much should they charge for popcorn served in the cylinder-shaped container and the box-shaped container to be fair to customers?

Fig. 2. A Standards-based textbook lesson

- Before looking at textbooks, clearly articulate the mathematics that should be addressed at each grade level. What central mathematical ideas in each content strand should be addressed in each grade level?
- Examine the mathematics addressed in the textbook series. How does the content align with the central mathematical ideas articulated in the curriculum framework?

- What types of activities are provided in the textbook? Are students challenged to think and develop understanding, or are they simply shown how to work some exercises, then asked to practice procedures? Will these activities engage students in mathematical thinking and activity?

- Is there a focus on mathematical thinking and problem solving? Are students expected to explain "why"? Are they asked to explore "what if" questions? Are they encouraged to offer and test conjectures?

- What support is necessary for teachers to learn to use this textbook? Are teachers prepared to implement the textbooks using the instructional strategies outlined? What professional development will be provided to support the use of the textbook?

Wise selection, coupled with supported implementation of high-quality textbooks, will contribute to teachers' ability to implement the vision of the Curriculum Principle and Learning Principle, enabling all students to learn mathematics with understanding.

REFERENCES

National Council of Teachers of Mathematics (NCTM). *Principles and Standards for School Mathematics*. Reston, Va.: NCTM, 2000.

Stigler, James W. "Understanding and Improving Classroom Mathematics Instruction." *Phi Delta Kappan* 79 (September 1997): 14–21.

U. S. Department of Education. "Introduction: *No Child Left Behind*." www.nclb.gov/next/overview/index.html. Retrieved March 6, 2003.

CLASSROOM DISCOURSE

High-quality mathematics tasks can have their intended impact only when a teacher is able to help students focus and articulate fundamental mathematical ideas. Effective discourse is essential in maximizing learning opportunities. John Holt (1970) posed the following question, which is still relevant today: "Who needs the most practice talking in school? Who gets it? Exactly. The children need it, the teacher gets it." No dialogue or interaction can occur if only one person is talking or asking the questions. The reader has probably been in a situation in which starting a discussion on a given topic was like playing tennis by oneself. You tossed a question over the net, but nothing was returned. Another hit over the net, no response. A similar outcome can happen when a teacher starts to use questioning in the mathematics classroom. Many articles in NCTM journals support teachers in this area. Here we have selected outstanding ones that can be used collectively or individually to support teachers who are attempting to improve student discussions in their classrooms. [Editorial comment: Readers interested in exploring the topic of classroom discourse further are referred to *Getting into the Mathematics Conversation: Valuing Communication in Mathematics Classrooms,* edited by Portia C. Elliott and Cynthia M. Elliott Garnett, published by NCTM (2008).]

COMMENTARY

Synopses of Articles

For the teacher who is trying to get started with classroom discussions and having trouble getting student participation, "Assessment and Accountability: Strategies for Inquiry-Style Discussions," by Beto (2004), is an excellent choice. The author shares specific strategies of how she presents a task, sets guidelines, incorporates communication strategies, and assesses students' performance on the task. The accessible tone and the practical examples in this piece make the process of using inquiry doable.

"Strategies for Advancing Children's Mathematical Thinking," by Fraivillig (2001), introduces a framework with three dimensions for supporting students' thinking—eliciting, supporting, and extending. The author discusses each dimension and offers strategies for each. This article is an outstanding comprehensive, yet specific, collection of ways to support student learning. It can be used as a tool in making classroom observations or viewing videotapes, as well as a tool for teachers establishing instructional goals for themselves.

Herbel-Eisenmann and Breyfogle (2005), in "Questioning Our Patterns of Questioning," contrast three questioning styles—initiation-response-feedback (IRF), funneling, and focusing—and offer vignettes describing each. This article is an excellent tool to help teachers think more deeply about their own patterns of questioning and ways to alter them to help students focus on essential mathematical ideas. The vignettes can be used prior to reading the article to prompt teachers to generate their own analyses of the different styles, and teachers can attempt to rewrite the funneling example, then read the article as a follow-up activity.

Another article that analyzes classroom discourse is Kazemi's (1998) "Discourse That Promotes Conceptual Understanding," which describes the "press for learning" in the classroom. "Press for learning" was measured by the degree to which the teacher emphasized students' effort, focused on understanding, supported students' autonomy, and emphasized reasoning. Classroom vignettes from a "low press" and a "high press" classroom are shared and contrasted, enabling the reader to consider subtle differences that make a difference in supporting students' thinking and learning.

Assessment and Accountability: Strategies for Inquiry-Style Discussions

Rachel A. Beto

Teachers' actions are what encourage students to think, question, solve problems, and discuss their ideas, strategies, and solutions. The teacher is responsible for creating an intellectual environment where serious mathematical thinking is the norm.

—Principles and Standards for School Mathematics *(NCTM 2000, p. 18)*

TEACHERS at my school recently watched a video of a model inquiry lesson, in which the instructor gathered her fifteen students on a rug to discuss fractions and share drawings and ideas. My neighbor whispered, "Sure, it's easy when you have just fifteen kids. Forget it! I've got thirty!" How does a teacher with a large class facilitate a mathematical discussion that produces important ideas, engages all students, and includes assessment? It may sound impossible, but many strategies can make inquiry-style discussions accessible to all teachers.

As mathematics instruction shifts from direct instruction in which students memorize algorithms to inquiry-style instruction in which students discover properties of numbers, teachers must modify the classroom environment and assessments. Susan Jo Russell defines computational fluency as "efficiency, accuracy, and flexibility" (p. 154). A strong understanding of the base-ten number system helps children choose strategies that demonstrate computational fluency, which requires both basic skills (to assist problem solving) and conceptual awareness (to justify answers). When students work on problems alone, share strategies, then practice the new strategies, they build flexibility from seeing one problem solved in multiple ways; accuracy arises from using these strategies to verify answers and justify solutions. In order to create efficiency and flexibility, I alternate mathematical discussions with practice using a variety of problems and hands-on investigations.

One way to create rich discussions, both in small and large groups, is to post a problem on large paper and gather the students in front of the problem with papers, pencils, and something to write on. Without teaching an algorithm, the teacher asks students to solve the problem using their own strategies and prior knowledge. Then students share their work and confirm or debate one another's reasoning. The emphasis is on how each student arrived at the answer. The teacher's main question and comment are the following:

1. How do you know?
2. Please label your pictures with numbers.

Asking this question when students are correct as well as incorrect is important because mistakes are an integral aspect of inquiry-based discussions. Students gain number sense when they make, discover, and analyze mistakes. **Figures 1 and 2** illustrate how a student's mistake can initiate discussion.

Teaching Children
Mathematics 10
(May 2004): 450–54

To make these mathematics discussions powerful opportunities for all students, teachers must foster a student-centered environment, make students accountable for participating, and take notes about student understanding. This article describes specific strategies that teachers can implement to create powerful participation and discusses how to assess students during mathematical discourse.

Setting the Scene

In inquiry-based instruction, students play the lead role while the teacher makes sure that students are listening to one another and building meaning from one another's work. Creating the discussion-centered classroom begins on the first day of school. Students must learn to pay attention to one another and be accountable for what others share with the class in all subjects.

The following three discussion "rules" provide a good base for the strategies discussed in this section:

1. Keep your eyes on the speaker.
2. Think about what others say.
3. Always be ready to explain another's ideas and/or offer your own.

Teachers should establish clear expectations that all students are aware of what has just been spoken and that students will ask questions when they do not hear or understand the speaker. One of the biggest mistakes that some teachers make is always repeating students' answers. This not only detracts from the authenticity of the students' answers but also places the teacher at center stage. Keeping a note card box that contains every student's name on three different note cards is one strategy that supports students' awareness of one another. At any point during the day, the teacher can randomly pull out a name and that student must either rephrase what another student has just said or ask a thoughtful question of the student. At the beginning of the year, this strategy is used heavily until the students begin to listen to one another naturally.

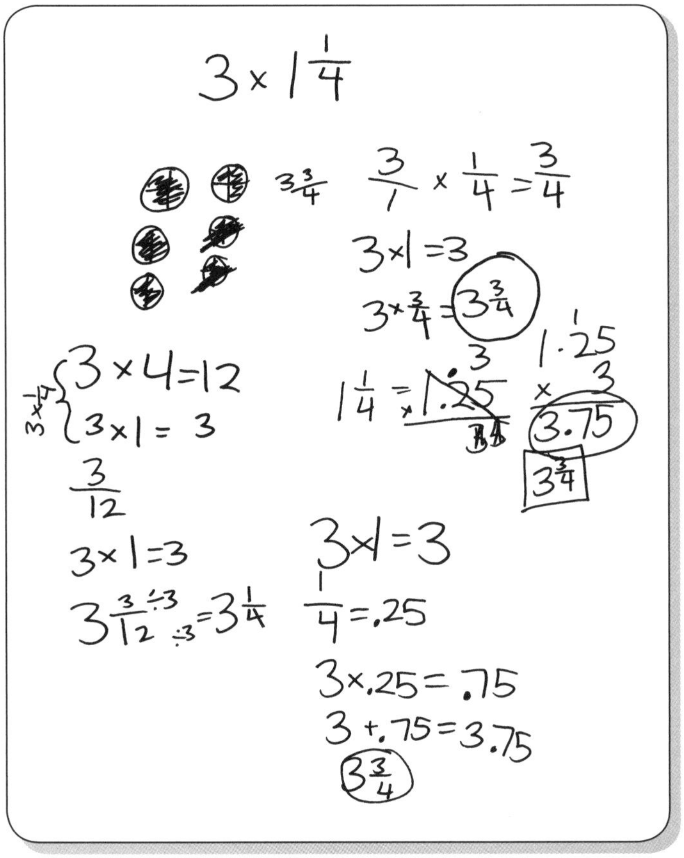

When students with quiet voices share ideas, the teacher can ask all students who heard that student's idea to raise their hands, then choose one student to repeat or rephrase the idea. The teacher asks again who knows the student's idea and chooses a new hand. Two to three rounds of this are usually necessary before the entire class understands the student's ideas. By that point, some students have heard the idea three times.

Students also must participate in discussions through attentive body language. When a student shares an answer, I announce, "Everyone put their eyes on [student's name], your new teacher." The class waits until all the students are looking at the speaker, while I stand behind or to the side of the class. Another attending strategy is to have students call on one another to share answers. Each year I choose a small (and soft) object, such as a stuffed animal or Koosh Ball, and students toss it to one another during mathematics discussions. Students are eager to receive the object and share an idea. Although I use this strategy often, I sometimes select students myself to make sure that a variety of ideas are shared by a diverse mix of students.

Fig. 1. Discussing a student's mistake

I wrote the problem in **figure 1** on a large piece of paper and told my fifth and sixth graders that we needed three batches of brownies and one of the ingredients is 1 1/4 cup of chocolate chips. The students had not yet learned how to multiply fractions. After working alone for three minutes, this is part of the discussion that followed.

Kendra. First I drew the 1 1/4. [She draws a circle, divides it into four pieces, and colors it in.] This is the 1. [She draws a circle beside it, divides it into four pieces, and colors one section in.] This is the 1/4. Since I was multiplying it by 3, I knew I needed to draw this two more times. [She draws two more identical pictures below the original.] So I knew I had three wholes [pointing to the three wholes] and I saw that I had 3/4. [She finishes by writing the numbers.]

Mark raised his hand to share next. I knew that his comments would be interesting because I had seen that he had the wrong answer.

Mark. I solved it the regular way. First I multiplied 3 × 4 and 3 × 1. Then I multiplied 3 times 1 again and I got 3 3/12.

Teacher. Mark, I'm confused. Why did you multiply 3 × 4 and 3 × 1? [I saw that he had actually multiplied 3/3 × 1/4 instead of 3/1 × 1/4, but I wanted the students to come to this realization on their own.]

Mark. Well, the 3 is the whole number, and I got the 4 and 1 from the fraction [pointing to the denominator and numerator]. [Several hands fly up.]

Teacher. Angel?

Angel. I did it the regular way, too, but Mark made a mistake. I got a different answer.

Teacher. Can you explain or show it? [Angel comes to the paper and writes "3 × 1 = 3."] Where did you get the 3 and 1 from?

Angel. They are the whole numbers [pointing]. Then I multiplied 3/1 × 1/4 and I got 3/4. I added this to the 3 and got 3 3/4. [Hands are up in the air again; students want to share and clarify.]

Teacher. Simon?

Simon. I know that Mark's answer doesn't make sense. 3/12 can be simplified to 1/4. I know the answer can't be 1/4 because if you multiply 3 × 1/4, you'll get more than 1/4. [Nods from classmates, including Mark.]

The discussion continued, with Simon changing the fractions into decimals and another student reiterating this, using the analogy of money.

Fig. 2. Sample dialogue

Keeping the Discussion Going

When a lull occurs in the discourse, teachers can help students generate ideas by posing a question and having students pair-share ideas. Using the name box to call on someone afterward ensures that all students are accountable for at least sharing what they discussed in pairs. The pair-share is also a good assessment of the students' knowledge: If most pairs do not have much to share, they do not have enough background knowledge to solve the problem. In this case, the teacher must either stop the lesson and revisit it after providing more experiences or directly teach a concept and revisit the challenge afterward.

Attaching meaning to the problem is another way to generate ideas during discussions. Sometimes I intentionally write numbers without context for the students to work with, because this can prompt a greater diversity of ideas. With more difficult concepts, such as multiplying a fraction by a whole number, I find that story problems work best. Having students write the problems brings their personal interests and experiences to the lesson, creating ownership. If the teacher decides to have students write scenarios for the mathematics problem, he or she can choose two or three scenarios that require different kinds of pictures in their solutions. In **figure 3,** students wrote story problems for 3 3/4 × 4. The first problem warrants drawing either round or rectangular cakes, then grouping them into one big whole. The candy-bar problem encourages students to draw individual bars, then find the total. For the third problem, students drew glasses and imagined them being filled with liquid. Although the answer to every problem is 15 units, the different situations encourage flexible thinking.

Making mathematics tools available is one more suggestion to generate discussion ideas. I have found this strategy most effective when students work in small groups before coming to the whole-class discussion. Instead of choosing manipulatives for the students, teachers should let them choose their own. This creates a wide range of solutions and piques the interest of students who like to experiment with different tools.

Assessment

Assessing students' knowledge is one of the biggest challenges during inquiry-style discussions in a large group. Most of the learning occurs inside students' heads as they make sense of one another's strategies, and because teachers cannot see the processes, we must create situations to make them visible.

I have found two strategies most helpful. The first is for students to record others' strategies and apply them to new mathematical problems. After the mathematics discussion, students choose four strategies that they are interested in trying on their own. They copy the work on a page I call "Multiple Strategies" (see **figs. 4** and **5**). Then I assign a new, yet similar, problem, and on the back of the page students apply three of the four strategies to the new problem. This homework assignment holds every child accountable for classroom discourse. When students know they need to use one another's strategies, they listen better and ask more questions of their peers. They also engage in a complex skill by applying the strategies to new situations. They build flexibility with numbers, as well as methods to check the accuracy of their work. I grade the students' work using a four-point scale:

4 = Exceeds standard

3 = At standard

2 = Approaching standard

1 = Not at standard

A second assessment strategy holds learners directly accountable for their participation during discourse. When facilitating a discussion, I carry a clipboard that includes all students' names beside blank lines (see **fig. 6**). At the top of the page, I write the concept for that day's lesson or the actual problem that we are discussing, then choose a colored pen and write the date in that color. I write all comments for that day in the same color, and beside my notes I assign points (four points are possible, using the same scoring system as above). As long as the lessons focus on that particular concept, I continue to record my observations on the same sheet, using different colors for subsequent days. This ensures that I reach all students by the end of the week. A benefit of using the clipboard is that students know I am recording notes on their understandings (as well as behavior), and they perk up when they see me with it.

Reflection

I have found great success using the strategies outlined in this article with my fifth- and sixth-grade classes, not only in mathematics but in all subjects. The name box makes students accountable for what is being said in class at any time; they learn to be ready to ask a question or make a comment, which keeps them engaged. When students practice attentive body language, the classroom develops a dynamic and caring atmosphere. Implementing "pair-sharing" ensures that all students participate in discussions, even though only a few students may stand in front of the class and explain a strategy.

Most important, when students call on one another instead of waiting for the teacher to call on them, they see themselves as the source of learning. This generates questions because students are not sure that information that comes from other students rather than from teachers is "correct." Instead of passively learning and memorizing information that a teacher feeds to them, students consider one another's ideas and discover even more about the concept. When students look to one another for the answers, diversity fills the classroom and students take ownership for building their own understandings.

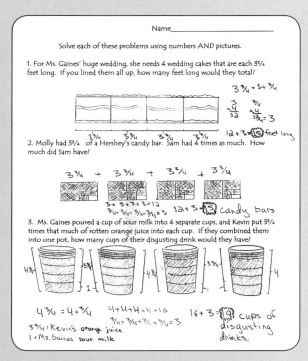

Fig. 3. Students' story problems

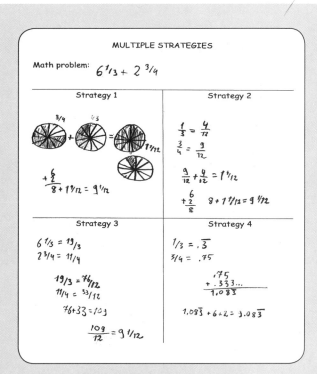

Fig. 4. Students try new strategies

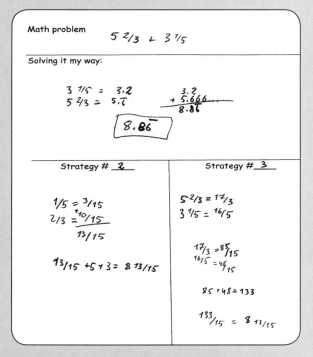

Fig. 5. Students' strategies

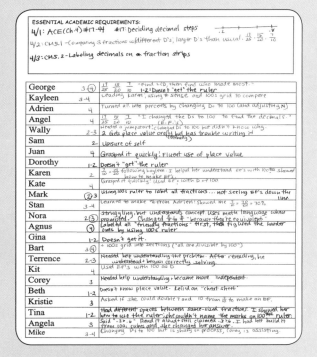

Fig. 6. Recording students' understanding

REFERENCES

Russell, Susan Jo. "Developing Computational Fluency with Whole Numbers." *Teaching Children Mathematics* 7 (November 2000): 154–58.

National Council of Teachers of Mathematics (NCTM). *Principles and Standards for School Mathematics.* Reston, Va.: NCTM, 2000.

Strategies for Advancing Children's Mathematical Thinking

Judith Fraivillig

This is a seminal article, very accessible to preservice and in-service teachers. The article helps teachers understand questions they might pose that will elicit, support, and extend students' thinking.

—*Rheta Rubenstein*

Ms. Smith, a first-grade teacher, asked her students how they could add thirty cents and twenty-seven cents. The students suggested several different solution methods. One student said that they could use the number grid to count. Ms. Smith asked where they should start, and the student said, "At thirty." Together with the student, Ms. Smith counted twenty-seven more squares on the grid. After finishing, Ms. Smith asked whether the students knew a shortcut. Another student said to start on 27 and count by tens to end at 57.

In this classroom example, the teacher encouraged the students to talk about how they might solve an addition problem, helped one student execute his method, and challenged students to consider alternative methods. This kind of interaction with students stands in sharp contrast with conventional mathematics teaching in which a teacher might ask a closed question, such as "If you have thirty cents and twenty-seven cents, how much do you have altogether?" Here, Ms. Smith engaged the children in mathematical thinking and generated mathematical discussion in the classroom. Her teaching exemplifies instructional ideas presented in the NCTM's *Standards* documents (1989, 1991, 2000).

What can we learn from this kind of teaching? How can teachers foster problem-solving skills in children? How can teachers advance children's thinking while students are engaged in mathematical inquiry? These questions have no easy answers. Fortunately, humans are naturally adept at learning from examples. By studying examples of effective instruction, we can begin to define instructional methods that have proved successful for other classroom teachers (Schifter 1996). To successfully apply lessons from these models, we must answer two principal questions: (1) What are the characteristics of effective teaching? (2) What general principles of instruction help children make sense of mathematics?

In this article, the practice of one teacher who masterfully engages students in mathematics learning serves as an example from which we can draw generalizations. The general principles of this kind of instruction are organized into a pedagogical framework that can guide other teachers as they move toward providing instruction that focuses on children's thinking. This "Advancing Children's Thinking" (ACT) framework can help teachers design and implement instruction to make mathematics personally meaningful for children. The ACT framework also establishes a structure for the often-complex interactions that occur when teachers and students grapple with real mathematics problems.

Background of the ACT Framework

The ACT framework was synthesized from an in-depth analysis of observed and reported data from one first-grade teacher who uses the *Everyday Mathematics* (*EM*) curriculum. This activity-oriented curriculum for the elementary grades draws on the child's "rich store of mathematics understanding and information" (Bell and Bell 1995, p. iii). The data used in this article were a subset of a larger study investigating the implementation of the *EM* curriculum among eighteen first-grade teachers and a

Teaching Children Mathematics 7 (April 2001): 454–59

longitudinal study charting the mathematics achievement of a group of students. Priscilla Smith, who allowed her real name to be used for this article, emerged as being distinct from the sample of teachers in the studies because of her ability to engage children in mathematical problem solving and to foster productive classroom discourse about complex mathematical issues.

What made Ms. Smith's instruction so effective? Three aspects of Ms. Smith's teaching contributed to her effectiveness: (1) her ability to elicit children's solution methods, (2) her capacity to support children's conceptual understanding, and (3) her skill at extending children's mathematical thinking. These related teaching components and the supportive classroom climate in which they were used form the basis of the ACT framework (see **fig. 1**). The following sections describe each of the framework's components and the underlying learning environment by articulating some of the effective teaching strategies used by Ms. Smith (see Fraivillig, Murphy, and Fuson [1999] for a fuller description of this study).

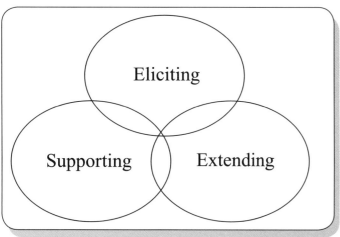

Fig. 1. Advancing Children's Thinking (ACT): A framework for providing learning opportunities in a safe environment

Eliciting children's solution methods

An important step in advancing children's thinking is to challenge children to describe and analyze their solution methods. The following paragraphs illustrate instructional strategies that Ms. Smith used to elicit children's solution methods (see **fig. 2**).

Elicit many solution methods for one problem. Rather than focus on a single answer to a mathematics problem, Ms. Smith attempted to foster discussion about how students solved a problem. She asked such questions as "Who did this problem another way?" "Did anyone do something else, that Allan did not do?" and "Can you use anything else besides your fingers and a number line to solve this problem?" By asking these types of questions, Ms. Smith encouraged children to share their ideas. Moreover, children in this classroom readily discovered that many approaches are available to solve problems.

Wait for, and listen to, students' descriptions of solution methods. A sense of calm and patience is needed to encourage children to express their ideas. Providing sufficient wait time and listening to ideas let children know that thoughtful explanations are more valuable than quick answers.

Encourage elaboration. Often, children need prompting to explain their thinking more completely. Even though Smith may have understood a child's response, she encouraged clarifications for the benefit of the entire class. On occasion, she assisted students in articulating their methods.

Use students' explanations for the lesson's content. Students' articulated ideas can furnish the content of class discussions. Ms. Smith treated the children's explanations of their solution methods and their mathematical thinking as the content of the lessons themselves. To support this goal, she usually listed the

Fig. 2. An elaboration of the eliciting component of ACT

students' methods on the board to help the class remember and refer to these methods in subsequent discussions.

Convey an attitude of acceptance toward students' errors and efforts. Ms. Smith accepted and often highlighted children's errors to turn them into "teachable moments." She explicitly told her students, "Don't worry about the answer yet," giving them time to explore and discuss various problem-solving strategies before evaluating the answer. The students in Smith's class realized that they would not necessarily be judged on the basis of their initial responses and enthusiastically contributed their thoughts. The students' eagerness to participate may be a result of Smith's accepting attitude.

Promote collaborative problem solving. Smith and her students exhibited a mutual respect for one another and worked as a team to solve problems. An atmosphere of intellectual excitement and team spirit pervaded the classroom when everyone was working on a problem. The students were eager to share their thinking.

Decide which students need opportunities to report. The art of facilitating a discussion requires a teacher to decide which students need opportunities to share their ideas. Sometimes this decision is based on prompting reports from a variety of students. Smith prompted many children to participate in the classroom conversation, but more important, she selected students who were able to contribute different solution methods to the discussion.

Supporting children's conceptual understanding

What should a teacher do after eliciting student's ideas? One possibility is to assist children in carrying out the solution methods that seem to mesh with their current cognitive abilities, or *zones of proximal development* (Vygotsky 1978). The supporting component of the ACT framework describes the instructional techniques used to support children's fragile understanding of their own solution methods, as well as to help them understand the ideas offered by peers. Examples of Smith's support of children's conceptual understanding during mathematics instruction are elaborated in the following paragraphs (see **fig. 3**).

Remind students of conceptually similar problem situations. To "jump-start" their thinking, children may need to be reminded that one problem is like another one that they have solved previously.

Review background knowledge. Reviewing necessary background knowledge with students is another effective support strategy. For example, Smith reviewed coin values for a student who was having trouble counting money.

Lead students through instant replays. Teachers can support the understanding of all children in the class by revisiting one child's elicited solution method in a slow and step-by-step fashion. This strategy is very different from that of a teacher who offers his or her own solution method as the only sanctioned method.

Write symbolic representations of each solution method on the board. Writing the symbols for the children's solution methods on the board helps children link the verbal descriptions of their thinking with the written mathematical marks.

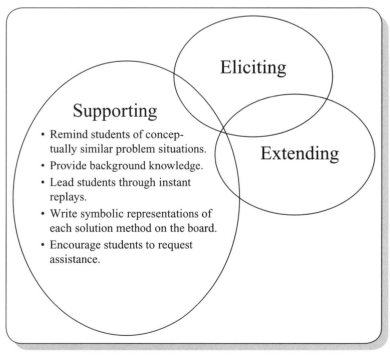

Fig. 3. An elaboration of the suppporting component of ACT

Smith noted another benefit of this strategy: "Recording on the board assists students in following the procedure. Some students must see the numbers. Constant review of this helps them write the digits."

Encourage students to request assistance. Ms. Smith expected children to request additional help when necessary. She attached no stigma to the requirement for extra help; on the contrary, students who requested assistance received extra time and attention from the teacher. This acceptance and expectation is an important aspect of a teacher's support of students' learning.

Extending children's mathematical thinking

Eliciting and supporting children's thinking alone do not challenge children's mathematical understanding. The extending component of the ACT framework describes teaching strategies that challenge children's thinking. In this regard, Smith's teaching was truly exceptional. The children in her classroom experimented with alternative solution methods, analyzed and compared solution methods, and generalized ideas across mathematical concepts. Some of the teaching strategies that she used to encourage this kind of learning are highlighted in the following paragraphs (see **fig. 4**).

Maintain high standards and expectations for all students. Ms. Smith asked all students to attempt to solve difficult problems. Students at all levels engaged in problem solving, although the complexity of their solution methods varied and they received different degrees of scaffolding from the teacher. All students, however, contributed to the classroom mathematics community.

Encourage students to draw generalizations. Smith challenged students to move beyond their initial problem-solving efforts and to generalize across mathematical concepts by modeling mathematical thinking. On one occasion, the teacher concluded a class discussion of the different number sentences that could be generated using the numbers 6, 2, and 8. As Smith began to distribute materials for a different activity, a student eagerly exclaimed, "It doesn't matter which way you put the numbers [6 and 2] together. It will always make the same answer as long as you use the same numbers." The teacher probed, "Is that true for addition *and* subtraction?" The first grader immediately began scribbling other number sentences on her paper, motivated by the teacher's challenge. Smith did not lose an opportunity to extend a student's thinking, even in the midst of performing management routines. She understood and accepted this student's expression of the commutative property, and she immediately led the student to extend the concept to test its generality.

List all solution methods on the board to promote reflection. During class discussions, Smith listed the different solution methods offered by the students. This strategy encourages reflection, chronicles the discussion for reference when a student's concentration wanders, and reinforces the classroom norm of valuing multiple ways of solving a problem.

Push individual students to try alternative solution methods. Smith challenged individual students to try solution methods that differed from their initial attempts. This strategy worked especially well for students who arrived at solu-

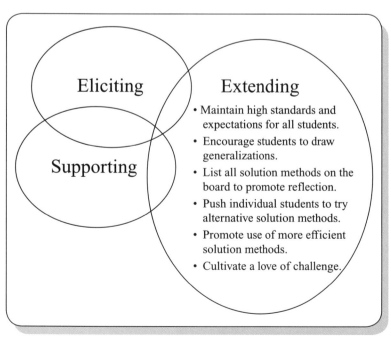

Fig. 4. An elaboration of the extending component of ACT

Eliciting

Supporting

Extending
- Maintain high standards and expectations for all students.
- Encourage students to draw generalizations.
- List all solution methods on the board to promote reflection.
- Push individual students to try alternative solution methods.
- Promote use of more efficient solution methods.
- Cultivate a love of challenge.

Growing Professionally

tion methods with ease. This type of challenge promotes flexibility in students' mathematical thinking. It also gave Smith a technique for managing the range of student achievement levels in her classroom.

Promote the use of more efficient solution methods. Even when a student was able to explain a solution method, Smith often asked whether the student could find a shorter way. She extended children's thinking beyond their first attempts. Smith's students began to understand that although many solution methods are valid, some methods are more efficient than others. Of course, this strategy requires the teacher to be sensitive to children's current and potential understanding so that students are not pushed beyond what they can do even with appropriate scaffolding.

Cultivate a love of challenge. The feeling of excitement in Smith's mathematics classroom was infectious. Smith cultivated this attitude by modeling her enthusiasm. She would respond to student-generated problems with such comments as "A challenge! I love it" or "Don't tell me. I want to figure it out myself." Smith encouraged students to challenge her and to challenge one another. The students in this class loved to pose difficult questions and to tackle complex problems.

Intersections between and among Components

Although eliciting, supporting, and extending describe elements of effective instruction, the art of teaching is much too complex to be captured by these three components. Classroom discussions and activities are interrelated and serve many functions. In fact, each classroom incident may be described by more than one of these components. For example, eliciting a response from one student may extend the thinking of other students. Smith demonstrated this interaction by following up students' explanations with such questions as "Does this rule apply in all situations?" or "How did you know that?" She also elicited and extended children's thinking at the same time by highlighting and discussing children's errors. For example, Smith pushed one student to examine his explanation of an incorrect solution method for determining the difference between 28 and 52. This strategy resulted in a productive whole-class mathematical discussion and one student's identification of a rule for using diagonals on the number grid. Using errors as learning opportunities was a hallmark of Smith's teaching.

Providing Learning Opportunities in a Safe Environment

The teaching practices described here could not be effective unless they occurred in a safe environment. Smith's efforts to elicit, support, and extend children's thinking were successful because she accepted children's ideas in a rational, nonthreatening manner and incorporated students' contributions into intellectual discussions. She established this environment by modeling respect for each child's thinking. She insisted that students listen to and respect other children's comments and questions. Smith explained this aspect of her philosophy:

> Students may not laugh at another student who is asking a question. That is the only time a student may be put out of the room. Without this rule, students would stop asking questions, which stops learning. Then students feel safe to explore, to do things. Then they are more available to learn and discuss.

The framework highlights the goal of helping children create their own meaningful understanding

Smith positively reinforced her students for expressing their ideas and emphasized each student's individual contributions to the classroom mathematics community.

Conclusion

Although no road map exists for reforming mathematics teaching, the ACT framework offers teachers a compass to guide their instruction. The framework highlights the goal of helping children create their own meaningful understanding of mathematical concepts. The eliciting component reminds teachers to consider how they might get children's thinking out in the open for discussion and build instruction on that thinking. The supporting component describes instructional strategies for assisting children at their current levels of understanding. The extending component prompts teachers to challenge children's thinking regardless of the student's initial efforts. Of course, the complex art of teaching for meaningful mathematical learning is highly personal; each teacher must incorporate these guidelines into his or her own instructional style. The ACT framework can serve teachers who wish to create classrooms in which all children's mathematical thinking is elicited, supported, and extended.

REFERENCES

Bell, Jean, and Max Bell. *First Grade Everyday Mathematics: Teacher's Manual and Lesson Guide.* 2nd ed. Chicago: Everyday Learning Corp., 1995.

Fraivillig, Judith, Lauren A. Murphy, and Karen C. Fuson. "Advancing Children's Mathematical Thinking in Everyday Mathematics Classrooms." *Journal for Research in Mathematics Education* 30 (March 1999): 148–70.

National Council of Teachers of Mathematics (NCTM). *Curriculum and Evaluation Standards for School Mathematics.* Reston, Va.: NCTM, 1989.

———. *Professional Standards for Teaching Mathematics.* Reston, Va.: NCTM, 1991.

———. *Principles and Standards for School Mathematics.* Reston, Va.: NCTM, 2000.

Schifter, Deborah, ed. *What's Happening in Math Class? Envisioning New Practices through Teacher Narratives.* Vol. 1. New York: Teachers College Press, 1996.

Vygotsky, L. S. *Mind in Society: The Development of Higher Psychological Processes.* Cambridge, Mass.: Harvard University Press, 1978.

The research reported in this article is based on a dissertation study and was supported by the National Science Foundation (NSF) under grant no. ESI 9252984. The opinions expressed in this article are those of the author and do not necessarily reflect the views of NSF. The author would like to recognize the support and cooperation of all participating teachers, particularly that of Priscilla Smith.

Questioning Our Patterns of Questioning

Beth A. Herbel-Eisenmann
M. Lynn Breyfogle

TEACHERS pose a variety of questions to their students every day. As teachers, we recognize that some questions promote deeper mathematical thinking than others (for more information about levels of questions, see Martens 1999, Rowan and Robles 1998, and Vacc 1993). For example, when asking, "Is there another way to represent or explain what you are saying?" students are given the chance to justify their thinking in multiple ways. The question "What did you do next?" focuses only on the procedures that students followed to obtain an answer. Thinking about the questions we ask is important, but equally important is thinking about the patterns of questions that are asked.

Although *Principles and Standards for School Mathematics* (NCTM 2000) highlights the importance of asking questions that challenge students, we conjecture that focusing only on the questions asked is not going far enough to help students to clarify and develop their mathematical thinking. When engaging students in discussion, consider what happens in the exchanges *after* an initial question is posed; in other words, examine the *interaction patterns* that occur. In some situations, the pattern of interaction encourages students to participate, shows that students' thinking is valued, and helps them clarify their thinking. In other situations, the interaction may hinder students from describing what they think. Early research on classroom interactions documented the Initiation-Response-Feedback (IRF) pattern (Mehan 1979) as the most prominent form of interaction that occurs between the teacher and learners. With this pattern, the teacher asks a question, a student provides a response, and the teacher offers evaluative feedback. This IRF pattern can be seen in the following example from an eighth-grade mathematics classroom.

Example 1

Teacher [Initiation]: What kind of mathematical relationship does this equation [$y = 2x + 5$] show?

Student [Response]: A linear relationship.

Teacher [Feedback]: Okay. It's a linear relationship [Herbel-Eisenmann 2000].

Although this form of interaction was identified and described over twenty-five years ago, it is still prevalent in classrooms today (Stigler and Hiebert 1999). Since this type of interaction has been shown to lead students through a predetermined set of information and does little to encourage students to express their thinking (Cazden 1988; Nystrand 1997), we offer alternative ways to broaden the interactions that occur.

We begin with another common form of interaction called "funneling" (Wood 1998), which limits the students' responses but not as much as the IRF described above. Next, we illustrate the open-ended "focusing" (Wood 1998) interaction, which draws on students' thinking, and then suggest ways to turn funneling into focusing interactions. The examples that we use come from two nontracked eighth-grade classrooms that are using curriculum materials from the Connected Mathematics Project (Lappan, Fey, Fitzgerald, Friel, and Phillips 1998a). After illustrating these two forms of interaction, we conclude

Ideas for using this article include these:

1. Before reading the text, share each of the vignettes, asking teachers to comment on the expectations of the teacher, the style of the discourse, and the opportunities for learning. Ask teachers to take on the task, described in the article, of turning a funneling example into a focusing example. The article is an excellent follow-up to this activity, and teachers can discuss what characteristics of discourse lead to a focusing pattern.

2. Similarly, this article can set the stage for developing unit plans. It is an excellent resource in helping teachers craft questions that "focus" students' thinking rather than "funnel" it. —*Karen Brannon*

3. This article can be used in study groups or discussion groups to discuss how to have a successful classroom discourse. It can follow a lesson study or shared observation, especially when the lesson includes some of the strategies recommended in the article. —*Phyllis Tam*

Mathematics Teaching in the Middle School 10 (May 2005): 484–89

1. Circle the name of the graph or graphs that show a linear relationship, and write their equations.
2. Explain how you can recognize a linear relationship from a graph.

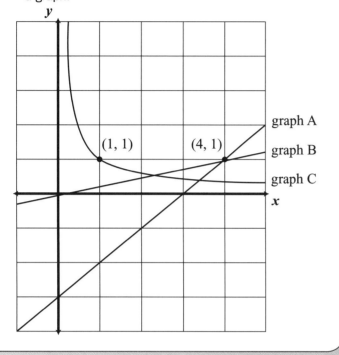

graph A
graph B
graph C

(1, 1)   (4, 1)

Fig. 1. Students find the equation for graph B.

by suggesting a plan for using these ideas to examine one's own classroom interactions by further encouraging students to share their thinking.

Funneling

Funneling occurs when the teacher asks a series of questions that guide the students through a procedure or to a desired end. In this situation, the teacher is engaged in cognitive activity and the student is merely answering the questions to arrive at an answer, often without seeing the connection among the questions. This pattern can be seen in example 2 below when the teacher directs students to find the equation for graph B in **figure 1** (Lappan, Fey, Fitzgerald, Friel, and Phillips 1998b, p. 63).

Example 2

Teacher: (0, 0) and (4, 1) [are two points on the line in graph B]. Great. What's the slope?

[Long pause—no response from students.]

Teacher: What's the rise? You're going from 0 on the *y* [axis] up to 1? What's the rise?

Students: 1.

Teacher: 1. What's the run? You're going from 0 to 4 on the *x* [axis]?

Students: 4.

Teacher: So the slope is _____?

Students: 0.25 [in unison with the teacher].

Teacher: And the *y*-intercept is?

Students: 0.

Teacher: So, $y = 1/4x$? Or $y = 0.25x$ would be your equation [Herbel-Eisenmann 2000].

When students do not respond to the teacher's initial question of "What's the slope?" the funneling pattern begins. The teacher walks through a series of steps with the students until they find the correct equation for the line. The students' attention is focused on subtracting the numbers that the teacher gives rather than on thinking about the relationship between points on a line and rise, run, and slope. With funneling, although the "teacher may intend that the child use strategies and learn about the relationships between numbers, the student needs to know only how to respond to the surface linguistic patterns to derive the correct answer" (Wood 1998, p. 172). The end result is that the teacher "funnels" the students' responses to include only the exact information that they were investigating. Only the teacher's thinking process is explicit; little is known about what the students were actually thinking.

Another interpretation of this example could be that the teacher is scaffolding the students' thinking by modeling the questions one would ask when finding a linear equation (given two points on a line). However, two important aspects need to occur in future interactions: (1) the teacher should discuss these particular questions and the purpose for attending to them, and (2) the questions need to be diminished and eventually removed. Students will not immediately understand the significance of this series of questions because they view asking questions as being characteristic of the teacher's role. To distinguish this set of questions as being *different* from other questions the teacher asks, it would be important to stop and discuss the purpose of these particular questions for finding a linear equation. Also, if the teacher continues to ask this same series over time, the questions are not serving a "scaffolding" purpose; the assistance would need to be gradually withdrawn until students had learned to ask themselves these questions independently (Cazden 1988).

Employing a funneling-interaction pattern limits what students are able to contribute because it directs their thinking in a predetermined path based only on how the teacher would have solved the problem. Students need more opportunities to articulate their thinking so that they can build on prior knowledge and make their ideas clear to the teacher and their classmates. An alternative interaction pattern that allows this situation to occur is called "focusing."

Focusing

As an alternative to funneling student responses, Wood (1998) suggests "focusing." A focusing-interaction pattern requires the teacher to listen to students' responses and guide them based on what the students are thinking rather than how the teacher would solve the problem. This pattern of interaction serves many purposes, such as allowing the teacher to see more clearly what the students were thinking or requiring the students to make their thinking clear and articulate so that others can understand what they are saying. This type of interaction values student thinking and encourages students to contribute in the classroom.

Example 3 below, from a different eighth-grade mathematics class, illustrates one way to focus the discussion. Students were given a point on a line (5, 9.5) and the slope for that line (1.5). They were asked to find the *y*-intercept so that they could write the linear equation. When one student, Becky, offered a novel way to use the graphing calculator to find the equation, the teacher asked questions or replied in ways that helped Becky articulate her thinking so that it would make sense to the teacher and the students in the class.

> A focusing interaction values student thinking and encourages students to contribute in the classroom

Example 3

Teacher: I don't know this [pointing to the *y*-intercept]. I've got to find it. How do I find it [the *y*-intercept] if this [the slope and one point on the line] is all I know?

Becky: You know what you can do? You can put an equation in the graph and just calculate it out.

Teacher: How?

Becky: If you put $y = 1.5x$ and then go to the table and find where 5 is.

Mark: Yeah, but then the starting point would be 0.

Becky: No, when 5 is *x*, you find whatever *y* is and then whatever the difference is between *y*, that *y* and your other *y*. . . .

Mark: 7.5 [is the *y*-value when *x* is 5].

Becky: . . . is the *y*-intercept.

Teacher: Help.

Sam: Come again?

Becky: You put in $y = 1.5x$ in the graphing calculator.

Teacher: Okay.

Becky: And you go to table [on the graphing calculator].

Teacher: Let's do that because she lost me after that. But if you're putting $1.5x$ into your calculator and you know it's crossing at 0, 0 . . . . So, you said put 1.5x in even though you know that's not the right equation?

Becky: Yeah.

Teacher: Okay, then you wanna do what?

Becky: Go to the table and look for where 5 is [the x-value]; it's 7.5, right?

Teacher: Yeah.

Becky: And then whatever the difference is between that one and 9.5 is your [inaudible] or your y-intercept.

Teacher: So, you're saying the y-intercept is 2?

Becky: Yep. And then you can [inaudible] 5x + 2 and where x is 5.

Keith: [Asks the teacher] can you say that in English so we can write that down?

Teacher: I'm not sure that I understand her. I'm going to ask if I'm correct. On her calculator—and you can look at it on yours—um, she said she put in [y =] 1.5x, even though she knew that wasn't right. That's assuming it crossed at (0, 0.) She went down to 5 and at that equation, it was at 7.5. So, then you took the difference between 9.5 and 7.5? And said the new y-intercept should be 2 [Herbel-Eisenmann 2000].

In this example, Becky is using the graphing calculator to figure out the equation of the line. The teacher recognizes that Becky's method is novel and has not appeared in the mathematical solutions in previous class sessions. Becky is asked to explain what she did to everyone in the class (including the teacher) when the teacher says, "Help." To assist Becky in articulating her strategy and to aid everyone else's sense-making, the teacher suggests that Becky go back through the process while everyone else follows along on their graphing calculators. At points when the teacher thinks Becky's strategy might be confusing, she asks questions (e.g., "So, you said put 1.5x in even though you know that's not the right equation?") and restates Becky's strategy, focusing student attention on what Becky did. This situation does not allow students' attention "to fade or change or be interrupted" (Wood 1998, p. 174). Rather than attempt to funnel Becky's strategy to the teacher or textbook's solution strategy, the teacher instead holds Becky responsible for articulating her thinking. The teacher "tries to anticipate what the other students might not understand and asks clarifying questions [and restates particular aspects of the solution] to keep attention focused on the discriminating aspects of the solution" (Wood 1998, p. 175). The classroom discussion then turned to figuring out why Becky's strategy worked and pursuing how changes in the slope and y-intercept in the equation effect the shifts in the line on the graph. Although funneling is a more common classroom interaction pattern, we maintain that the long-term benefits of focusing make it imperative that mathematics teachers "focus" more often.

When a strategy might be confusing, the teacher asks questions

Turning "Funneling" into "Focusing"

We now return to example 2 and discuss how this classroom interaction *could* have "focused" students' ideas rather than "funneled" them. The first two italicized lines are the same as those in the original example 2; the remainder of the dialogue explores how a funneling pattern was changed to a focusing pattern. In this revised version, the teacher helps students make conceptual connections and draws out students' thinking.

Example 2 (Revised)

Teacher: (0, 0) and (4, 1) [are two points on the line in graph B]. Great. What's the slope?

[Long pause—no response from students.]

Teacher: What do you think of when I say slope?

Kara: The angle of the line.

Teacher: What do you mean by the angle of the line?

Kara: What angle it sits at compared to the *x*- and *y*-axis.

[Pause for students to consider.]

Teacher: What do you think Kara means?

Sam: I see what Kara's saying, sort of like when we measured the steps in the cafeteria and the steps that go up to the music room—each set of steps went up at a different angle.

Teacher: How did we know they went up at a different angle?

Sam: The music room steps were steeper than the cafeteria steps.

Teacher: How did we decide that the music room steps were steeper?

Lana: We measured how far up the step went and then we measured how far back the step went and then we divided the numbers.

Teacher: Lana, could you draw us an example of what you mean?

Lana: Hmm. Yea. [She draws stair steps on the board where the height is 12 inches and the depth is 12 inches.] So here the steepness is 1, because 12 ÷ 12 is 1.

Teacher: Okay. Let's say the height was 10 inches and the depth was 12 inches—which set of stairs is steeper? Jennifer?

Jennifer: I would say the first set, because you are going up as much as going forward, but in the second set you aren't going up as much as forward.

Teacher: Tom, do you agree?

Tom: Yes, because I think the steepness of the second is 10/12, which is not as big as 1.

Teacher: So, let's consider what Jennifer and Tom are saying. If I were to lean a board against the two sets of stairs, the 12-by-12 steps have a steepness, or slope, of 1 and are steeper than the second set of steps, which have a slope of 10/12. Is this right?

[Class nods and says "yes."]

Teacher: So, let's go back to our original problem and think through it again. This time I need to think about leaning a board against the points (0, 0) and (4, 1). How steep would it be—or what is its slope?

Jennifer: Well, we would go up 1 and over 4.

Teacher: Okay, so how could we determine the value of the slope?

Lana: We have to divide the numbers.

Teacher: How do we divide them?

[Students respond with both 1/4 and 4/1.]

Lana: I would say that it's 4, because you should do 4 divided by 1.

Jennifer: But 4 is bigger than 1/4 and 4 would be steeper than the 12 by 12 we looked at, so to me that would mean that we went up 4 and over 1, not up 1 and over 4.

Tom: Right, I say its 1/4.

Teacher: Tom, why do you say it is 1/4?

Tom: Because like we talked about with the music stairs, it's the amount we go up or down divided by the amount that we went over. It was 10/12, not 12/10.

Teacher: Lana, what do you think about what Tom and Jennifer are saying?

Lana: Yes, I agree, it makes sense what they said—steeper would mean up more than over. And, the slope of 4 would be much steeper than the slope of the 12 by 12, but this line is not as steep as that.

Is the
interaction
pattern allowing
the discussion
to achieve the
goals of the
lesson?

Teacher: Now, I would like you to consider the points (–1, 3) and (2, 5) and write down the value of the slope and what you thought about to arrive at your answer.

In this revised example, the teacher thinks that the pause indicates that students are not sure about what is being asked; they may not remember what the slope is or how to find it. Acting on this assumption, the teacher requires the students to articulate what the slope is and refer back to a previous problem they solved. Students are often asked what they mean and to decide if they agree or disagree with others' ideas. The teacher repeats important information and keeps students focused on the components of slope, not only valuing the language that students use ("steep") but also subtly offering the more mathematically appropriate language ("slope") (Herbel-Eisenmann 2002). In this revised example, the teacher does not do the thinking for the students. Instead, students are helped to make connections and articulate their thinking by using their contributions to probe further and by referring back to common activities that occurred in the classroom. Not only does this action value and draw out student thinking but it also supports two of the teacher's goals: (1) to help students make connections and (2) to encourage multiple representations (by capturing a visual image of the slope of a line and its relationship with two points on the line).

In sum, managing classroom interactions needs to include paying attention to how an initial question is followed up and how it relates to the goals of the lesson. After a question is asked, a teacher might only offer evaluative feedback, which does little to further the thinking about the mathematical content. Two classroom interactions are "funneling" and "focusing." When funneling, the student is still guided toward a predetermined solution strategy. The teacher takes over the thinking for the students, who may be paying more attention to language cues rather than the mathematical topics at hand. An alternative to consider following initial questions, and one that we suggest is applicable to most lessons, is to "focus" student solutions. In this situation, the teacher points out salient features of the students' solution strategies by asking them to explain what they mean, then restating what students have said. This interaction pattern helps students articulate their own thinking to one another and encourages students to make sense of one another's strategies and reasoning.

Examining Management Strategies

Principles and Standards for School Mathematics (NCTM 2000) challenges teachers to "encourage students to think, question, solve problems, and discuss their ideas, strategies, and solutions" (p. 18). Getting students to articulate their thinking is difficult and must include looking beyond the initial questions that are posed. To help with the transition to focusing more often, we have seen how important it is to get a broader view on one's own teaching by audio- or videotaping a classroom session (Breyfogle and Herbel-Eisenmann 2004). By watching segments of classroom discussions, it is easy to identify what kinds of interaction patterns are taking place.

An important question to consider when investigating one's own teaching practice is this: Is the interaction pattern allowing the discussion to achieve the goals of the lesson? It is then important to examine whether the pattern is helping students' articulate their thinking or is mainly providing feedback (as in the IRF) or funneling students to use only the strategy we want them to use. Once we identify our current interaction patterns, we can then try to modify them to focus student thinking more often so that students contribute more frequently and can see that we value their thinking. For a way to use these ideas to reflect on your own classroom interactions, we suggest the following reflective process:

- When students are prepared to discuss a "worthwhile mathematical task," use an audio- or videorecorder to capture the conversation that takes place.

- Listen to the interaction that took place and attend carefully to both the initial question that was asked and (more important) how that question was followed up. Write down the series of questions that were asked and try to identify when an IRF pattern was being used, when funneling was occurring, and when students' thinking was being focused. Then pinpoint instances when student's thinking could have been focused rather than using an IRF or a funneling pattern. Make a list of questions that could have helped in understanding or valuing the student's thinking in a "focusing" manner.

- Find another worthwhile mathematical task to use with students. When planning, try to anticipate multiple solution strategies that students might offer as well as areas that might be confusing for some students. Use that information to decide what kinds of questions to ask to focus student thinking.

- Audio- or videotape the implementation of this task. Repeat the reflection process to see if students' were helped to focus their thinking.

REFERENCES

Breyfogle, M. Lynn, and Beth A. Herbel-Eisenmann. "Focusing on Students' Mathematical Thinking." *Mathematics Teacher* 97 (April 2004): 244–47.

Cazden, Courtney B. *Classroom Discourse: The Language of Teaching and Learning.* Portsmouth, N.H.: Heinemann, 1988.

Herbel-Eisenmann, Beth A. "How Discourse Structures Norms: A Tale of Two Middle School Mathematics Classrooms." Unpublished Ph.D. diss., Michigan State University, 2000.

———. "Using Student Contributions and Multiple Representations to Develop Mathematical Language." *Mathematics Teaching in the Middle School* 8 (October 2002): 100–105.

Lappan, Glenda, James Fey, William Fitzgerald, Susan Friel, and Elizabeth Phillips. *The Connected Mathematics Project.* Palo Alto, Calif.: Dale Seymour Publications, 1998a.

———. *Thinking with Mathematical Models.* Menlo Park, Calif.: Dale Seymour Publications, 1998b.

Martens, Mary L. "Productive Questions: Tools for Supporting Constructivist Learning." *Science and Children* (May 1999): 24–27; 53.

Mehan, Hugh. *Learning Lessons.* Cambridge, Mass.: Harvard University Press, 1979.

National Council of Teachers of Mathematics (NCTM). *Principles and Standards for School Mathematics.* Reston, Va.: NCTM, 2000.

Nystrand, Martin. "Dialogic Instruction: When Recitation Becomes Conversation." In *Opening Dialogue: Understanding the Dynamics of Language and Learning in the English Classroom.* New York: Teachers College Press, 1997.

Rowan, Thomas E., and Josepha Robles. "Using Questions to Help Children Build Mathematical Power." *Teaching Children Mathematics* 4 (May 1998): 504–9.

Stigler, James W., and James Hiebert. *The Teaching Gap.* New York: The Free Press, 1999.

Vacc, Nancy N. "Implementing the *Professional Standards for Teaching Mathematics:* Questioning in the Mathematics Classroom." *Arithmetic Teacher* 41 (October 1993): 88–91.

Wood, Terry. "Alternative Patterns of Communication in Mathematics Classes: Funneling or Focusing?" In *Language and Communication in the Mathematics Classroom,* edited by Heinz Steinbring, Maria G. Bartolini Bussi, and Anna Sierpinska, pp. 167–78. Reston, Va.: NCTM, 1998.

Discourse That Promotes Conceptual Understanding

Elham Kazemi

A S MATHEMATICS teachers, we want students to understand mathematics, not just to recite facts and execute computational procedures. We also know that allowing students to explore and have fun with mathematics may not necessarily stimulate deep thinking and promote greater conceptual understanding. Tasks that are aligned with the NCTM's curriculum standards (NCTM 1989) and that are connected to students' lives still may not challenge students to build more sophisticated understandings of mathematics. The actions of the teacher play a crucial role.

This article presents highlights from a study that demonstrates what it means to "press" students to think conceptually about mathematics (Kazemi and Stipek 1997), that is, to require reasoning that justifies procedures rather than statements of the procedures themselves. This study assessed the extent to which twenty-three upper elementary teachers supported learning and understanding during whole-class and small-group discussions. "Press for learning" was measured by the degree to which teachers (1) emphasized students' effort, (2) focused on learning and understanding, (3) supported students' autonomy, and (4) emphasized reasoning more than producing correct answers. Quantitative analyses indicated that the higher the press in the classroom, the more the students learned.

Like researchers in other studies (e.g., Fennema et al. [1996]), we observed that when teachers helped students build on their thinking, student achievement in problem solving and conceptual understanding increased. To understand what press for learning looks like in classrooms, we studied in depth two classes with higher scores for press and two classes with lower scores, and we looked closely at mathematical activity and discourse in the classes. The high-press classroom of Ms. Carter is contrasted with the low-press classroom of Ms. Andrew.

Students in Ms. Carter's and Ms. Andrew's classes were exploring the concept of equivalence and the addition of fractions. They worked on fair-share problems, such as the following:

> I invited 8 people to a party (including me), and I had 12 brownies. How much did each person get if everyone got a fair share? Later my mother got home with 9 more brownies. We can always eat more brownies, so we shared these out equally too. This time how much brownie did each person get? How much brownie did each person eat altogether? (Corwin, Russell, and Tierney 1990, 76)

Similarities between Classrooms: Social Norms

In both Ms. Carter's and Ms. Andrew's classes, we saw students huddled in groups, materials scattered about them, figuring out how to share a batch of brownies equally among a group of people. The students seemed to be engaged in and enjoying their work. Often each group found a slightly different strategy to solve the problem. After moving from group to group, listening to and joining student conversations, both teachers stopped group activity to ask students to share their work and explain how they solved the problem.

The NCTM *Standards* document supports the view that social norms—practices such as explaining thinking, sharing strategies, and collaborating that we see in both

Teaching Children Mathematics 4 (March 1998): 410–14

classrooms—afford opportunities for students to engage in conceptual thinking. Many teachers establish those social norms in their classrooms quite readily. But social norms alone may not advance students' conceptual thinking.

Differences between Classrooms: Sociomathematical Norms

Although Ms. Andrew and Ms. Carter both valued problem solving and established the same social norms in their classrooms, important differences were seen in the quality of their students' engagement with the mathematics. To understand those differences, we looked more closely at the norms that guide the quality of mathematical discourse, the *sociomathematical* norms (Yackel and Cobb 1996). Teachers and students actively negotiate the sociomathematical norms that develop in any classroom. Sociomathematical norms identify what kind of talk is valued in the classroom, what counts as a mathematical explanation, and what counts as a mathematically different strategy. In the brownie problem, for example, students grapple with ideas of equivalence, part-whole relations, and the addition of fractional parts. Sociomathematical norms help us understand the ways in which fraction concepts are supported within the context of sharing and explaining strategies.

Through our study of the four classrooms, we identified four sociomathematical norms that guided students' mathematical activity and helped create a high press for conceptual thinking:

- Explanations consisted of mathematical arguments, not simply procedural summaries of the steps taken to solve the problem.
- Errors offered opportunities to reconceptualize a problem and explore contradictions and alternative strategies.
- Mathematical thinking involved understanding relations among multiple strategies.
- Collaborative work involved individual accountability and reaching consensus through mathematical argumentation.

Other norms may also contribute to a high press, but these norms captured the major differences in the way that mathematics was treated by the high- and low-press teachers.

Explaining strategies

The following examples illustrate some of the differences in the two classrooms. First, in Ms. Carter's class, explanations were not limited to descriptions of steps taken to solve a problem. They were always linked to mathematical reasons. In the following example, Ms. Carter asked Sarah and Jasmine to describe their actions and to explain why they chose particular partitioning strategies.

Sarah: The first four we cut them in half.

[Jasmine divides squares in half on an overhead transparency. See **fig. 1.**]

Ms. Carter: Now as you explain, could you explain why you did it in half?

Sarah: Because when you put it in half, it becomes four . . . four . . . eight halves.

Ms. Carter: Eight halves. What does that mean if there are eight halves?

Sarah: Then each person gets a half.

Ms. Carter: Okay, that each person gets a half. [Jasmine labels halves 1 through 8 for each of the eight people.]

Sarah: Then there were five boxes [brownies] left. We put them in eighths.

Ms. Carter: Okay, so they divided them into eighths. Could you tell us why you chose eighths?

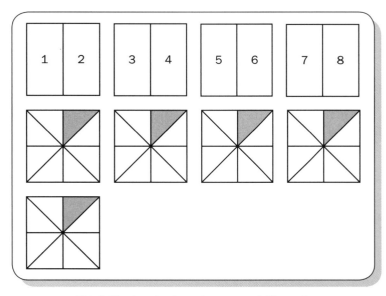

Fig. 1. Sharing nine brownies among eight people

Sarah: It's easiest. Because then everyone will get . . . each person will get a half and [addresses Jasmine] "How many eighths?"

Jasmine: [Quietly] Five-eighths.

Ms. Carter: I didn't know why you did it in eighths. That's the reason. I just wanted to know why you chose eighths.

Jasmine: We did eighths because then if we did eighths, each person would get each eighth, I mean one-eighth out of each brownie.

Ms. Carter: Okay, one-eighth out of each brownie. Can you just, you don't have to number, but just show us what you mean by that? I heard the words, but . . . .

[Jasmine shades in one-eighth of each of the five brownies that were divided into eighths.]

Jasmine: Person one would get this . . . [points to one-eighth].

Ms. Carter: Oh, out of each brownie.

Sarah: Out of each brownie, one person will get one-eighth.

Ms. Carter: One-eighth. Okay. So how much did they get if they got their fair share?

Jasmine and Sarah: They got a half and five-eighths.

Ms. Carter: Do you want to write that down at the top, so I can see what you did?

[Jasmine writes $1/2 + 1/8 + 1/8 + 1/8 + 1/8 + 1/8$ at the top of the overhead transparency.]

The exchange among Sarah, Jasmine, and Ms. Carter highlighted the conceptual focus of the lesson on fair share. Ms. Carter asked Sarah to explain the importance of having eight halves and why the partitioning strategy using eighths made sense. After Jasmine gave a verbal justification, Ms. Carter continued to press her to link her verbal response to the appropriate pictorial representation—by shading the pieces—and to the symbolic representation—by writing the sum of the fractions.

The same degree of press did not exist in Ms. Andrew's classroom. Ms. Andrew's students engaged in the same social practice of sharing their strategies with the class, but the mathematical content of classroom conversations was different. Students shared solutions by giving procedural summaries of the steps they took to solve the problem, as demonstrated by the following exchange, in which Raymond described his solution for sharing twelve brownies among eight people. Ms. Andrew had drawn twelve squares on the chalkboard.

[Raymond divides four of the brownies in half.]

Ms. Andrew: Okay, now would you like to explain to us what . . . loud . . . .

Raymond: Each one gets one, and I give them a half.

Ms. Andrew: So each person got how much?

Raymond: One and one-half.

Ms. Andrew: One-half?

Raymond: No, one and one-half.

Ms. Andrew: So you're saying that each one gets one and one-half. Does that make sense? [After a chorus of "yeahs" comes from students, Ms. Andrew moves on to another problem.]

Unlike Ms. Carter, Ms. Andrew did not ask her students to justify why they chose a particular partitioning strategy. Instead, Ms. Andrew often asked questions that required a show of hands or yes-no responses, such as "How many people agree?" "Does this make sense?" or "Do you think that was a good answer?" Ms. Andrew wanted to engage her students in the activity and to see if they understood, but the questions she asked yielded general responses without revealing specific information about the students' thinking.

Reacting to mathematical errors

By emphasizing mathematical reasons for actions, Ms. Carter created opportunities for her students to prove that their solutions were correct. She resisted telling students that an answer or reason was wrong, and she invited others to respond to incorrect solutions. Ms. Carter modeled the kinds of questions that may help students think through their own confusion by using their existing knowledge. Those questions usually involved graphical representations of the fractions. In small groups, students challenged one another when they disagreed on a solution and helped one another find errors.

The interaction among Ms. Carter, Jasmine, and Sarah continued with the following conversation.

[Jasmine writes 1/2 + 1/8 + 1/8 + 1/8 + 1/8 + 1/8 at the top of the overhead transparency.]

Ms. Carter: Okay, so that's what you did. So how much was that in all?

Jasmine: It equals 1 1/8 or 6/8.

Ms. Carter: So she says it can equal 6 and 6/8? [She misheard Jasmine.]

Jasmine: No, it can equal 6/8 or it can equal 1 1/8.

Ms. Carter: Okay, so you have two different answers. Could you write them down so people can see it? And boys and girls, I'd like you to respond to what they've written up here. She says it either could equal 6/8 or 1 1/8.

Ms. Carter: Matthew had his hand up and was thinking about it. Someone from team 5. Anybody from team 6 that has a response? Right now. I'm just going to let you look a minute. See if anyone has a response. Andrew, you had your hand up, is that right? [waits] Still only have four hands up. I wonder if you're all looking up here and seeing? She's given us two answers here, 6/8 or it can equal 1 1/8. Okay, could those four people right now . . . do you agree with both answers?

Students: No . . . .

Ms. Carter: Do you have a reason why you don't agree? Don't explain it to me, but do you have a reason? Raise your hand if you have a reason why you don't agree. [hands] One, two, three, four, five, six . . . okay. Would those six people please stand? Okay. Would you please, you're going to be in charge of explaining why you don't agree to your team. [She assigns those six students to teams.] Right now, if you don't agree, would you please tell them what you think the answer is and why you don't agree. Go ahead. Explain.

Ms. Carter could have stepped in and pointed out why 6/8 and 1 1/8 are not equal. Instead, her response to this mistake was to encourage her students to explore the error by providing the conceptual reasons for why 6/8 and 1 1/8 are not equal. She engaged the entire class in thinking about which solution was correct instead of talking with only the two presenters or correcting their mistake herself, and she created an opportunity for her students to practice articulating their thinking.

The mistake also created an opportunity for the entire class to explore contradictions in the solution and to build an understanding of fractional equivalence and the

Ms. Carter created opportunities for her students to prove that their solutions were correct

addition of fractions by using an area model. This type of activity and discourse was typical in Ms. Carter's classroom. In a whole-class discussion, each group shared its proof that 1 1/8 was correct. Neither the students nor Ms. Carter belittled, penalized, or discredited anyone who made a mistake. The atmosphere of mutual respect between the students and Ms. Carter allowed the class to think about and build conceptual understandings eagerly.

Ms. Andrew treated errors differently. Note how she provides the mathematical reasoning when three boys explained their solution for sharing five brownies among six people.

Ms. Andrew: They got 1/2, you already said that. And then 1/6 and then another sixth. So, how many sixths did they get?

Anthony: One, two.

Ryan: One, two.

Joe: 1/12.

Ms. Andrew: What did you say? [to Joe] They got two . . .

Ryan: Sixths.

Anthony: 2/12.

Joe: 2/6.

Ms. Andrew: 2/6 [confirming the right answer]. Why did you say 2/12? Because there are twelve parts altogether?

Anthony: Yeah.

Ms. Andrew: Okay, be sure not to get confused. Because there are two brownies not one. Perfect. Good, good job.

At first, the boys appeared to be guessing the answer to Ms. Andrew's question. She focused on Joe once he stated the right answer. Although she predicted accurately why Anthony said 2/12, she did not ask him to think about why his answer did not work. Instead, she asked and answered the question herself and did not press Anthony to sort out his confusion. Her statement "Because there are two brownies, not one" was left unexplained. As this example illustrates, limited opportunity was available for the members of the group to engage in conceptual thinking about what 1/6 and 1/12 signify and how the graphical representation is linked to the numeric representation.

Both Ms. Carter and Ms. Andrew allowed students to make mistakes. That social norm, however, was not enough to press students to examine their work conceptually. Both teachers wanted their students to learn from their mistakes, but Ms. Andrew often supplied the conceptual thinking for her students. In Ms. Carter's class, inadequate solutions served as entry points for further mathematical discussion.

Comparing strategies

Students in both classrooms worked together, shared their strategies, and were praised for their efforts. Students in both classrooms attended to nonmathematical similarities between shared solutions, such as the layout of the paper or the uses of color. In Ms. Andrew's class, strategies were typically offered one after the other, with discussion limited to nonmathematical aspects of students' work. For example, a pair of students noted that they cut the paper brownies and pasted the pieces under stick-figure illustrations. Another pair had drawn lines from the fractional parts of the brownies to the individuals that received them. Although the partitioning strategy in both was the same, students viewed the strategies as different because the representations were different. Ms. Carter, however, pressed her students to go beyond their initial observations and reflect on the mathematical similarities and differences between strategies.

Ms. Carter pressed her students to reflect on the mathematical similarities and differences between strategies

Accountability and consensus

In inquiry-based classrooms, students often work together to share interpretations and solutions and construct new understandings. Important differences arose between Ms. Andrew's and Ms. Carter's classes in the way in which they emphasized individual accountability and consensus. Ms. Carter required her students to make sure that each person contributed to, and understood the mathematics involved in, the group's solution. If students disagreed about an answer, she encouraged them to prove their answers mathematically and to work until they arrived at a consensus. If she noticed that students were not listening to others during an activity, she reminded them that they had to prove their solutions and that each group member must be prepared to discuss the reasons for the solution in front of the class. As a result, the distribution of work was more equitable. Students listened to one another's ideas and evaluated their appropriateness before using them.

Ms. Andrew did not describe and discuss collaboration beyond the general directive to "work with a partner" or "remember to work together." Neither individual accountability nor consensus emerged as topics of discussion in whole-class activity. Typically, only one person would be in control of group work at any particular time and would complete most of the work.

Conclusion

We saw a consistently higher press for conceptual thinking in Ms. Carter's class. She took her students' ideas seriously as they engaged in building mathematical concepts. In both whole-class discussions and small-group work, all students were accountable for participating in an intellectual climate characterized by argument and justification. Four sociomathematical norms governed mathematical discourse in Ms. Carter's classroom: explanations were supported by mathematical reasons, mistakes created opportunities to engage further with mathematical ideas, students drew mathematical connections between strategies, and each student was accountable for the work of the group.

When teachers create a high press for conceptual thinking, mathematics drives not only the activities but the students' explanations as well. As a result, student achievement in problem solving and conceptual understanding increases.

All students were accountable for participating

Action Research Ideas

- Over time, listen for differences in the number of times that you that interrupt a student's explanation, restate a student's explanation, or provide a solution strategy. By keeping a daily log, notice any changes in the nature and quantity of your responses.
- *(a)* Identify the social norms and the sociomathematical norms that characterize your classroom. *(b)* Discuss the issue of sociomathematical norms with a colleague. Share your goals and the problems that you expect to encounter. Continue to discuss your progress with your colleague over time. Encourage your colleague to engage in a similar program to create a higher press. *(c)* Observe and discuss each other's teaching.
- *(a)* Reflect on the discourse associated with a problem recently discussed in your classroom. Using a four-point scale from 0 (low press) to 4 (high press), rate the discourse according to each of the sociomathematical norms that characterize Ms. Carter's classroom. *(b)* Set personal goals for each of the sociomathematical norms. Use such questions as the following to help establish a high press: "How can you prove that your answer is right? Can you prove it in more than one way?

How is your strategy mathematically different from, or mathematically like, that of [another student]? Do you agree or disagree with [another student's] solution? Why? Why does [strategy *x*] work? Why does [strategy *y*] not work?" *(c)* After four weeks, reevaluate your classroom, using the same scale and the same socio-mathematical norms. Note your areas of improvement, and set new goals for the next four weeks.

REFERENCES

Corwin, R. B., S. J. Russell, and C. C. Tierney. *Seeing Fractions: Representations of Wholes and Parts. A Unit for the Upper Elementary Grades*. Sacramento, Calif.: Technical Education Research Center, California Department of Education, 1990.

Fennema, Elizabeth, Thomas P. Carpenter, Megan L. Franke, Linda Levi, Victoria R. Jacobs, and Susan B. Empson. "A Longitudinal Study of Learning to Use Children's Thinking in Mathematics Instruction." *Journal for Research in Mathematics Education* 27 (July 1996): 403–34.

Kazemi, E., and D. Stipek. "Pressing Students to Be Thoughtful: Promoting Conceptual Thinking in Mathematics." Paper presented at the annual meeting of the American Educational Research Association, Chicago, 1997.

National Council of Teachers of Mathematics (NCTM). *Curriculum and Evaluation Standards for School Mathematics*. Reston, Va.: NCTM, 1989.

Yackel, Erna, and Paul Cobb. "Sociomathematical Norms, Argumentation, and Autonomy in Mathematics." *Journal for Research in Mathematics Education* 27 (July 1996): 458–77.

COMMENTARY

As educators, we need to strike a balance between the constraints of children's development and what we want them to learn about the structure, or "big ideas," of mathematics (Fosnot and Dolk 2001). Part of challenging students is giving them tasks that have the potential to challenge them (see the "Meaningful Mathematics Tasks" section for more on this topic), but equally important is that students have the opportunity to struggle, attempt various strategies, and behave like mathe-maticians be-have. Literacy educator Carolyn Burke cautions, "Don't step in front of the struggle" (1995). This ad-vice may counter teach-ers' instincts to help their students. The articles in this section help readers think deeply about this di-lemma of challenging and supporting students.

Synopses of Articles

Buschman's (1994) classic article "Sometimes Less Is More" is a provocative discussion about the use of examples to teach multiplication. Specifically, the author shares a study in which second graders did worse on multiplication tasks when an example of how to solve the problem was given than when they were left to their own devices. He concludes that students thought more creatively when no example was given. Although this scenario took place in a second-grade class, it applies to all grades, and the short article is an excellent device to prompt discussion about instructional practices that support or hinder learning.

"Isn't That Interesting!" also by Buschman (2005), focuses on supporting students in a problem-based classroom as they struggle with solving challenging problems. The author notes that rather than point out students' misconceptions, when he observes students' strategies, right or wrong, he responds using the title of this article and asks students to explain their thinking. "Reflect and discuss" questions prepared by the *Teaching Children Mathematics* editorial panel accompany this article.

"Is a Rectangle a Square? Developing Mathematical Vocabulary and Conceptual Understanding," by Renne (2004), shares a classroom discussion on the topic of geometry, describing how the discussion led to descriptions, then to definitions and deeper understanding. This quality article makes helpful suggestions on challenging students' understanding and also on facilitating discourse.

A good complement to the discourse articles that offers numerous examples of challenging tasks that lead to rich discussions is Maylone's (2000) "Using Counterintuitive Problems to Promote Student Discussion." Six tasks are shared for which the answers appear to "fly in the face of common sense." Each task is briefly discussed. One way to use this article is to explore these tasks in a professional development setting, then hold a debriefing session about how to facilitate the discussions.

Sometimes Less Is More

Larry Buschman

This article is an eye-opener for preservice teachers. Within a larger program that helps them understand the value of a reasoning and problem-centered curriculum, this article helps them understand the disadvantages of "showing" students how to "do" mathematics.

—Rheta Rubenstein

THIS is a true story. Not long ago, on a cool, chilly day in November, a teacher named Mr. B. gave his second-grade students some problems similar to those found in most mathematics textbooks (**fig. 1**). Since these students would be "problem solving," Mr. B. reminded them that they could use the classroom calculators and would be working in cooperative-learning pairs.

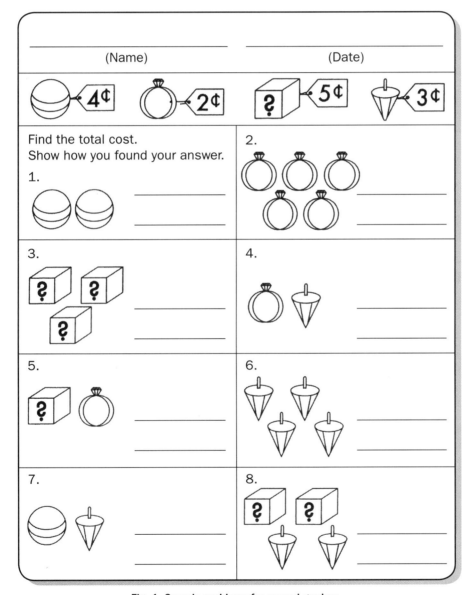

Fig. 1. Sample problems for second graders

Although the students had not been introduced to multiplication, Mr. B. was confident that they could complete this activity successfully. His confidence was based on the

Arithmetic Teacher 41 (March 1994): 378–80

fact that all the students had participated in the school's problem-a-day program since the beginning of the school year. This program supplemented the mathematics textbook used in the classroom. Daily problem-solving activities required students to solve problems that required a solution and that had no apparent or immediate answer. To ensure that the students practiced their problem-solving skills and strategies, Mr. B. decided not to include a solution example for the first problem.

The students completed the paper with little difficulty after engaging in long and exuberant discussions over how to solve the problems and the accuracy of their solutions. Mr. B. was pleased with the students' performance during the activity and with their scores. These scores are summarized in **table 1**.

TABLE 1

Number of Students to Answer Incorrectly without a Solution Example

0–1 problem	2–3 problems	4–5 problems
23	3	1

Later in the school year, as part of the regular textbook curriculum on multiplication, Mr. B. again passed out the same paper previously given to the students in November. However, this time he followed the practice of most textbook publishers and included a solution example for the first problem.

Since many of the students completed the paper in record time without the usual noisy discussions that accompanied most classroom problem-solving activities, Mr. B. assumed that students had remembered the solutions to the problems from their previous exposure. Expecting the students' performance to be even higher, Mr. B. was shocked to learn that their scores were actually lower. The scores appear in **table 2**.

TABLE 2

Number of Students to Answer Incorrectly with a Solution Example

0–1 problem	2–3 problems	4–5 problems
7	17	3

Including a solution to the first problem had a profound impact on the students' performance. When given an example to follow, that is exactly what these second-grade students did. After seeing the solution example, it was as if the students no longer felt the need to apply higher-order-thinking skills to the solution of the problems and instead relied on lower-order thinking by simply following directions. Perhaps they had come to the conclusion that the solution example indicated that someone else had done their thinking for them and they were free to fill in the blanks and perform the calculations mechanically.

Determined to unravel the mystery, Mr. B. performed an experiment the following school year. He selected four pages from the second-grade mathematics textbook that he could use as problem-solving activities. The students completed each page twice, first with the solution example *removed* from the paper, and second, with the solution example *included* on the paper one-to-two months later. When given the paper the second time, most students did not recognize it as a paper they had completed two months before. The results of this experiment are summarized in **table 3**.

Although these scores indicate that including a solution to the first problem on each paper influenced students' performance, the scores do not indicate *why* student perfor-

Table 3

Number of Students to Answer Incorrectly

	0 – 1 problem	2 – 3 problems	4 – 5 problems
Page A			
Without example	23	5	0
With example	11	15	2
Page B			
Without example	18	7	3
With example	9	16	3
Page C			
Without example	13	11	4
With example	5	19	4
Page D*			
Without example	6	5	2
With example	3	6	4

* The results for page D represent the students' scores for a slightly different version of the experiment conducted by Mr. B. In this version, the class was divided into two groups, each with comparable ability levels. Both groups solved the problems at the same time; half the students completed the page with examples, and the other half completed the page without examples.

mance was affected so dramatically. Better to understand this question, the classroom observations conducted by Mr. B. lent some insight into the thinking processes used by these students. These observations were documented through (1) tape recordings of students' conversations as they worked in cooperative-learning pairs, (2) notes kept by Mr. B. of questions students asked during the activity, and (3) notes kept by Mr. B. of interviews conducted with students after the activity. The following summaries are results of these classroom observations.

When a solution example was not provided, the following behavior occurred:

- Students seemed to assume more ownership for solving the problems and actively explored various solution strategies.
- Students' thinking was generally more divergent in nature and focused on all aspects of the process they used to arrive at an answer.
- Discussions within and between pairs of students were more numerous and lasted longer. These discussions centered on figuring out what to do, clarifying and organizing information, determining a strategy for solving the problem, and checking to make sure the solution was correct and reasonable. (The tone of these discussions was, for the most part, accepting in nature and avoided sharp criticism of other students' ideas, problem solutions, or answer.)
- The most frequently used phrase by students during the activity was "I think …."
- Although the students initially seemed confused by the lack of an example to follow, they displayed a great deal of confidence in their solution strategies.
- Students checked their problem-solving strategies and answers with other groups.
- Students adopted a more individualized and flexible approach to the activity. They had a legitimate problem and searched for a solution.

Students thought more creatively when no examples were given

When a solution example was provided, students exhibited the following behavior:

- Many students relied on the example as "the" correct solution to all the problems on the paper.
- Students' thinking was generally more convergent in nature and focused almost entirely on the answers produced using the example as a guide.
- Discussions within and between pairs were infrequent and brief. They centered primarily on how to apply the solution example to each of the problems and whether the answers themselves were correct. (Of greatest concern to Mr. B. was the tone in which these discussions were carried out. Exchanges between students were often very critical and harsh, especially when anyone was chastised for straying from "the way" shown in the example.)
- The most frequently used phrase by students during the activity was "But the paper [the example solution] says . . . . "
- Although the solution example allowed the students to begin working on the problems immediately, they gradually became more confused and started to distrust their own thinking when it became evident that applying the example was yielding some very questionable answers.
- Students rarely checked their answers with other groups or attempted to validate their thinking process by consulting with others.
- Students tended to impose on the activity a rigid structure that they seemed obligated to follow. They had a legitimate solution and searched for ways to make it fit the problems on the paper.

These results can be represented by the following axiom: *When learning how to solve problems having no apparent or immediate answer, it is often better not to supply students with solution examples but rather to encourage them to develop their own solutions through applying higher-order thinking skills and problem-solving strategies.* This procedure will help to ensure that students do their own thinking instead of relying on the thinking of someone else as represented by the solution example. It also helps to ensure that students will think more about the problems and their solution, since they have no preconceived "right way" of doing the problems. In a shortened version, this axiom can be stated thus: *Less can be more because sometimes the less information you give students, the more thinking you get from them.* When examples are not included, students are required to take full responsibility for their own thinking, which is a necessary and essential ingredient for successful problem solving.

As more problem-solving activities are incorporated into the mathematics curriculum, it is our responsibility to ensure that these activities promote the development of students' thinking skills rather than direction-following skills. When students are learning how to solve problems, we need to examine carefully how and when solution examples are included on students' assignments. We also need to be aware of the possible negative effects that these examples can have on students' creativity, thinking processes, and performance.

> **Sometimes the less information you give students, the more thinking you get from them**

Isn't That Interesting!

Larry E. Buschman

*P*RINCIPLES *and Standards for School Mathematics* (NCTM 2000) states, "Students should have frequent opportunities to formulate, grapple with, and solve complex problems that require a significant amount of effort and should then be encouraged to reflect on their thinking" (p. 52). The *Standards* go on to say, "Mathematics teaching in the lower grades should encourage students' strategies and build on them as ways of developing more-general ideas and systematic approaches" (p. 76). In short, the Standards expect children to solve complex problems and in ways that make sense to them by crafting personal solutions to problems. When young children solve problems using their own individualized styles of sense making, however, their solutions may not make sense to others. This can create awkward moments in the classroom that are mathematically challenging and emotionally uncomfortable for both teachers and children.

I Don't Get It!

Elementary school teachers often face a difficult dilemma when asked by a child to comment on a drawing the child has made. This may not seem like much of a problem to most adults, but teachers of young children realize that commenting on a child's drawing is not a simple matter, especially if you cannot tell what the child has drawn. Not wanting to embarrass the child (or themselves) by making a comment such as "What is it?" teachers use a more respectful response, such as "Tell me about your drawing."

A similar difficulty arises in mathematics classrooms when a problem-solving approach is used for mathematics instruction. In this type of classroom, situations often arise that children and teachers find perplexing or confusing. This situation is in sharp contrast to classrooms that use a traditional drill-and-practice approach, in which the procedures for completing routine exercises are clear and unambiguous—at least to the teacher.

But when children solve problems in ways that make sense to them instead of practicing procedures that the teacher has shown them, things can become messy both mathematically and socially. When these situations occur, children and even teachers have a tendency to make comments such as "I don't get it!" I have found a better response to these situations: "Isn't that interesting!"

The following examples demonstrate how mathematical problems can be challenging for both children and teachers in some unexpected ways.

Young Children Often Do Not Think Like Adults

The pencil problem

Like many other teachers, I have a tendency to assume that children understand the meaning of commonly used words that find their way into mathematics lessons. One such word is *each*. Most young children seem to understand the meaning of this word as it applies to situations in which *each* person in a group gets something, such as a cookie. When I presented the following problem to first and second graders, however, I was surprised by some of the responses.

> Jason has 2 pencils in each hand. How many pencils does he have in all?

Teaching Children Mathematics 12 (August 2005): 34–40

Several children said the answer was 2 pencils, and at first I wanted to say, "That's not correct. Perhaps you should try the problem again—only this time read the problem more carefully and think about the answer more thoughtfully." Instead I said to the children, "Isn't that interesting! Tell me how you solved the problem."

A typical description of the children's solution process involved modeling the problem for me, since it was apparent to these children that it would take more than a verbal explanation to correct my lack of understanding. These children would place two pencils on their desk, and then they would pick up one pencil in each hand and say, "See, Mr. B.? Two pencils—in each hand. Just like it says in the problem." At first I found it hard to understand what the children were trying to say, since I knew the answer to the problem should be 4 pencils. After several attempts to explain to these children that "2 pencils in each hand" meant 2 pencils in *both* hands, I realized that my explanations were not going to change the minds of these children. Therefore, I did something that some teachers might find strange—I simply ended the discussion by saying, "This is a really interesting problem. Let's all think about it some more and come back to it at a later time." (For the resolution of this misunderstanding, see the Author's Note at the end of this article.—*Ed.*)

It has taken me a long time to learn that not all children are ready to acquire certain concepts on the day I decide to present them. The traditional teacher inside me wants to ensure that every child understands everything, every day. In reality, however, some children need more time than others to digest all the information teachers ask them to learn, or they may need more experiences to help fine-tune working definitions of words and the concepts they represent.

Howard Gardner has noted that children form "primitive beliefs" about most things in their world at an early age (Gardner 1993). These beliefs are highly resistant to change and are modified very slowly over time as children construct new understandings based on experiences that broaden and deepen their knowledge base. Gardner has found that although children can be trained to give the correct answers to questions teachers commonly ask in mathematics and science classrooms (such as "What are your chances of winning the lottery?" or "What causes the seasons of the year?"), children frequently continue to hold on to their primitive beliefs well into adulthood, even though they are aware that their beliefs are in direct conflict with the correct answers they have been taught in school.

> **Not all children are ready to acquire certain concepts on the day I decide to present them**

The circus problem

Confusion over the word *each* arose on another occasion when I asked the first graders to solve this problem:

> Jennifer, her twin brothers, and her parents went to the circus. How much did it cost for the whole family to go to the circus?

> **Circus Tickets**
> Children $2
> Adults $3

Several first graders said the answer to this problem was $5. At first I thought the children had simply added the numbers on the sign in the problem, but in fact when I questioned them about their answer they had a solution to the problem that I found interesting. A typical explanation went like this: "The tickets for the kids [meaning *all* the children—not *each* child] is $2 and the tickets for the adults [again, meaning *all* the adults—not *each* adult] is $3. So 2 + 3 is $5." Although most of the children in the classroom realized that the amounts shown on the sign were for *each* child and *each* adult, several young children in the classroom were unaware of this convention because they had never been to a circus and thought that the prices reflected an admissions policy that allowed all children to enter the circus for $2 and all adults for $3.

Unlike with the misunderstanding of the Pencil problem, this time I easily clarified the children's mistake. I have found that when children misinterpret information because of confusion over a social convention (as in this example), they respond favorably to direct instruction and more readily accept a teacher's explanation. When the confusion is over a misunderstanding of a mathematical idea or concept (as in the Pencil problem), however, children are less willing to abandon their misconceptions. Instead of simply accepting information from others, children appear to need to construct this kind of conceptual understanding for themselves.

Older Children Think More Like Adults, but Confusion Still Exists

The following problems were challenging for the children and me because they resulted in dilemmas that as a mathematical community we were unable to resolve to everyone's satisfaction.

The Ghost problem

This problem generated an interesting discussion in our classroom:

> Julia made some paper ghosts for a Halloween party. It took her 2 1/2 minutes to make each ghost. How many ghosts did she make in 1 hour?

Two of the third graders each solved the problem in a different manner, as shown in **figure 1**. Both solutions seem to make sense mathematically, yet each solution produces a different answer: Darlene's solution yields an answer of 24 ghosts, and Austin's solution yields an answer of 25 ghosts.

When these two third graders presented their solution methods to the rest of the class, the resulting discussion did not go as I had anticipated. Instead of focusing on the mathematical validity of each solution, the children engaged in a heated debate over the number of answers that are possible for particular types of problems.

All the children in the classroom were familiar with open-ended problems, which can have more than one correct answer. The children also realized that some problems—such as the one under discussion—should have only one mathematically correct answer. Because they could not determine which solution for this problem was in error, however, they decided that a third category of problems existed and offered this hypothesis: Perhaps this was a "very, very, very

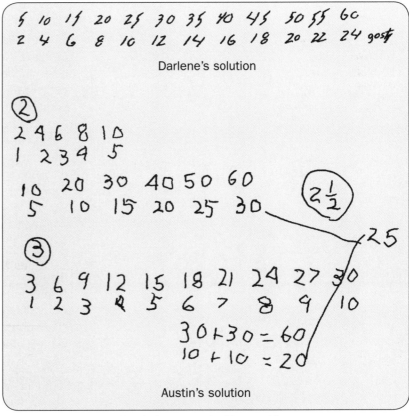

Fig. 1. Answers to the Ghost problem

special problem that even though it should have only one answer, somehow it has two answers." They thought they had made a "great discovery" and we should let the "president and the guys who do mathematics" know about this new kind of problem.

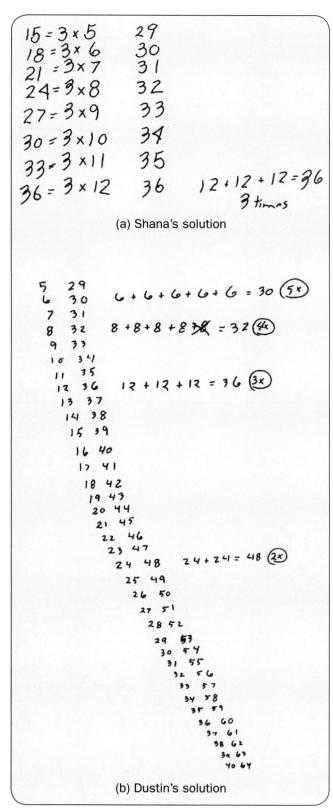

(a) Shana's solution

(b) Dustin's solution

Fig. 2. Answers to the Age problem

At the conclusion of the children's discussion, all eyes in the room were on me, and it was apparent that the students expected me to confirm or refute their hypothesis about the new kind of problem. I was fairly certain that the correct answer was 24 ghosts, but I had no explanation for why Austin's answer (25 ghosts) was in error. So I said to the class, "Isn't this interesting! I think the answer is probably 24 ghosts, but Austin's solution seems to make sense. This is a problem we all need to take some time and really think about."

Although I continue to think about this problem, it has stretched my mathematical understanding to the limit, and I must admit that I am still unsure why Austin's solution is incorrect. If anyone can offer an explanation that both young children and I can understand, I would appreciate your help.

The Age problem

Unlike the previous problem, which presented a mathematical dilemma, this problem resulted in a conclusion that was not only interesting but also somewhat humorous:

When Jason was 5 years old, his father was 29 years old. How old will Jason be when his father is 3 times his age?

A typical student solution for this problem is shown in **figure 2a**. One child, however, noted a pattern that the other children had not seen and decided to continue the list (see **fig. 2b**). This child reasoned that because at one point Jason's father was 5 times as old as Jason, and later he was 4 times as old as Jason, and still later he was 3 times as old as Jason, then there should come a time when Jason is 1 times as old as his father—that is, they are the same age. The children found this supposition to be very interesting, and I found it to be quite amusing.

Later I commented to a colleague that this problem raised all kinds of interesting questions, such as "If Jason's father is 5 times as old as Jason today, isn't he 5 times as old as Jason tomorrow?" and if this is not true (which the solution to the problem seems to indicate), then "Is Jason's father aging slower than Jason, or is Jason aging faster than his father?" My colleague jokingly responded, "Perhaps this is what people mean when they say that as you get older you *slow down*." To which I replied, "Or perhaps it is what people mean when they say that children are growing up *faster*."

The Egg Carton problem

A common manipulative in our classroom is egg cartons, which the children use for counting, sorting activities, and games, and as problem-solving aides. Some of these egg cartons have been cut into sections with 2, 4, 6, 8, or 10 cups. One day while using the egg cartons, a child made the following comment to another child in the classroom: "I just don't get it. You got this small egg carton [see **fig. 3a**] and this other egg carton [see **fig. 3b**]. But the small one is like 1/4 smaller than the other one [see **fig. 4**]. But the big one is 1/3 bigger than the small one [see **fig. 5**]. It's like they should be the same (1/3 or 1/4) but they're not—like one is 1/3 bigger and the other is 1/4 smaller." The other child replied, "I don't get it either, but it's kind of interesting. Maybe we should think about it."

As I listened to these two girls discuss their dilemma, several questions entered my mind: How could I help them gain a deeper understanding of the fractional concepts represented in this situation? Should I let them attempt to resolve the situation on their own or should I intervene? If I did intervene, should I use a direct instructional approach or should I attempt to guide them using leading questions? If I did not intervene, should I have the girls present the problem to the rest of the children in the class and see if the other students could offer any insights into the situation? These and other questions confront a teacher on a daily basis when children solve problems instead of completing drill-and-practice worksheets. These are not easy questions because the answers often depend on the age and mathematical abilities of children as well as the nature and level of difficulty of the problems.

Final Thought

Principles and Standards for School Mathematics (NCTM 2000) states, "By allowing time for thinking, believing that young students can solve problems, listening carefully to their explanations, and structuring an environment that values the work that students do, teachers promote problem solving and help students make their strategies explicit" (p. 119). During the activities described in this article, I allowed time for thinking; I believed in my students and their abilities as budding mathematicians; I listened attentively to their explanations; I created a learning environment in which children took risks and shared their original thinking; and I helped children make their strategies explicit. In doing all the things recommended by the *Standards*, I also created situations that challenged my abilities and exposed my weaknesses as a teacher of mathematics.

I discovered that a teacher of problem solving must have a deep understanding of mathematical ideas and concepts and a thorough working knowledge of the children in the classroom: how they interpret the meanings of words, their misconceptions about mathematical concepts, and what kinds of problems challenge children without being too complex or overwhelming. The whole experience of using a problem-solving approach for teaching mathematics is very *interesting* indeed.

REFERENCES

Gardner, Howard. *The Unschooled Mind: How Children Think and How Schools Should Teach*. New York: Basic Books, 1993.

National Council of Teachers of Mathematics (NCTM). *Principles and Standards for School Mathematics*. Reston, Va.: NCTM, 2000.

Author's Note

In classrooms where accountability and high-stakes testing drive the curriculum, teachers have a tendency to look for a "quick fix" to correct children's misconceptions.

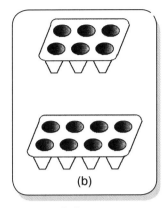

Fig. 3. Egg cartons

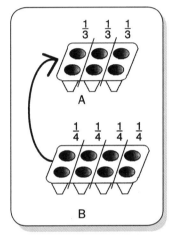

Fig. 4. If egg carton B is divided into fourths, then egg carton A appears to be 1/4 smaller than egg carton B.

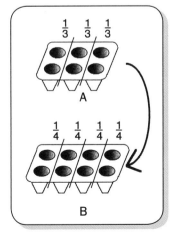

Fig. 5. If egg carton A is divided into thirds, then egg carton B appears to be 1/3 larger than egg carton A.

Reflective teaching is a process of self-observation and self-evaluation. It means looking at your classroom practice, thinking about what you do and why you do it, and then evaluating whether it works. By collecting information about what goes on in our classrooms, then analyzing and evaluating this information, we identify and explore our own practices and underlying beliefs.

The following questions related to "Isn't That Interesting!" by Larry E. Buschman are suggested prompts to aid you in reflecting on the article and on how the author's idea might benefit your own classroom practice. You are encouraged to reflect on the article independently as well as discuss it with your colleagues.

- **What are some other common words for your grade level that students might misinterpret during a mathematics lesson? Here are just a few additional terms that Buschman has found to give primary-grade students difficulty: *each, altogether, probably, total, alike, every other, change, end to end, round trip, day*. What can you do to help resolve student misconceptions related to language?**

- **As a classroom teacher, what would you do to help the two girls described in this article resolve the apparent contradiction in fractional quantities represented by the Egg Carton problem?**

- **What can you do to convince your students that it is worth their time and effort to solve challenging problems that require patience and perseverance when children grow up in a society in which instant gratification has become the norm?**

I have found that giving children time to ponder a mathematical problem and discuss the problem with others provides children with the opportunity to correct their misconceptions on their own and is often more effective than my attempts to teach them the correct answer. Because this advice is also true for adults, later that day I continued to think about the children's responses to the Pencil problem, and I decided to discuss the matter with a colleague. She suggested rewording the problem: *Jason has 1 pencil in each hand. How many pencils does he have in all?*

The next day I gathered all the first and second graders together and presented them with the reworded version of the problem. When the children finished solving the problem, I asked them to share how they solved the problem, and encouraged the other children to ask questions about each solution or to tell why they agreed or disagreed with the answer. Some of the children who had experienced difficulty the previous day immediately became aware of the dilemma raised by this new version of the problem and were able to self-correct their misconception. Other children refused to let go of their misconceptions and solved the problem by tossing one pencil back and forth between their hands, or they broke the pencil in half so they could hold "one" pencil in each hand. By the end of the children's discussion, they all agreed that the answer for this problem was 2 pencils. Although all the children said they agreed with the answer, I knew that some children would need to solve other problems containing the word *each* before they would correct all their misconceptions about this term.

Is a Rectangle a Square?
Developing Mathematical Vocabulary and Conceptual Understanding

Christine G. (Chris) Renne

INSTEAD of asking the usual "Is a square a rectangle?" I asked my thirty fourth grad- ers, "Is a rectangle a square?" Expecting a resounding "No," I was quite surprised when more than half the students replied, "Yes!" We were working with basic geometric terms and concepts including parallel lines, naming angles, and congruency in relationship to simple shapes. That the students could readily identify various two- dimensional shapes such as squares, triangles, and rectangles quickly became appar- ent; however, they were unable to use mathematical terms and concepts to describe the shapes. They did not have a conceptual understanding of the properties or the vocabu- lary to express distinguishing characteristics and compare shapes in a systematic manner (NCTM 2000; see **fig. 1**).

Technically, a square is a particular rectangle, and in the unique case that all sides are of equal length, a rectangle is a square. Because my students had difficulty going beyond simple identification, my immediate purpose became to give them experiences that required them to describe and then compare basic attributes of squares and rectangles. We needed to develop the initial conceptual understanding and appropriate vocabulary so that we had a common language to later explore the special relationships of squares, rectangles, and other geometric shapes and solids.

The Challenge

To learn to better encourage students' thinking and conceptual development, I decided to purposefully examine the discourse during their exploration (NCTM 1991; see **fig. 2**). Following the lead of several mathematics educa- tors (Cobb, McClain, and Whiteneck [1997]; Cobb et al. [1992]), I examined my students' responses in depth and reflected on my own role as their teacher and discussion leader. I tape-recorded a whole-class discussion to capture their often rapid exchanges and reveal the discourse of teaching and learning mathematics, specifically how the students were using mathematical vocabulary (see Lampert and Blunk [1998]; Silver [1996]). Instead of accepting the brief replies that are so common in classrooms (Cazden 2001), I wanted the students to explain and justify their thinking. I hoped to avoid inaccurate assumptions about what the students knew and understood.

The students' first challenge was to identify similarities and differences between squares and rectangles using descriptive and appropriate geometric terminology. As an example, I asked the students to describe a circle. Students suggested that a circle does

Instructional programs from prekindergarten through grade 12 should enable all students to analyze characteristics and properties of two- and three-dimensional geometric shapes and develop mathematical arguments about geomet- ric relationships.

Expectations (selected)

In grades 3–5, all students should—

- identify, compare, and analyze attributes of two- and three-dimensional shapes and de- velop vocabulary to describe the attributes;
- classify two- and three-dimensional shapes according to their properties and develop definitions of classes of shapes such as triangles and pyramids;
- explore congruence and similarity;
- make and test conjectures about geometric properties and relationships and develop logical arguments to justify conclusions. (NCTM 2000, p. 164)

Fig. 1. Geometry Standard for grades 3–5

Teaching Children Mathematics **10** (January 2004): **258–63**

The teacher of mathematics should orchestrate discourse by—

- posing questions and tasks that elicit, engage, and challenge each student's thinking;
- listening carefully to students' ideas;
- asking students to clarify and justify their ideas orally and in writing;
- deciding what to pursue in depth from among the ideas that students bring up during a discussion;
- deciding when and how to attach mathematical notation and language to students' ideas;
- deciding when to provide information, when to clarify an issue, when to model, when to lead, and when to let a student struggle with a difficulty;
- monitoring students' participation in discussions and deciding when and how to encourage each student to participate.

(NCTM 1991, p. 35)

Fig. 2. Standard 2: The Teacher's Role in Discourse

not have any sides or angles. After demonstrating how to construct a chart, I asked small groups of about three students each to create a chart that described squares and rectangles and compared the similarities and differences. I gave students paper packets that contained a square and two rectangles to provide concrete examples. About thirty minutes later, the students reconvened as a total class to make a composite chart from their small-group work.

Creating the Composite Comparison Chart

The first contribution to the chart was that a square is similar to a rectangle because it has four angles. Gina stated that a square has congruent sides. When I asked her what she meant by congruent sides, she could not explain. Scott suggested, "Two sides like on a square; the top, the line, the top side, and the bottom side go the same way and the sides, the two sides go the same way." Michael said that they are "two sides that never touch." At this point, it was obvious that the concept the students were describing was not *congruent* but *parallel*. I had believed that the students had a grasp of *parallel* from the previous work we had done. I stopped to ask what term describes lines that go on forever and never touch. Michael answered, "Parallel lines." He knew the term *parallel* even though he was the one who had suggested that congruent sides never touch. When I asked them if congruent sides had to be parallel, the students did not respond. I dropped the discussion about congruence, thinking that we would return to it later, and continued to ask about similarities between squares and rectangles. Jennifer said that both have four sides.

The students observed that the two shapes have right angles. Julie stated, "A rectangle is much longer." I asked if that is the case for all rectangles. Cathy noted, "All four sides on the rectangle are not all the same size." The students went on to identify that rectangles have top and bottom lengths that are the same but different from the two side lengths, which are also the same.

After more discussion about sides being the same, Julie stated, "A square has all congruent sides." I asked what she meant by *congruent*. "All the same length," she said. I followed up by asking if that condition is special only to a square. Willie said it is not. When I asked the students what makes it a square and not a rectangle, Joey responded, "'Cause it's not long, it's short." Michael stated, "It has all four equal sides and four right angles." I asked the students if that worked, and many agreed.

I went back to the original question: "Is a rectangle a square?" The students' responses included the following:

Joey. It is not a square because all the sides are not the same but it has right angles.

Richard. It is a square. It's just a little longer.

Gina. I think a rectangle is a square because they all have four sides, they all have congruent sides, they all have length.

Michael. It's not a square because the rectangle has two equal sides—the opposite sides—and it has to have four equal sides.

Julie. It's not a square because on a square all the sides are the same length, and on a rectangle they aren't.

Lacy. No, because they all have to be four equal sides and right angles.

Just as I was thinking that the students were beginning to accurately describe the similarities and differences between rectangles and squares, the conversation took an unexpected turn. While the students explained their stances, Joey challenged Steven's description of why a rectangle is a square.

Teacher. Some of you are saying yes, a rectangle is a square; some of you are saying no, a rectangle is not a square. What is it?

Steven. It's a square because [coming to the board] these two sides are equal [pointing to the two vertical sides of the rectangle]; they're the same; and those two sides are just the same [pointing to the horizontal sides of the rectangle].

Teacher. OK.

Joey. Not all the sides are the same.

Steven. They are.

Joey. Nuh-uh. See those two. You're saying those short sides are as long as the longer sides?

Steven. Uh, no.

Teacher. OK. We still seem to have some disagreement here. Let's see what Amber has to say.

Amber. It is a square; it's just that a rectangle's sides are stretched out.

Joey and Steven continued their discussion independently while other students began to talk about whether a rectangle is still a square if it is "stretched out." What eventually evolved from Joey and Steven's separate discussion was the notion that the particular rectangle I drew on the board was simply two squares put together, and therefore a square (see **fig. 3**). If Joey had not challenged Steven on this point, I am not sure I would have picked up on this curious interpretation and misconception. Variations of thinking about a rectangle as two or more squares put together continued to arise periodically in later discussions and writings.

Fig. 3. Some students erroneously concluded that
because this rectangle is made of two squares, it is
therefore a square.

After we had constructed the chart (see **fig. 4**), I asked if a rectangle is a square. The replies still varied.

	Square	**Rectangle**
What is the same?	Four angles, congruent sides, four sides	Four angles, congruent sides, four sides
What is different?	All four sides equal and all four right angles	Top and bottom lengths are the same but different from the two sides, which are the same

Fig. 4. Chart of similarities and differences

The Discussion Continues

The following day, I asked the class to write descriptions of the two shapes followed by specific definitions of them. As a class, the students designed a chart (see **fig. 5**). Students then broke into groups of four or five and I asked them to come to agreement on whether a rectangle is a square. They were to create convincing statements supporting their viewpoints that they would present to the class.

	Square	Rectangle
Description	· It has equal sides. · It has four sides. · It has four right angles. · It has four congruent sides.	· It has four sides. · It does not have four equal sides. · Top line and bottom line are equal. · Side lines are not equal to the top and bottom lines. · It has four right angles. · Two pairs of sides are congruent.
Definition	Right angles and congruent sides and four equal sides	Right angles, congruent sides, two sides the same and the other two sides the same

Fig. 5. Chart of descriptions and definitions

After we came back together as a class, each group reported that a rectangle is not a square and gave its reasons. The students individually wrote their response to "Is a rectangle a square?" and I asked them to be specific about why they thought what they did. Only five of the twenty-nine students in the class said yes, compared with more than half the students when I asked this question the first day. Interestingly, one of the students who said yes had res-ponded for his group that a rectangle is not a square. He was not convinced after all.

Some of the reasons the students gave for their answers were surprising. One usu-ally eloquent student said that a square and a rectangle are like the ones in the textbook, so a square is a square and a rectangle is a rectangle. Other students used some of the vocabulary I had presented and expressed more in-depth reasoning (see **fig. 6**). For

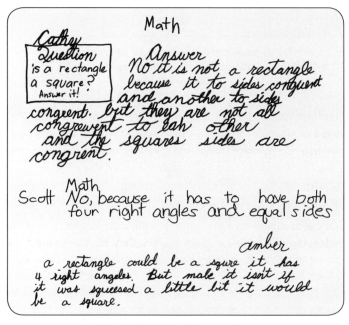

Fig. 6. Student writing samples

example, Cathy accurately used the term *congruent* in her justification of why a rectangle is not a square. Other explanations ranged from Scott's very succinct statement to Amber's tentative (and correct) suggestion that a rectangle could be a square—but maybe it is not.

Hindsight and Insight

Looking back through the transcript and at the written work provides hindsight and insight into new practices, including limitations, decisions, missed opportunities, and mathematical discourse.

Limitations

The convenience of having materials "ready to go" may have actually limited the students' opportunities to construct their own examples and possibly their understanding. One of the sample rectangles was about the size of two side-by-side squares, which may have inadvertently contributed to the misconception that if two squares are put together, the resulting rectangle is actually a square. In addition to giving students preassembled packets of squares and rectangles, I could have encouraged them to create their own shapes and then discuss their creations with a partner or in a small group to challenge and justify their thinking. The few examples in the packets should have been starting points rather than the sole evidence for the exploration.

Decisions

Knowing when and to what extent to intervene in a discussion often is difficult. I was thrilled when Richard offered his answer to whether a rectangle is a square because he rarely participated in whole-class discussions. At that moment, the social dimension of accepting his participation took precedence over asking for further clarification. I knew that the students would be doing additional work in small groups and individually writing responses, so encouraging and supporting Richard's involvement was an on-the-spot judgment call. What may appear to be a failure to challenge a student's mathematical thinking actually was a deliberate decision to accept the student's participation.

Establishing connections between relationships of mathematical concepts and terminology is essential. Incorporating all the possible variations at the same time is extremely difficult, especially when the exploration is a first attempt to have students analyze attributes and develop vocabulary to describe squares and rectangles. Instead of focusing on the finer nuances of all squares being rectangles and on a rectangle with equal sides being a square, during this initial experience I made the decision to emphasize development of the concepts of congruency, right angles, parallel, and so on, in the context of describing basic geometric shapes. I knew that we would continue our study of geometry and I would have multiple opportunities in the weeks ahead to enhance and deepen students' knowledge and understanding of special cases and relationships.

> Establishing connections between relationships of mathematical concepts and terminology is essential

Missed opportunities

Not responding directly to Richard's statement ("It is a square. It's just a little longer") or to Gina's subsequent observation ("I think a rectangle is a square because they all have four sides, they all have congruent sides, they all have length") were missed opportunities to potentially uncover misconceptions or extend the students' growing understanding. When Julie explained that congruent sides meant that they all had the same length, I asked if that is special only to a square. Willie replied no, but I did not ask him to explain his answer.

When Joey challenged Steven's conjecture that a rectangle comprised of two squares actually is a square, his comments were part of a side conversation that I overhead. Although it is true that multiple squares sometimes comprise a rectangle, the conclusion that a rectangle is therefore a square is incorrect. In hindsight, I probably should have stopped and explored this misconception as a whole class at that point.

If the lesson format had been direct instruction rather than a more constructivist approach, however, the misconception probably would not have surfaced. It is much simpler to list the attributes of squares and rectangles and tell the students that the two shapes share several characteristics and relationships; for example, a square is a rectangle and some rectangles are squares. Using this approach, a misunderstanding of a basic geometric tenant would go undiscovered and unchallenged, potentially causing future difficulties in building accurate conceptual understanding.

Mathematical discourse

In examining the students' use of appropriate mathematical terminology, one instance in particular illustrates the tenuous learning that takes place. When they worked with the concept of *parallel* in the days prior to this exploration, the students appeared to have a solid understanding of lines that continue forever and never touch. When they used both *congruent* and *parallel* as descriptors of geometric shapes, however, it became apparent that their understanding was not fully developed. Building conceptual knowledge requires time and multiple experiences so that students can expand, apply, and refine appropriate and accurate vocabulary in various contexts.

Teachers must constantly make decisions and monitor students' learning and behavior. Although they are efficient, teacher questions that elicit short one- or two-word answers (such as "Yes" or "No," "Parallel lines," and "Right angles") are not always sufficient to determine what students actually know. Orchestrating engaging whole-class discourse with thirty students, however, is demanding and difficult. When the students were justifying their reasoning, the whole-class discussion sometimes broke into side conversations, often at important times. For example, when Joey challenged Steven and then they broke off into a separate discussion, the class was at a highly engaged point. The whole-class discourse dissolved into several small-group conversations that were engaging but often difficult to follow or monitor. The teacher must decide whether to continue or change the immediate format of interaction.

> Engaging whole-class discourse with thirty students is demanding and difficult

Concluding Thoughts

The students need to continue developing mathematical vocabulary and deeper conceptual understanding. Intentionally intermingling small-group work, whole-class discussions, and individual writing created multiple opportunities for students to explore, conjecture, display, and clarify their thinking and understanding.

Although I hoped that most of the students would adopt more sophisticated usage of appropriate vocabulary after completing this exploration, the students' progress was inconsistent. In comparing the charts that the class designed (see **figs. 4** and **5**), the students were more specific and accurate in their descriptions on the second chart. The charts included some of the vocabulary, such as *congruent,* but few of the students' individual written responses did. Overall, however, the students did demonstrate a stronger ability to adequately describe the basic properties of squares and rectangles.

A close review of the mathematical discourse in my classroom revealed some areas for attention and consideration. In most cases, I asked students to explain their thinking; however, I missed a few important opportunities. Balancing verbal participation at the whole-class level with the engagement and enthusiasm that contributed to multiple side conversations was a constant judgment call. Creating opportunities for misconceptions to surface is essential, but knowing what to do when they unexpectedly arise is yet another on-the-spot decision. Because of the often unpredictable nature of classroom discourse, attaining the NCTM Standards for geometry and the teacher's role in discourse remain challenging goals.

REFERENCES

Cazden, Courtney. *Classroom Discourse: The Language of Teaching and Learning.* 2nd ed. Portsmouth, N.H.: Heinemann, 2001.

Cobb, Paul, Ada Boufi, Kay McClain, and Joy Whitenack. "Reflective Discourse and Collective Reflection." *Journal for Research in Mathematics Education* 28 (May 1997): 258–77.

Cobb, Paul, Terry Wood, Erna Yackel, and B. McNeal. "Characteristics of Classroom Mathematics Traditions: An Interactional Analysis." *American Educational Research Journal* 29 (1992): 573–604.

Lampert, Magdalene, and Merrie L. Blunk, eds. *Talking Mathematics in School.* Cambridge, England: Cambridge University Press, 1998.

National Council of Teachers of Mathematics (NCTM). *Professional Standards for Teaching Mathematics.* Reston, Va.: NCTM, 1991.

———. *Principles and Standards for School Mathematics.* Reston, Va.: NCTM, 2000.

Silver, Edward. "Moving beyond Learning Alone and in Silence: Observations from the QUASAR Project Concerning Communication in Mathematics Classrooms." In *Innovations in Learning: New Environments for Education,* edited by Leona Schauble and Robert Glaser, pp. 127–59. Mahwah, N.J.: Lawrence Erlbaum, 1996.

Using Counterintuitive Problems to Promote Student Discussion

Nelson J. Maylone

T HE NCTM's *Curriculum and Evaluation Standards* document (1989) has sparked a revolution in curriculum and textbook design and, to some extent, the evaluation of students. The Standards encourage problem solving in context; increasing opportunities for students to discuss mathematics; and actively involving students in exploring, conjecturing, analyzing, and applying mathematics. Recommended instructional techniques include promoting group work, making presentations, conducting experiments, and working with manipulatives.

Using counterintuitive mathematics problems helps to keep adolescent students actively involved in their mathematics education. Counterintuitive problems can be defined as answers and solutions that may not, at first, seem right to students or adults. A classic example is the Birthday Match problem:

> What is the minimum number of people, randomly chosen, who need to be assembled in a room so that the chances are at least 50 percent that two of them have exactly the same birthday? (Month and day, not year)

Typical guesses include 365, which is the number of days in most years, and 183, which is about half of 365. The correct answer is surprising, but more on this problem is found at the end of this article.

Counterintuitive problems are not necessarily trick questions; rather, before discussion, their answers appear to fly in the face of common sense. Therein lies one of their chief benefits: these problems promote momentary confusion; then debate; and, finally, understanding. An identifying characteristic of counterintuitive problems is that even after a solution has been verified, one may have difficulty ignoring that little voice that keeps whispering, "But that answer can't be right!" That dissonance is part of the fun of using counterintuitive problems, since adolescents tend to be passionate about defending that which they see as right. Even reticent students will find themselves being drawn in by these challenges. Motivated students not only will pay attention but will think and reflect, be eager to contribute, and perhaps most important, continue to want to contribute even after being proved "wrong."

Any middle school teacher can confirm that adolescents enjoy debating. Discussion and debate engage students and promote new thinking, reinforce learning, and offer important social erudition opportunities. Verbal interaction can help teach students how to listen, be patient, respond appropriately to others' comments and opinions, and be concise and persuasive. Using counterintuitive problems can also provide opportunities to incorporate students' intuitive solutions, thereby increasing learning. Furthermore, if students are enthusiastically engaged, discipline problems will be minimized.

Several problems follow. Some are variations of ancient classics; some are original—all are counterintuitive. The middle-grades-mathematics teacher will want to share them with students and to allow the subsequent debate to rage in the classroom.

Any middle school teacher can confirm that adolescents enjoy debating

Mathematics Teaching in the Middle School 5 (April 2000): 542–46

The Busy Lawyers

The classic Busy Lawyers problem is shown in **figure 1**. Please read it and answer the question at its end before proceeding.

The answer is yes, Writ could have had a better all-year winning percent than Barr in spite of Barr's having the better first half *and* the better second half.

Intuition loudly screams that Writ could not possibly have had a better yearlong winning percent than Barr. When first confronted with this problem, one seventh grader asserted that "Barr beat Writ twice. There is no way that two wins could be a 'loss.'" Barr beat Writ in the first half, Barr beat Writ in the second half; case closed.

Although those ideas are good opening arguments from students, the teacher should challenge them to push their thinking beyond the obvious. If students do not bring up the point on their own, the teacher should ask them whether the number of cases that each lawyer handled is relevant. (It is.) This question may spark some students to examine the situation more closely. Here are other questions that the teacher may want to pose if students are struggling with this issue:

1. Barr's yearlong winning percent will fall between what two other percents? (Answer: Barr's first-half and Barr's second-half winning percents.)
2. Using the same logic, where will Writ's yearlong winning percent lie?
3. Will Barr's and Writ's yearlong winning percents always be *exactly* halfway between their half-year winning percents? If not, what factor might determine to *which half-year percent* each lawyer's yearlong percent is closest?

The teacher's purpose here is to encourage students to rely on their intuition, which says no, each lawyer's yearlong winning percent may *not* lie exactly halfway between their half-year percents. This revelation may open up the possibility that Barr may have the better yearlong winning percent.

Table 1 describes a possible scenario in which Barr wins the individual halves of the year but Writ, nonetheless, becomes the yearlong champion. Note that the winning percent is given in baseball-statistics style, that is, as a decimal shown to the thousandths place.

An everyday example may help to clarify the issue for students:

> Terry is given only one mathematics test during the first semester, and he earns a score of 100 percent. During the second semester, twelve tests are administered, and Terry consistently earns a score of 80 percent on each one. Assuming that the teacher assigns equal weight to all thirteen tests, is it fair to take the first-semester score of 100 percent and average it with the consistent second-semester score of 80 percent for a yearly average of 90 percent?

Terry might wish it so, but that approach is bad mathematics.

Table 1

	Lawyer	Wins from Cases Tried	Winning Percent
First half of the year	Barr	60 out of 100	0.600
	Writ	29 out of 50	0.580
Second half of the year	Barr	40 out of 50	0.800
	Writ	79 out of 100	0.790
Entire year	Barr	100 out of 150	0.667
	Writ	108 out of 150	0.720

> Attorneys Barr and Writ are two highly competitive injury lawsuit lawyers. They are so competitive that twice a year, at the end of June and December, they compare their trial won-lost records, with the "winner" treating the other lawyer to dinner.
>
> For the first half of a recent year, Barr's winning percent, the number of cases won divided by the total number of cases handled, was better than Writ's winning percent. For the second half of that year, Barr again had a better winning percent than Writ.
>
> Is it possible that attorney Writ had a better winning percent than attorney Barr for that entire year? If so, how? If not, why not?

Fig. 1. The Busy Lawyers problem

The net worth of Pat's Candy Company at the start of a particular year was $32,000.00. Because of declining sales, Pat's company began losing 10 percent of its worth each month. At the end of the year, the company was still worth $9,037.74. How can that be? Wouldn't losing 10 percent per month for just ten consecutive months, let alone twelve, leave Pat with a worth of $0?

Fig. 2. The Why Isn't Pat Broke? problem

To raise money for a new outdoor school sign, the principal decided to hold a raffle. Exactly 1000 tickets would be sold for $1 each. One winning ticket would be chosen, the prize being free school lunches for a year. The first 600 tickets were sold to 600 separate individuals, and the remaining 400 were all sold to Jesse Bernhoot.

Question: Who will likely win the raffle?

Fig. 3. The Raffle problem

Students should walk away from this activity with the understanding that, generally, *one should not average averages*. As one student observed, "I think denominators matter."

Why Isn't Pat Broke?

Discussing the problem shown in **figure 2** will help clarify the truth that 10 percent of a dwindling amount is an ever-dwindling amount. Ten percent of something remains a consistent size only if the original amount under consideration does not vary. The teacher may want to ask students this series of questions:

1. "Reduce $20 by 50 percent. What do we have?" Most students will be able to give the correct response of $10.
2. "Now reduce that $10 by 50 percent." A moment of hesitation is likely as students see that this new reduction takes them to $5.
3. "We have done two reductions of 50 percent. Shouldn't two reductions of 50 percent take us to $0?" Most students will be able to state that, no, we should not now be at zero; we took "50 percent off" two different amounts. The hope is that students will be able to recognize that the same phenomenon is at work as Pat's company loses money.

The teacher's goal is to lead students to state in their own words the phenomenon noted through the questions. One sixth-grade student commented, "When you chop off 10 percent of what you have, you have something smaller, and 10 percent of *it* will be smaller, too." That observation is praiseworthy. It indicates that the student has understood the essence of this mathematical phenomenon.

The Raffle

Consider the problem shown in **figure 3**. Answer the question before proceeding. The answer is that we cannot say who will win for sure, but it probably will *not* be Jesse! Students will object to this statement. In one sixth-grade classroom, one student stated, "I know that Jesse might not win, but Jesse has lots more tickets than anyone else, so that person will win *probably*." Although it is correct that Jesse is 400 times as likely to win as anyone else, it is still true that someone else will probably win. Here is why.

The sold tickets can be seen as forming two blocks, one of 400, another of 600 tickets. When those 1000 tickets are thoroughly mixed together in a drum, neither the drum nor the ticket drawers know that Jesse has purchased so many. It is more likely, although by no means certain, that the winning ticket will come from the block of 600 than from Jesse's block of 400.

This problem is counterintuitive because of the following true, but apparently contradictory, statements:

1. Jesse is 400 times as likely as anyone else to win.
2. Someone other than Jesse is still more likely to win.

Why? Because Jesse is indeed 400 times more likely than any other *individual* ticket holder to win the raffle, but Jesse is not 400 times more likely than *all* the other 600 ticket holders to win.

A good classroom followup to this problem would be to discuss the mathematical wisdom of playing regional and state lotteries. Perhaps students have read about people who have spent their life savings buying thousands of tickets for one day's drawing, their thinking being that if they buy 5,000 tickets, and no one else has bought more than, say, 20 for that drawing, they are sure to win. After all, 5,000 is many times more than 20.

Like Jesse, however, that individual will probably not win. Although he or she may have purchased 5,000 tickets, far more than anyone else, total sales might equal 1,000,000 tickets, nearly assuring that someone else will win.

Some lottery advertisements use a slogan, such as, "*Someone's* gonna win!" Truth in advertising would require the follow-up line "And it probably won't be you!" As a popular bumper sticker points out, "The lottery is a tax on the mathematically challenged."

Rolling the Block

The interesting problem shown in **figure 4** is for students who know that the circumference of a circle is pi times the diameter. They are likely to propose an answer of 31.4 inches, the circumference of one log. Students should be given credit for knowing and applying the circumference formula and for demonstrating an understanding of the meaning of pi. They should also be tactfully informed that they are incorrect. An explanation follows.

Rolling over the tops of 10-inch-diameter logs as they make one turn will move the block 31.4 inches. This situation fails to take into account that the logs themselves also move forward by 31.4 inches. The forward progress of the block is combined with the forward motion of the logs themselves, so the block will roll 31.4 + 31.4 inches, or 62.8 inches, for each log.

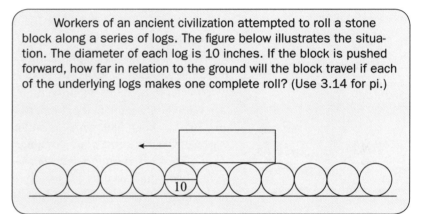

Workers of an ancient civilization attempted to roll a stone block along a series of logs. The figure below illustrates the situation. The diameter of each log is 10 inches. If the block is pushed forward, how far in relation to the ground will the block travel if each of the underlying logs makes one complete roll? (Use 3.14 for pi.)

Fig. 4. The Rolling the Block problem

The Careless Cat

Many students will want to make a sketch of the problem in **figure 5**. Encourage this tactic, as it is an appropriate problem-solving strategy. Students' sketches may look something like **figure 5**. The zigzags represent the cat's upward, then downward, progress and regression.

One eighth grader who was working on this problem stated, "Going up two feet in a minute, then going down one foot the next is like going up one foot every two minutes." At that rate, the cat would emerge from the six-foot hole at 12:12 p.m.

However, once the cat reaches the top of the hole at 12:09 it no longer falls back. Continuing with the zigzag pattern suggested in **figure 5** may clarify this problem for skeptical students.

Birthday Match

Here are some comments on the Birthday Match problem mentioned previously. The teacher may want to backtrack for a moment and ask, "How many people would need to be in a room for it to be a *certainty* that two of them have the exact same birthday in a nonleap year?" Many students will quickly give the correct answer of 366. The teacher can then present the Birthday Match problem:

> What is the minimum number of people, randomly chosen, who need to be assembled in a room so that the chances are at least 50 percent that two of them have exactly the same birthday?

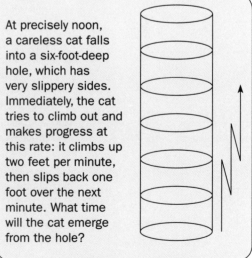

At precisely noon, a careless cat falls into a six-foot-deep hole, which has very slippery sides. Immediately, the cat tries to climb out and makes progress at this rate: it climbs up two feet per minute, then slips back one foot over the next minute. What time will the cat emerge from the hole?

Fig. 5. The Careless Cat problem

Students may make a first guess of 183—half of 366. The teacher can expect other wild guesses in excess of the correct answer of 23.

Even students who have developed a sense of caution when approaching such questions as these may state that the number 23 simply cannot be right. "Twenty-three is hardly any people compared with 365" was how one eighth grader put it. One way to convince them is to have them check several other classrooms in the school for common birthdays.

Even after empirical verification of the answer, students will not be clear as to why 23 is correct. Here is a brief explanation, offered with the assumption that students have at least some basic prior knowledge of probability.

What are the chances that a birthday match will *not* occur if *one* person stands alone in a room? The chances are 100 percent—no one is in the room to match up with! That probability can also be shown as 365/365; 365 possibilities divided by 365 matches that are sure not to happen.

What is the no-match probability if *two* people are in the room? Since 364 non-matching possibilities exist, we have chances of 365/365 × 364/365, multiplying the individual probabilities together, giving about .997, or near certainty.

With *three* people in the room, the no-match chances become 365/365 × 364/365 × 363/365, or about .989, which is still near certainty, but not as near. Continuing in this way, we would at some point compute a no-birthday-match probability of approximately .5, or 50 percent, meaning that the chances of having a match are *also* about 50 percent. In fact, we would need to go "twenty-three fractions deep" for the probability to hit approximately 50 percent. The computation can easily be verified with a calculator.

Use of counter-
intuitive problems
can promote
meaningful thought
in middle-grades
classrooms

Summary

Use of counterintuitive problems such as those presented here can promote meaningful thought in middle-grades classrooms and can provide a framework for substantive, intelligent discussion. The broader theme of course is student engagement, in keeping with the NCTM's Standards.

BIBLIOGRAPHY

Blum, Ray, and Jeff Sinclair. *Math Tricks, Puzzles and Games*. Northampton, Mass.: Sterling Publications, 1998.

Dewdney, A. K. *200 Percent of Nothing: An Eye-Opening Tour through the Twists and Turns of Math Abuse and Innumeracy*. New York: John Wiley & Sons, 1993.

Dudeney, Henry Ernest. *536 Curious Problems and Puzzles*. New York: Barnes & Noble Books, 1995.

Gardner, Martin. *Mathematical Carnival*. New York: Alfred A. Knopf, 1975.

Kordensky, Boris A. *The Moscow Puzzles: 359 Mathematical Recreations*. New York: Dover Publications, 1992.

Maylone, Nelson. *That Can't Be Right! Using Counterintuitive Problems in the Classroom*. Lancaster, Penn.: Technomic Publishing Co., 1999.

National Council of Teachers of Mathematics (NCTM). *Curriculum and Evaluation Standards for School Mathematics*. Reston, Va.: NCTM, 1989.

Paulos, John Allen. *Innumeracy*. New York: Vintage Books, 1988.

Vecchione, Glen, and Nina Zottoli. *Challenging Math Puzzles*. Northampton, Mass.: Sterling Publications, 1998.

COMMENTARY

"Some books are to be tasted, others to be swallowed and some few to be chewed and digested" (Bacon 1909). A similar continuum holds true for mathematical ideas and concepts. Although the timing of topics within the curriculum for any given school system will vary, certain big ideas in mathematics need to be examined in depth, or as Bacon suggests, "digested." We know that the shallow understanding that results from merely memorizing procedures and formulas limits students to being able to solve only basic and sometimes trivial problems. Liping Ma (1999) states that teachers' comprehensive knowledge of a topic contributes to students' opportunities to learn it. Many articles in NCTM journals focus on a deeper treatment of content. We have limited our selections here to span some fundamental concepts in the curriculum. Each article is heavily focused on developing that concept in a meaningful way, and therefore may help teachers deepen their own content knowledge as they consider instruction that will deepen their students' understanding.

"HEY, WAIT A MINUTE! JUST YESTERDAY, SHE SAID X WAS EQUAL TO _FIVE!_"

Synopses of Articles

"Relational Understanding and Instrumental Understanding," a classic article by Skemp (1978), contrasts the disparate goals of teaching for a robust understanding of mathematics with the more common classroom goal of seeking proficiency at the exclusion of understanding (what Skemp calls "rules without reason"). This timeless article is appropriate for teachers at all grade levels and experience levels.

Buchholz (2004) discusses the importance of focusing on strategies when teaching the basic facts in her article "Learning Strategies for Addition and Subtraction Facts: The Road to Fluency and the License to Think." The many strategies available and how she uses them have the potential to have a strong positive influence on teachers who practice traditional rote memorization of facts.

Cotter's (2000) article "Using Language and Visualization to Teach Place Value" offers numerous models for helping students develop a deep understanding of place value, including the ten frame, tallies, a double-sided abacus, and base-ten blocks. She also addresses the language considerations in learning place value. She shares a study indicating that these strategies led to improved learning by students.

This collection includes several articles on the important topic of computation. First, a classic among classics, "Meaning and Skill—Maintaining the Balance," by William Brownell (1956), articulately argues the existence of levels of understanding that go beyond knowing how to compute. This article is as relevant today as it was fifty years ago. The fact that it is more than fifty years old underscores how difficult it is to change the way we teach mathematics, and therefore the critical importance of professional learning opportunities to enable such fundamental change.

More recently, Wu (2001) shares models for making sense of multiplication in "Multiplying Fractions." The article includes two examples, one an area model and one a length model. This short piece can support teachers' growth in knowledge of multiplication of fractions and instructional strategies to support student understanding of the topic.

Lannin (2003), in "Developing Algebraic Reasoning through Generalization," lists and discusses strategies that students use to study patterns. A knowledge of these strategies can support teachers in understanding their students' algebraic reasoning.

"Three Balloons for Two Dollars: Developing Proportional Reasoning," by Langrall and Swafford (2000), offers four developmental levels gathered from the literature that follow how students reason proportionally (or nonproportionally). The authors elaborate to discuss essential components of teaching proportional reasoning.

Relational Understanding and Instrumental Understanding

Richard R. Skemp

I**N** *this article, the author defines* relational *and* instrumental understanding. *He then explains the impact he feels these two disparate goals have on the attitudes and understanding of students. We believe the reader will find his ideas about the teaching and learning of mathematics remarkably contemporary and thought-provoking.*

Faux Amis

Faux amis is a term used by the French to describe words which are the same, or very alike, in two languages, but whose meanings are different. For example:

French word	Meaning in English
histoire	story, not history
libraire	bookshop, not library
chef	head of any organization, not only chief cook
agrément	pleasure or amusement, not agreement
docteur	doctor (higher degree) not medical practitioner
médecin	medical practitioner, not medicine
parent	relations in general, including parents

One gets *faux amis* between English as spoken in different parts of the world. An Englishman asking in America for a biscuit would be given what we [Englishmen] call a scone. To get what we [Englishmen] call a biscuit, he would have to ask for a cookie. And between English as used in mathematics and in everyday life there are such words as field, group, ring, ideal.

A person who is unaware that the word he is using is a *faux ami* can make inconvenient mistakes. We expect history to be true, but not a story. We take books without paying from a library, but not from a bookshop; and so on. But in the foregoing examples there are cues which might put one on guard: difference of language, or of country, or of context.

If, however, the same word is used in the same language, country and context, with two meanings whose difference is non-trivial but as basic as the difference between the meaning of (say) 'histoire' and 'story', which is a difference between fact and fiction, one may expect serious confusion.

Two such words can be identified in the context of mathematics; and it is the alternative meanings attached to these words, each by a large following, which in my belief are at the root of many of the difficulties in mathematics education to-day.

One of these is 'understanding'. It was brought to my attention some years ago by Stieg Mellin-Olsen, of Bergen University, that there are in current use two meanings of this word. These he distinguishes by calling them 'relational understanding' and 'instrumental understanding'. By the former is meant what I have always meant by understanding, and probably most readers of this article: knowing both what to do and why. Instrumental understanding I would until recently not have regarded as understanding

Reprinted from the December 1976 issue of Mathematics Teaching, *the journal of the Association of Teachers of Mathematics, Great Britain. All rights reserved. Also reprinted in the* Arithmetic Teacher, *November 1978, pp. 9–15.*

A word on the editorial approach to reprinted articles: Obvious typographical errors have been silently corrected. Additions to the text for purposes of clarification appear in brackets. No effort has been made to reproduce the layouts or designs of the original articles, although the subheads are those that first appeared with the text. The use of words and phrases now considered outmoded, even slightly jarring to modern sensibilities, has likewise been maintained in an effort to give the reader a better feel for the era in which the articles were written.—Ed.

Mathematics Teaching in the Middle School 5 (September 2006): 88–95

at all. It is what I have in the past described as 'rules without reasons', without realising that for many pupils *and their teachers* the possession of such a rule, and ability to use it, was what they meant by 'understanding'.

Suppose that a teacher reminds a class that the area of a rectangle is given by $A = L \times B$. A pupil who has been away says he does not understand, so the teacher gives him an explanation along these lines. "The formula tells you that to get the area of a rectangle, you multiply the length by the breadth." "Oh, I see," says the child, and gets on with the exercise. If we were now to say to him (in effect) "You may think you understand, but you don't really," he would not agree. "Of course I do. Look; I've got all these answers right." Nor would he be pleased at our de-valuing of his achievement. And with his meaning of the word, he does understand.

We can all think of examples of this kind: 'borrowing' in subtraction, 'turn it upside down and multiply' for division by a fraction, 'take it over to the other side and change the sign', are obvious ones; but once the concept has been formed, other examples of instrumental explanations can be identified in abundance in many widely used texts. Here are two from a text used by a former direct-grant grammar school, now independent, with a high academic standard.

Multiplication of fractions To multiply a fraction by a fraction, multiply the two numerators together to make the numerator of the product, and the two denominators to makes its denominator.

E.g.

$$\frac{2}{3} \text{ of } \frac{4}{5} = \frac{2 \times 4}{3 \times 5} = \frac{8}{15}$$

$$\frac{3}{5} \times \frac{10}{13} = \frac{30}{65} = \frac{6}{13}$$

The multiplication sign $\times$ is generally used instead of the word 'of'.

Circles The circumference of a circle (that is its perimeter, or the length of its boundary) is found by measurement to be a little more than three times the length of its diameter. In any circle the circumference is approximately 3.1416 times the diameter which is roughly 3 1/7 times the diameter. Neither of these figures is exact, as the exact number cannot be expressed either as a fraction or a decimal. The number is represented by the Greek letter π (pi).

$$\text{Circumference} = \pi d \text{ or } 2\pi r$$
$$\text{Area} = \pi r^2$$

The reader is urged to try for himself this exercise of looking for and identifying examples of instrumental explanations, both in texts and in the classroom. This will have three benefits. (i) For persons like the writer, and most readers of this article, it may be hard to realise how widespread is the instrumental approach. (ii) It will help, by repeated examples, to consolidate the two contrasting concepts. (iii) It is a good preparation for trying to formulate the difference in general terms. Result (i) is necessary for what follows in the rest of the present section, while (ii) and (iii) will be useful for the others.

If it is accepted that these two categories are both well-fitted, by those pupils and teachers whose goals are respectively relational and instrumental understanding (by the pupil), two questions arise. First, does this matter? And second, is one kind better than the other? For years I have taken for granted the answers to both these questions: briefly, 'Yes; relational.' But the existence of a large body of experienced teachers and of a large number of texts belonging to the opposite camp has forced me to think more about why I hold this view. In the process of changing the judgement from an intuitive to a reflective one, I think I have learnt something useful. The two questions are not entirely separate, but in the present section I shall concentrate as far as possible on the first: does it matter?

The problem here is that of a mismatch, which arises automatically in any *faux ami* situation, and does not depend on whether A or B's meaning is 'the right one'. Let us imagine, if we can, that school A sends a team to play school B at a game called 'football', but that neither team knows that there are two kinds (called 'association' and 'rugby'). School A plays soccer and has never heard of rugger, and vice versa for B. Each team will rapidly decide that the others are crazy, or a lot of foul players. Team A in particular will think that B uses a mis-shapen ball, and commit one foul after another. Unless the two sides stop and talk about what game they think they are playing at, long enough to gain some mutual understanding, the game will break up in disorder and the two teams will never want to meet again.

Though it may be hard to imagine such a situation arising on the football field, this is not a far-fetched analogy for what goes on in many mathematics lessons, even now. There is this important difference, that one side at least cannot refuse to play. The encounter is compulsory, on five days a week, for about 36 weeks a year, over 10 years or more of a child's life.

Leaving aside for the moment whether one kind is better than the other, there are two kinds of mathematical mis-matches which can occur.

1. Pupils whose goal is to understand instrumentally, taught by a teacher who wants them to understand relationally.
2. The other way about.

The first of these will cause fewer problems *short-term* to the pupils, though it will be frustrating to the teacher. The pupils just 'won't want to know' all the careful ground-work he gives in preparation for whatever is to be learnt next, nor his careful explanations. All they want is some kind of rule for getting the answer. As soon as this is reached, they latch on to it and ignore the rest.

If the teacher asks a question that does not quite fit the rule, of course they will get it wrong. For the following example I have to thank Mr. Peter Burney, at that time a student at Coventry College of Education on teaching practice. While teaching area he became suspicious that the children did not really understand what they were doing. So he asked them: "What is the area of a field 20 cms by 15 yards?" The reply was: "300 square centimetres". He asked: "Why not 300 square yards?" Answer: "Because area is always in square centimetres."

To prevent errors like the above the pupils need another rule (or, of course, relational understanding), that both dimensions must be in the same unit. This anticipates one of the arguments which I shall use against instrumental understanding, that it usually involves a multiplicity of rules rather than fewer principles of more general application.

There is of course always the chance that a few of the pupils will catch on to what the teacher is trying to do. If only for the sake of these, I think he should go on trying. By many, probably a majority, his attempts to convince them that being able to use the rule is not enough will not be well received. 'Well is the enemy of better,' and if pupils can get the right answers by the kind of thinking they are used to, they will not take kindly to suggestions that they should try for something beyond this.

The other mis-match, in which pupils are trying to understand relationally but the teaching makes this impossible, can be a more damaging one. An instance which stays in my memory is that of a neighbour's child, then seven years old. He was a very bright little boy, with an I.Q. of 140. At the age of five he could read *The Times*, but at seven he regularly cried over his mathematics homework. His misfortune was that he was trying to understand relationally teaching which could not be understood in this way. My evidence for this belief is that when I taught him relationally myself, with the help of Unifix, he caught on quickly and with real pleasure.

A less obvious mis-match is that which may occur between teacher and text. Suppose that we have a teacher whose conception of understanding is instrumental, who for one reason or other is using a text which aim is relational understanding by the pupil. It will take more than this to change his teaching style. I was in a school which was using my own text[1], and noticed (they were at Chapter 1 of *Book 1*) that some of the pupils were writing answers like

'the set of {flowers}'.

When I mentioned this to the teacher (he was head of mathematics) he asked the class to pay attention to him and said: "Some of you are not writing your answers properly. Look at the example in the book, at the beginning of the exercise, and be sure you write your answers exactly like that."

Much of what is being taught under the description of 'modern mathematics' is being taught and learnt just as instrumentally as were the syllabi which have been replaced. This is predictable from the difficulty of accommodating (restructuring) our existing schemas[2]. To the extent that this is so, the innovations have probably done more harm than good, by introducing a mis-match between the teacher and the aims implicit in the new content. For the purpose of introducing ideas such as sets, mappings and variables is the help which, rightly used, they can give to relational understanding. If pupils are still being taught instrumentally, then a 'traditional' syllabus will probably benefit them more. They will at least acquire proficiency in a number of mathematical techniques which will be of use to them in other subjects, and whose lack has recently been the subject of complaints by teachers of science, employers and others.

Near the beginning I said that two faux amis could be identified in the context of mathematics. The second one is even more serious; it is the word 'mathematics' itself. For we are not talking about better and worse teaching of the same kind of mathematics. It is easy to think this, just as our imaginary soccer players who did not know that their opponents were playing a different game might think that the other side picked up the ball and ran with it because they could not kick properly, especially with such a mis-shapen ball. In which case they might kindly offer them a better ball and some lessons on dribbling.

It has taken me some time to realise that this is not the case. I used to think that maths teachers were all teaching the same subject, some doing it better than others. I now believe that *there are two effectively different subjects being taught under the same name, 'mathematics'*. If this is true, then this difference matters beyond any of the differences in syllabi which are so widely debated. So I would like to try to emphasise the point with the help of another analogy.

Imagine that two groups of children are taught music as a pencil-and-paper subject. They are all shown the five-line stave, with the curly 'treble' sign at the beginning; and taught that marks on the lines are called E, G, B, D, F. Marks between the lines are called F, A, C, E. They learn that a line with an open oval is called a minim, and is worth two with blacked-in ovals which are called crotchets, or four with blacked-in ovals and a tail which are called quavers, and so on—musical multiplication tables if you like. For one group of children, all their learning is of this kind and nothing beyond. If they have a music lesson a day, five days a week in school terms, and are told that it is important, these children could in time probably learn to write out the marks for simple melodies such as *God Save the Queen* and *Auld Lang Syne,* and to solve simple problems such as

'What time is this in?' and 'What key?', and even 'Transpose this melody from C major to A major'. They would find it boring, and the rules to be memorised would be so numerous that problems like 'Write a simple accompaniment for this melody' would be too difficult for most. They would give up the subject as soon as possible, and remember it with dislike.

The other group is taught to associate certain sounds with these marks on paper. For the first few years these are audible sounds, which they make themselves on simple instruments. After a time they can still imagine the sounds whenever they see or write the marks on paper. Associated with every sequence of marks is a melody, and with every vertical set a harmony. The keys C major and A major have an audible relationship, and a similar relationship can be found between certain other pairs of keys. And so on. Much less memory work is involved, and what has to be remembered is largely in the form of related wholes (such as melodies) which their minds easily retain. Exercises such as were mentioned earlier ('Write a simple accompaniment') would be within the ability of most. These children would also find their learning intrinsically pleasurable, and many would continue it voluntarily, even after O-level or C.S.E.

For the present purpose I have invented two non-existent kinds of 'music lesson', both pencil-and-paper exercises (in the second case, after the first year or two). But the difference between these imaginary activities is no greater than that between two activities which actually go on under the name of mathematics. (We can make the analogy closer, if we imagine that the first group of children was initially taught sounds for the notes in a rather half-hearted way, but that the associations were too ill-formed and unorganised to last.)

The above analogy is, clearly, heavily biased in favour of relational mathematics. This reflects my own viewpoint. To call it a viewpoint, however, implies that I no longer regard it as a self-evident truth which requires no justification: which it can hardly be if many experienced teachers continue to teach instrumental mathematics. The next step is to try to argue the merits of both points of view as clearly and fairly as possible; and especially of the point of view opposite to one's own. This is why the next section is called *Devil's Advocate*. In one way this only describes that part which puts the case for instrumental understanding. But it also justifies the other part, since an imaginary opponent who thinks differently from oneself is a good device for making clearer to oneself why one does think that way.

Devil's Advocate

Given that so many teachers teach instrumental mathematics, might this be because it does have certain advantages? I have been able to think of three advantages (as distinct from situational reasons for teaching this way, which will be discussed later).

1. Within its own context, *instrumental mathematics is usually easier to understand;* sometimes much easier. Some topics, such as multiplying two negative numbers together, or dividing by a fractional number, are difficult to understand relationally. 'Minus times minus equals plus' and 'to divide by a fraction you turn it upside down and multiply' are easily remembered rules. If what is wanted is a page of right answers, instrumental mathematics can provide this more quickly and easily.

2. *So the rewards are more immediate, and more apparent.* It is nice to get a page of right answers, and we must not under-rate the importance of the feeling of success which pupils get from this. Recently I visited a school where some of the children describe themselves as 'thickos'. Their teachers use the term too. These children need success to restore their self-confidence, and it can be argued that they can achieve this more quickly and easily in instrumental mathematics than in relational.

3. Just because less knowledge is involved, *one can often get the right answer*

more quickly and reliably by instrumental thinking than relational. This difference is so marked that even relational mathematicians often use instrumental thinking. This is a point of much theoretical interest, which I hope to discuss more fully on a future occasion.

The above may well not do full justice to instrumental mathematics. I shall be glad to know of any further advantages which it may have.

There are four advantages (at least) in relational mathematics.

1. *It is more adaptable to new tasks.* Recently I was trying to help a boy who had learnt to multiply two decimal fractions together by dropping the decimal point, multiplying as for whole numbers, and re-inserting the decimal point to give the same total number of digits after the decimal point as there were before. This is a handy method if you know why it works. Through no fault of his own, this child did not; and not unreasonably, applied it also to division of decimals. By this method 4.8 ÷ 0.6 came to 0.08. The same pupil had also learnt that if you know two angles of a triangle, you can find the third by adding the two given angles together and subtracting from 180°. He got ten questions right this way (his teacher believed in plenty of practice), and went on to use the same method for finding the exterior angles. So he got the next five answers wrong.

I do not think he was being stupid in either of these cases. He was simply extrapolating from what he already knew. But relational understanding, by knowing not only what method worked but why, would have enabled him to relate the method to the problem, and possibly to adapt the method to new problems. Instrumental understanding necessitates memorising which problems a method works for and which not, and also learning a different method for each new class of problems. So the first advantage of relational mathematics leads to:

2. *It is easier to remember.* There is a seeming paradox here, in that it is certainly harder to learn. It is certainly easier for pupils to learn that 'area of a triangle = 1/2 base × height' than to learn why this is so. But they then have to learn separate rules for triangles, rectangles, parallelograms, trapeziums; whereas relational understanding consists partly in seeing all of these in relation to the area of a rectangle. It is still desirable to know the separate rules; one does not want to have to derive them afresh everytime. But knowing also how they are inter-related enables one to remember them as parts of a connected whole, which is easier. There is more to learn—the connections as well as the separate rules—but the result, once learnt, is more lasting. So there is less re-learning to do, and long-term the time taken may well be less altogether.

Teaching for relational understanding may also involve more actual content. Earlier, an instrumental explanation was quoted leading to the statement 'Circumference = πd'. For relational understanding of this, the idea of a proportion would have to be taught first (among others), and this would make it a much longer job than simply teaching the rules as given. But proportionality has such a wide range of other applications that it is worth teaching on these grounds also. In relational mathematics this happens rather often. Ideas required for understanding a particular topic turn out to be basic for understanding many other topics too. Sets, mappings and equivalence are such ideas. Unfortunately the benefits which might come from teaching them are often lost by teaching them as separate topics, rather than as fundamental concepts by which whole areas of mathematics can be inter-related.

3. *Relational knowledge can be effective as a goal in itself.* This is an empiric fact, based on evidence from controlled experiments using non-mathematical material. The need for external rewards and punishments is greatly reduced, making what is often called the 'motivational' side of a teacher's job much easier. This is related to:

4. *Relational schemas are organic in quality.* This is the best way I have been able to formulate a quality by which they seem to act as an agent of their own growth. The

connection with 3 is that if people get satisfaction from relational understanding, they may not only try to understand relationally new material which is put before them, but also actively seek out new material and explore new areas, very much like a tree extending its roots or an animal exploring a new territory in search of nourishment. To develop this idea beyond the level of an analogy is beyond the scope of the present paper, but it is too important to leave out.

If the above is anything like a fair presentation of the cases for the two sides, it would appear that while a case might exist for instrumental mathematics short-term and within a limited context, long-term and in the context of a child's whole education it does not. So why are so many children taught only instrumental mathematics throughout their school careers? Unless we can answer this, there is little hope of improving the situation.

An individual teacher might make a reasoned choice to teach for instrumental understanding on one or more of the following grounds.

1. That relational understanding would take too long to achieve, and to be able to use a particular technique is all that these pupils are likely to need.
2. That relational understanding of a particular topic is too difficult, but the pupils still need it for examination reasons.
3. That a skill is needed for use in another subject (e.g. science) before it can be understood relationally with the schemas presently available to the pupils.
4. That he is a junior teacher in a school where all the other mathematics teaching is instrumental.

All of these imply, as does the phrase 'make a reasoned choice', that he is able to consider the alternative goals of instrumental and relational understanding on their merits and in relation to a particular situation. To make an informed choice of this kind implies awareness of the distinction, and relational understanding of the mathematics itself. So nothing else but relational understanding can ever be adequate for a teacher. One has to face the fact that this is absent in many who teach mathematics; perhaps even a majority.

Situational factors which contribute to the difficulty include:

1. *The backwash effect of examinations.* In view of the importance of examinations for future employment, one can hardly blame pupils if success in these is one of their major aims. The way pupils work cannot but be influenced by the goal for which they are working, which is to answer correctly a sufficient number of questions.

2. *Over-burdened syllabi.* Part of the trouble here is the high concentration of the information content of mathematics. A mathematical statement may condense into a single line as much as in another subject might take over one or two paragraphs. By mathematicians accustomed to handling such concentrated ideas, this is often overlooked (which may be why most mathematics lecturers go too fast). Non-mathematicians do not realise it at all. Whatever the reason, almost all syllabi would be much better if much reduced in amount so that there would be time to teach them better.

3. *Difficulty of assessment* of whether a person understands relationally or instrumentally. From the marks he makes on paper, it is very hard to make valid inference about the mental processes by which a pupil has been led to make them; hence the difficulty of sound examining in mathematics. In a teaching situation, talking with the pupil is almost certainly the best way to find out; but in a class of over 30, it may be difficult to find the time.

4. *The great psychological difficulty for teachers of accommodating (re-structuring) their existing and longstanding schemas,* even for the minority who know they need to, want to do so, and have time for study.

From a recent article[3] discussing the practical, intellectual and cultural value of a mathematics education (and I have no doubt that he means relational mathematics!) by Sir Hermann Bondi, I take these three paragraphs. (In the original, they are not consecutive.)

> So far my glowing tribute to mathematics has left out a vital point: the rejection of mathematics by so many, a rejection that in not a few cases turns to abject fright.
>
> The negative attitude to mathematics, unhappily so common, even among otherwise highly-educated people, is surely the greatest measure of our failure and a real danger to our society.
>
> This is perhaps the clearest indication that something is wrong, and indeed very wrong, with the situation. It is not hard to blame education for at least a share of the responsibility; it is harder to pinpoint the blame, and even more difficult to suggest new remedies.

If for 'blame' we may substitute 'cause', there can be small doubt that the widespread failure to teach relational mathematics—a failure to be found in primary, secondary and further education, and in 'modern' as well as 'traditional' courses—can be identified as a major cause. To suggest new remedies is indeed difficult, but it may be hoped that diagnosis is one good step towards a cure. Another step will be offered in the next section.

A Theoretical Formulation

There is nothing so powerful for directing one's actions in a complex situation, and for co-ordinating one's own efforts with those of others, as a good theory. All good teachers build up their own stores of empirical knowledge, and have abstracted from these some general principles on which they rely for guidance. But while their knowledge remains in this form it is largely still at the intuitive level within individuals, and cannot be communicated, both for this reason and because there is no shared conceptual structure (schema) in terms of which it can be formulated. Were this possible, individual efforts could be integrated into a unified body of knowledge which would be available for use by new-comers to the profession. At present most teachers have to learn from their own mistakes.

For some time my own comprehension of the difference between the two kinds of learning which lead respectively to relational and instrumental mathematics remained at the intuitive level, though I was personally convinced that the difference was one of great importance, and this view was shared by most of those with whom I discussed it. Awareness of the need for an explicit formulation was forced on me in the course of two parallel research projects; and insight came, quite suddenly, during a recent conference. Once seen it appears quite simple, and one wonders why I did not think of it before. But there are two kinds of simplicity: that of naivity; and that which, by penetrating beyond superficial differences, brings simplicity by unifying. It is the second kind which a good theory has to offer, and this is harder to achieve.

A concrete example is necessary to begin with. When I went to stay in a certain town for the first time, I quickly learnt several particular routes. I learnt to get between where I was staying and the office of the colleague with whom I was working; between where I was staying and the university refectory where I ate; between my friend's office and the refectory; and two or three others. In brief, I learnt a limited number of fixed plans by which I could get from particular starting locations to particular goal locations.

As soon as I had some free time, I began to explore the town. Now I was not wanting to get anywhere specific, but to learn my way around, and in the process to see what I might come upon that was of interest. At this stage my goal was a different one: to construct in my mind a cognitive map of the town.

These two activities are quite different. Nevertheless they are, to an outside observer, difficult to distinguish. Anyone seeing me walk from A to B would have great difficulty in knowing (without asking me) which of the two I was engaged in. But the most important thing about an activity is its goal. In one case my goal was to get to B, which is a physical location. In the other it was to enlarge or consolidate my mental map of the town, which is a state of knowledge.

A person with a set of fixed plans can find his way from a certain set of starting points to a certain set of goals. The characteristic of a plan is that it tells him what to do at each choice point: turn right out of the door, go straight on past the church, and so on. But if at any stage he makes a mistake, he will be lost; and he will stay lost if he is not able to retrace his steps and get back on the right path.

In contrast, a person with a mental map of the town has something from which he can produce, when needed, an almost infinite number of plans by which he can guide his steps from any starting point to any finishing point, provided only that both can be imagined on his mental map. And if he does take a wrong turn, he will still know where he is, and thereby be able to correct his mistake without getting lost; even perhaps to learn from it.

The analogy between the forgoing and the learning of mathematics is close. The kind of learning which leads to instrumental mathematics consists of the learning of an increasing number of fixed plans, by which pupils can find their way from particular starting points (the data) to required finishing points (the answers to the questions). The plan tells them what to do at each choice point. And as in the concrete example, what has to be done next is determined purely by the local situation. (When you see the post office, turn left. When you have cleared brackets, collect like terms.) There is no awareness of the overall relationship between successive stages, and the final goal. And in both cases, the learner is dependent on outside guidance for learning each new 'way to get there'.

In contrast, learning relational mathematics consists of building up a conceptual structure (schema) from which its possessor can (in principle) produce an unlimited number of plans for getting from any starting point within his schema to any finishing point. (I say 'in principle' because of course some of these paths will be much harder to construct than others.)

This kind of learning is different in several ways from instrumental learning.

1. The means become independent of particular ends to be reached thereby.
2. Building up a schema within a given area of knowledge becomes an intrinsically satisfying goal in itself.
3. The more complete a pupil's schema, the greater his feeling of confidence in his own ability to find new ways of 'getting there' without outside help.
4. But a schema is never complete. As our schemas enlarge, so our awareness of possibilities is thereby enlarged. Thus the process often becomes self-continuing, and (by virtue of 3) self-rewarding.

Taking again for a moment the role of devil's advocate, it is fair to ask whether we are indeed talking about two subjects, relational mathematics and instrumental mathematics, or just two ways of thinking about the same subject matter. Using the concrete analogy, the two processes described might be regarded as two different ways of knowing about the same town; in which case the distinction made between relational and instrumental understanding would be valid, but not that between instrumental and relational mathematics.

But what constitutes mathematics is not the subject matter, but a particular kind of knowledge about it. The subject matter of relational and instrumental mathematics may be the same: cars travelling at uniform speeds between two towns, towers whose heights are to be found, bodies falling freely under gravity, etc. etc. But the two kinds of knowledge are so different that I think that there is a strong case for regarding them as different kinds of mathematics. If this distinction is accepted, then the word 'mathematics' is for many children indeed a false friend, as they find to their cost.

The State of Play

This is already a long article, yet it leaves many points awaiting further development. The applications of the theoretical formulation in the last section to the educational problems described in the first two have not been spelt out. One of these is the relationship between the goals of the teacher and those of the pupil. Another is the implications for a mathematical curriculum.

In the course of discussion of these ideas with teachers and lecturers in mathematical education, a number of other interesting points have been raised which also cannot be explored further here. One of these is whether the term 'mathematics' ought not be used for relational mathematics only. I have much sympathy with this view, but the issue is not as simple as it may appear.

There is also research in progress. A pilot study aimed at developing a method (or methods) for evaluating the quality of children's mathematical thinking has been finished, and has led to a more substantial study in collaboration with the N.F.E.R. as part of the TAMS continuation project. A higher degree thesis at Warwick University is nearly finished; and a research group of the Department of Mathematics at the University of Quebec in Montreal is investigating the problem with first and fourth grade children. All this will I hope be reported in due course.

The aims of the present paper are twofold. First, to make explicit the problem at an empiric level of thinking, and thereby to bring to the forefront of attention what some of us have known for a long time at the back of our minds. Second, to formulate this in such a way that it can be related to existing theoretical knowledge about the mathematical learning process, and further investigated at this level and with the power and generality which theory alone can provide.

ENDNOTES

1 R. R. Skemp: *Understanding Mathematics* (U.L.P.)

2 For a fuller discussion see R. R. Skemp: *The Psychology of Learning Mathematics* (Penguin 1972) pp. 43–46

3 H. Bondi: *The Dangers of Rejecting Mathematics* (*Times Higher Education Supplement,* 26.3.76)

Learning Strategies for Addition and Subtraction Facts: The Road to Fluency and the License to Think

Lisa Buchholz

TEACHING the basic facts seemed like the logical thing to do. Wouldn't a study of the basic facts make mathematics computation much easier for my students in the future? How could I help my students memorize and internalize this seemingly rote information? How could I get rid of finger counting and move on to mental computation? As I embarked on my first year of teaching second grade following many years of teaching first grade, these questions rolled through my head.

I had spent the summer poring through the curriculum for grade 2. After familiarizing myself with the mathematics concepts that I would be teaching, I decided to begin the school year with an intense study of strategies for learning and remembering the basic addition and subtraction facts. I looked to *Principles and Standards for School Mathematics* (NCTM 2000) for guidance. I was even more excited about my choice of topics after reading the following quote: "As children in prekindergarten through grade 2 develop an understanding of whole numbers and the operations of addition and subtraction, instructional attention should focus on strategies for computing with whole numbers so that students develop flexibility and computational fluency. Students will generate a range of interesting and useful strategies for solving computational problems, which should be shared and discussed" (NCTM 2000, p. 35). I believed that I was definitely on the right track by focusing on strategies. My next step was to research effective strategies for learning basic facts. *Facts That Last: A Balanced Approach to Mathematics* by Larry Leutzinger (1999a, 1999b) was an invaluable resource. After reading these wonderful books, I set a goal for my students to memorize the basic facts and "move on," but I was in for quite a surprise. I was completely unaware of the impact that this experience would have on my students and their number sense.

The Journey Begins

The first week of school, I began the unit with great excitement. I chose the "doubles" for a starting point. We worked with this concept for a few days; I wanted to be sure that everyone understood it. The first day, we used manipulatives to create equations showing doubles. The next day, we illustrated and wrote about everyday situations in which we can see doubles, such as "5 fingers plus 5 fingers equals 10 fingers" or "2 arms plus 2 legs equals 4 limbs." On the third day, the children wrote their own definitions for "doubles" in their mathematics journals. They also created equations that showed doubles. My students of average to lower-average ability had equations such as $3 + 3 = 6$ and $7 + 7 = 14$. My higher-ability students had equations such as $175 + 175 = 350$ and $324 + 324 = 648$. These students particularly enjoyed the open-ended structure of the mathematics journal assignment, and their enthusiasm ran high. We had begun to really learn and apply doubles instead of merely recognizing them. If I had been paying closer attention, I would have seen the beginning of our number-sense explosion.

Teaching Children Mathematics 10 (March 2004): 362–67

After making sure that the students had a full grasp of the doubles, we started working with combinations related to "Doubles Plus One." Using manipulatives, illustrations, and journal writing, we began to train our brains to see "5 + 6" but think "5 + 5 + 1." (I say "we" because I was learning as well. I was taught not to think about equations but to memorize their answers. This "journey" was an awakening for me too.) Soon my students began to compute Doubles Plus One equations mentally, with very little effort. Even my lowest-ability students were solving the equations with ease.

In an attempt to integrate home and school and to share our mathematics excitement and learning, I devised homework for the Doubles Plus One strategy. The assignment asked the children to write an explanation of the strategy and to create problems that could be categorized as Doubles Plus One. The homework also included a parent information sheet, which explained the strategy and gave specific examples of its use. The children were excited to share this homework and show off their new "math brains," a term they had invented.

During our Doubles Plus One exploration, I recalled seeing the same strategy in the district-adopted textbook, which I was not using. The book devoted only one page to the concept of Doubles Plus One. Devoting only one page to this strategy is like expecting someone to successfully ride a two-wheeler after one experience. The sharpest, most capable child might attempt it once and understand all its nuances and complexities, but the majority of children need time and experience to learn and apply the strategy. The page in the book would have presented the strategy, but would the students internalize and apply it or would they only recognize it? In *Elementary and Middle School Mathematics: Teaching Developmentally,* Van de Walle (2001) advises not to expect students to have an understanding of an introduced strategy after just one activity or experience. He states that students should use a strategy for several days so they can internalize it. I wholeheartedly agree.

Falling into the Routine

After we studied Doubles Plus One, my lesson plans for the strategies seemed to fall into a consistent pattern. My "mini unit" for each strategy followed this sequence of events:

- Introduce and explore the strategy using manipulatives.
- Create illustrations of the strategy.
- Use mathematics journals: Write a definition of the strategy and create problems that match its criterion.
- Assign homework including a Parent Information Page.

Additionally, each day I read a story problem aloud and asked students to come up with an answer mentally. I called on one student to answer the problem and explain how he or she solved it. Then I asked if anyone had a different way of solving the problem. I called on volunteers until we had heard five or six different ways to solve the problem. If the children had trouble understanding someone's "strategy," I wrote the student's explanation on the board step by step. A chorus of "Oh, I get it now" or "That makes sense" usually followed this illustration. We named this time of day "Mental Math." This was a daily chance for the children to apply their strategies. Over time, I featured two-digit numbers in the story problems. The children simply carried over their knowledge of the strategies to the new problem.

Many of the strategies we studied are featured in Leutzinger (1999a, 1999b) and Van de Walle (2001). We tailored some of these strategies to meet our needs and generated some strategies ourselves. With the combination of the "strategy" lessons (and related activities) and Mental Math sessions, the children demonstrated an amazing command of the world of numbers. They actively used the strategies I had taught them (see **figs. 1** and **2**) and explained them with confidence and conviction, both orally and in written form (see **figs. 3–5**). The children even began to make up their own strategies and explain them with enthusiasm and pride. They named their strategies "Jenna's Strategy" or "Jack's Favorite Strategy." Every day, the knowledge base that we were building became stronger. It was as if I had given my students a license to think. To my students, equations were not just equations anymore; they were numbers that they could

Addition Strategies	Our Interpretations and Descriptions
Doubles	Adding two of the same number together, such as 5 + 5 or 7 + 7
Doubles Plus One	Finding "hidden" doubles in expressions where one addend is one more than the other, such as 5 + 6 (thinking 5 + 5 + 1)
Doubles Plus Two	Finding "hidden" doubles in expressions where one addend is two more than the other, such as 5 + 7 (thinking 5 + 5 + 2)
Doubles Minus One	Locating doubles in expressions where one addend is one more than the other, such as 5 + 6 (but thinking 6 + 6 − 1 versus 5 + 5 + 1)
Doubles Minus Two	Locating doubles in expressions where one addend is two more than the other, such as 5 + 7 (but thinking 7 + 7 − 2)
Combinations of Ten	We learned to recognize expressions equaling 10 such as 6 + 4 and 7 + 3 for use in other strategies; we would picture our ten fingers.
Counting Up	This strategy was used only when adding 1 or 2 to a given number; we would see 9 + 2 and think, "9 . . . 10, 11."
Add One to Nine	Used when adding 9 to any number. This was how we "primed" ourselves for the Make Ten strategy; we would see 6 + 9 and think 6 + 10 − 1.
Make Ten	Turning more difficult expressions into expressions equaling 10 and then adding the "leftovers"; we would see 7 + 4 and think 7 + 3 + 1.
Adding Ten	Adding 10 to any number increases the digit in the tens place by one: 5 + 10 = 15, 12 + 10 = 22.
Commutative Property	Any given addends have the same sum regardless of their order: 8 + 7 = 7 + 8.

Fig. 1. Addition strategies that the students used

Subtraction Strategies	Our Interpretations and Descriptions
Counting Back	Beginning with the minuend, count back the number you are subtracting; we would see 9 − 3 and think, "9 . . . 8, 7, 6" for an answer of 6.
Counting Up	Beginning with the number you are subtracting, count up to the other number; we would see 12 − 9 and think, "9 . . . 10, 11, 12." Our answer would be 3 because we counted three numbers.
Doubles	We would see 14 − 7 and think 7 + 7 = 14.
Think Addition	We learned to think of related addition problems when confronted with subtraction facts; we would see 7 − 5 and think 5 + 2 = 7.
Fact Families	Similar to Think Addition above, we would think of the fact family to recall the "missing number." For a problem such as 8 − 5, we would recall 5 + 3 = 8, 3 + 5 = 8, 8 − 5 = 3, 8 − 3 = 5.
Subtracting from Ten	In equations with 10 as a minuend, we would mentally picture 10 (10 fingers, 10 frames, and so on) to learn what remained when some were taken away.

Fig. 2. Subtraction strategies that the students used

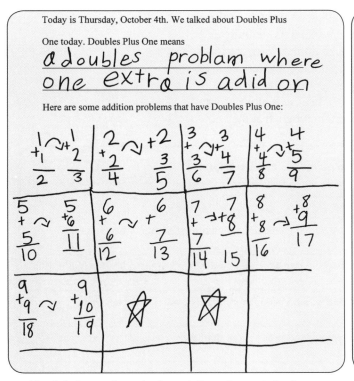

Today is Thursday, October 4th. We talked about Doubles Plus

One today. Doubles Plus One means

<u>a doubles problem where
one extra is a did on</u>

Here are some addition problems that have Doubles Plus One:

Fig. 3. In her mathematics journal, Tessa demonstrates how a Doubles equation is found in a Doubles Plus One equation.

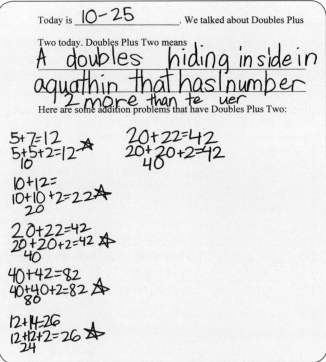

Today is ____10-25____. We talked about Doubles Plus

Two today. Doubles Plus Two means

<u>A doubles hiding inside in
aquathin that has l number
2 more than te uer</u>

Here are some addition problems that have Doubles Plus Two:

Fig. 4. Mary Caroline shows that a Doubles equation is found in a Doubles Plus Two equation.

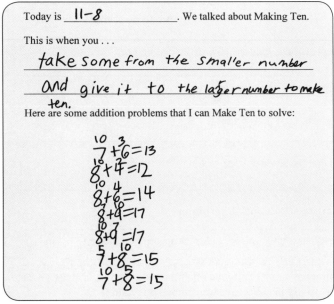

Today is ____11-8____. We talked about Making Ten.

This is when you . . .

<u>take some from the smaller number
and give it to the larger number to make
ten.</u>

Here are some addition problems that I can Make Ten to solve:

Fig. 5. Sarah shows how the Make Ten strategy works. She begins by creating a problem with two addends equaling more than ten when combined. Then she subtracts from one number to make the other number equal ten.

manipulate in any way that made sense to them. The following dialogue occurred during a Mental Math session. I have noted the strategies that the students used.

Me. Sarah had 9 fish. Her mother gave her 8 more fish. How many fish did Sarah now have?

Evan. Seventeen, because 9 + 9 is 18. One less would be 17. *[Doubles Minus One]*

Jenna. I know 9 + 9 is 18 and 8 + 8 is 16. . . . In the middle is 17. *[This strategy later became known as Jenna's Favorite Strategy or The Doubles Sandwich.]*

Jack. I took 2 from the 9 and gave it to the 8, which made 10. Then I took the 7 left over and put it with the 10, which gave me 17. *[Make Ten]*

Peter. Okay, I took 100 plus 100, which gave me 200. I took 200 – 400, which gave me negative 200. Then I took negative 200 + 209, which gave me 9. Then I added 1, which made 10. That's the Make Ten strategy. Then I added the 7 that were left, which gave me 17. *[Peter loved to work his way away from an answer and come back to it. We named his method the Walk All around the World Just to Cross the Street strategy.]*

Julian. I just knew the answer. *[Fluency]*

Hannah. I started on 9 and counted 8 more. *[Counting Up]*

Mark. I took 8 + 8 and got 16. Then I added 1 more to make 17. *[Doubles Plus One]*

The more strategies we learned, the longer our Mental Math time took. Every minute was worth it. My students seemed to be picturing one another's strategies mentally. This combination of an intense study of strategies and a daily opportunity for practice added up to success. Even now, well after our study of the strategies has ended (although we still review and practice each day during Mental Math), my students use their solid base of number sense to embrace every new mathematics challenge that comes along.

Another Discovery

Many positive things came out of our journey, such as the number sense and mental mathematics fluency I have already mentioned. I made yet another discovery, however: insight into my students and their abilities. I learned that I could not judge a student by others' impressions of them. Some students who seemed "lower ability" to their first-grade teachers actually were my best thinkers. They were able to help others understand concepts. Conversely, some students who were named top mathematics students in the previous year had difficulty. These few students were very good at performing algorithms or were already proficient with the addition and subtraction facts. They could not, however, answer the "how" and "why" questions that I posed daily as part of their mathematics journaling, such as "How did you get your answer?" and "Why did this work?" Simply learning what to do is a much easier task than is learning why to do it (Burns 1992).

As the year progressed and we explored new concepts such as renaming, these students' shaky foundation became more apparent. When solving problems involving regrouping, they could not explain why they crossed out a number and renamed it. They did not really understand mathematics; they understood algorithms that they had been taught and had quick recall of the basic facts, showing that they were good at memorizing information. I had uncovered a gap in the mathematics education of these students, a gap that desperately needed to be filled with a better understanding of numbers. A teacher who uses only the mathematics textbook might have a different idea of his or her top mathematics students than would a teacher who probes deeper and expects students to think about their strategies.

One student, who I will refer to as Steve, fell particularly hard into this mathematics textbook "void." On our "Meet the Teacher" day before school began, Steve's mother informed me that Steve had done third-grade mathematics the previous year. She also told me that he could "borrow" and "carry" to five or six digits. Two weeks into school, she saw Steve struggling and learned that my "lower ability" mathematics students were helping him with his assignments. Steve kept asking, "Can we just do borrowing and carrying? When will we get to borrowing and carrying?" His mother admitted that she wondered what kind of "weird math" I was teaching that had her son so confused. She began to blame his confusion on me and my methods until she saw the hole in her son's mathematics education filled with concepts and strategies. At the end of the year, she stopped in my classroom and said, "Steve has come a long way in understanding numbers. I think he's ready to handle third-grade math now." I firmly believe that if Steve had been taught with a mathematics textbook and a "dabble at the surface" approach, he really would have difficulty later when his ability to perform algorithms was no longer enough.

Final Thoughts

This experience was every teacher's dream. Not only did I grow in my own understanding of numbers but I now have a fresh enthusiasm for teaching mathematics. I have transformed from a "page a day" mathematics teacher to a facilitator of mathematics and its concepts.

This adventure into number sense took us about two months. In today's crowded curriculum, that is a considerable amount of time to invest in just one concept. My students emerged from this study, however, as amazing thinkers ready to take on any challenge that comes their way. The experience helped my students become faster and more accurate with mathematics flash cards and timed assessments, but the real gain was in number sense.

REFERENCES

Burns, Marilyn. *About Teaching Mathematics*. Sausalito, Calif.: Math Solutions Publications, 1992.

Leutzinger, Larry. *Facts That Last (Addition): A Balanced Approach to Memorization*. Chicago: Creative Publications, 1999a.

———. *Facts That Last (Subtraction): A Balanced Approach to Memorization*. Chicago: Creative Publications, 1999b.

National Council of Teachers of Mathematics (NCTM). *Principles and Standards for School Mathematics*. Reston, Va.: NCTM, 2000.

Van de Walle, *John A. Elementary and Middle School Mathematics: Teaching Developmentally*. New York: Addison Wesley Longman, 2001.

Using Language and Visualization to Teach Place Value

Joan A. Cotter

FOR years, a slow weight gain for premature infants was accepted as normal. Recently, however, better understanding of the special nutritional needs of these infants has resulted in a substantial increase in weight gains. In the same way, mathematics education research is giving us insights into helping young children learn substantially more mathematics.

Because international mathematics studies show that Asian children perform better than their age mates in the United States, I conducted a research project to study the effects of two essential elements of learning—language patterns and visualization—that are found in Japanese primary classrooms. I incorporated these elements into the mathematics instruction in a first-grade classroom in Minnesota. The following discussion presents the rationale behind these two elements and describes the positive results of the instruction.

Language Patterns

Young children seek patterns as they learn about the world. The fact that patterns are predictable allows children to make great advances in their learning. For example, young children know that an *s* added to a word means *more than one* and that a *d* sound at the end of a word indicates past tense, without having studied any English grammar.

Discovering the patterns for counting in English is difficult. The quantity ten has three names: *ten, -teen,* and *-ty.* The quantity three has another name, thir-, as in third, thirteen, and thirty; likewise, five's alias, fif-, appears in fifth, fifteen, and fifty. Also confusing are the numbers 11–19; the words eleven and twelve seem arbitrary, and for 13–19, the word order is reversed, with the ones stated before the tens. Besides blurring the pattern, these inconsistencies obscure the tens groupings.

Contrast the inconsistencies in English with the predictable patterns in most Asian languages, which follow the ancient Chinese method of number naming. The numbers 1–10 are single-syllable words. From 11–19, the names are *ten 1, ten 2, ten 3,* and so forth. The number 20 is named *2-ten,* and 21 is *2-ten 1.*

Counting to 100 in an Asian language requires learning the sequence 1–10 plus the word for 100, a total of just eleven words. In contrast, counting to 100 in English requires learning the words for 1–19 plus the decade names for 20 through 90, plus the word for 100, a total of twenty-eight words.

Korean is an interesting example for language study. The Korean language uses two distinct systems of number words—an everyday system with irregularities and the system described above, sometimes called the *academic system.* Song and Ginsburg (1988) studied how far Korean and American preschool children could count. They found that at four years of age, the average Korean child could count only to 8 in both systems; whereas the average American child could count to 22. At five years old, the average Korean child counted to 29 in the academic system and to 23 in the everyday system, and the American child, to 45. At six years old, after the Korean child learned the academic system in school, the average number reached in counting jumped to 91 in the academic system and to 61 in the everyday system. Meanwhile, at six years old,

Teaching Children Mathematics 7 (October 2000): 108–14

the average for American children was 72, showing approximately the same increase between four and five years old and between five and six years old. Korean children learn the counting sequence by recognizing counting patterns, whereas American children learn by rote memory.

Standard 13 in the NCTM's *Standards* (1989) recommends that children study patterns. The most important pattern is the base-ten number system. What better way to learn it than through a counting system that highlights that pattern?

Visualization

Visualization is an important part of the Japanese primary school curriculum. Mental images of quantities are necessary to work with these quantities mentally. To imagine or visualize a quantity, a person must be able to subitize it, that is, to recognize it immediately without counting. Subitizing has long been recognized as an important skill for developing number sense (Clements 1999).

Very young children can subitize small quantities. Wynn (1992) found that five-month-old babies can distinguish among one, two, and three objects. Researchers know that healthy babies will look longer at an object that is novel than one that is familiar. For example, Wynn showed a baby one teddy bear, which was then hidden behind a screen. Next, she showed the baby a second teddy bear and placed it behind the screen. A teddy bear was then added or removed from behind the screen when the baby's attention was focused elsewhere. When the screen was removed, Wynn measured the baby's viewing time. Babies showed significant differences in viewing times of correct and incorrect collections; it was found that they looked longer at incorrect collections. Strauss and Curtis (1981) found that half of twelve-month-old babies could distinguish up to four objects.

To help a young child associate the correct number with a small collection, we should refer to the whole collection by its number and avoid having children focus on individual objects through the counting ritual. In counting, the child loses the idea of the whole and assumes that we are tagging, or naming, each object. When we point and count four objects with a young child, then ask, "Show me the four," the child will often point to the fourth object rather than indicate the whole collection.

People cannot, however, recognize and visualize quantities of six to ten without some type of grouping. Try to imagine eight identical apples in a row without any grouping; the task is virtually impossible. Next imagine that five of those apples are red and that three are green. You can most likely visualize the eight. The Japanese use the five grouping for quantities of six to ten. The Romans constructed their numerals in groups of fives 2500 years ago; originally, they wrote IIII to represent 4, V for 5, VI for 6, and VIIII for 9.

Manipulatives for Visualizing Number and Place Value

Manipulatives should enhance young children's abilities to visualize number through appropriate groupings. Sometimes dominoes are used for quick recognition, but dominoes have a serious limitation because they are not additive. For example, adding one dot to the five-dot pattern on a domino does not result in the six-dot pattern. A ten-frame, composed of a five-by-two grid with counters or dots placed in adjacent rectangles (see **fig. 1**), is a grouping that can be determined immediately without counting (Wirtz 1980). Ten-frames become cumbersome, however, for quantities greater than about thirty.

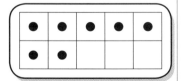

Fig. 1. Representing 7 with a ten-frame

Japanese teachers emphasize visualization through their choice of manipulatives. The teachers use few manipulatives but require them to help children practice visualization. The manipulatives used in this study include fingers and tally sticks; the AL abacus, which is a special double-sided abacus with the beads grouped in fives through color (shown in figs. **3–6**); and place-value cards.

Fingers and tally sticks

The most basic manipulative for children, especially young children, exists naturally on their hands—their fingers, and they are already grouped in fives! Use fingers for naming quantities, not counting. For quantities of one to five, children should use their left hands, to foster reading in the usual left-to-right sequence. For quantities of six to ten, children should use five on the left hand plus the amount over five on the right hand.

Tally sticks, or craft sticks, are an inexpensive tool for the next step in representing quantities. The children represent quantities of one to four by placing the sticks vertically about two centimeters apart. To represent five, children place a fifth stick horizontally across the four sticks; this action introduces the concept of grouping (see **fig. 2**). To emphasize the importance of ten, children are instructed to start a new row after each group of ten.

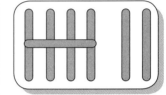

Fig. 2. Representing 7 with tally sticks

The AL abacus

For representing quantities to one hundred, the children use the AL abacus, which has two groups of five beads in contrasting colors strung on each of ten wires (see **fig. 3**). The children enter quantities by moving beads to the left and reading the quantities from left to right. The colors are reversed after five rows, helping the children subitize the number of tens. To develop number sense, children must operate on quantities in terms of tens and ones. The quantity 7-ten 4 (74), for example, is simply seven rows of beads and four beads in the next row. Children can enter and visualize any quantity from one to one hundred without counting. The children can also construct hundreds by stacking abacuses. For example, to represent the quantity three hundred, three abacuses can be stacked.

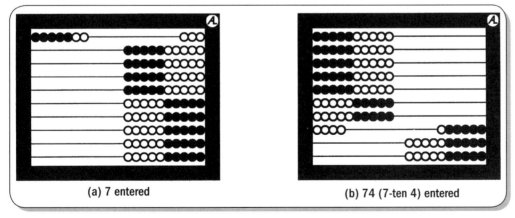

(a) 7 entered

(b) 74 (7-ten 4) entered

Fig. 3. Representations on the AL abacus

This abacus configuration has several advantages over rods of varying lengths and colors that are often used to represent quantities:

- Young children frequently regard each rod, regardless of length, as a single unit.
- Eight percent of the population has some color deficiency and cannot see ten distinct colors.

- For rods representing the quantity five or greater, only the color, not the quantity, can be visualized.
- Combining two rods does not give the immediate sum; the child must either compare with a third rod or count. On the abacus, the result is seen immediately.
- When the sum is over 10, the rods do not reveal the tens structure. On the abacus, the resulting sum is seen immediately as a ten and ones.
- Quantities to one hundred can be subitized and visualized.
- The "counters" are self-contained, allowing more class time to be spent on concepts and less time on management of small pieces.

Place-value cards

To connect the physical representations with the pattern of written numbers, the children use place-value cards (shown in **fig. 4**), sometimes called *expanded-notation* cards. The set of place-value cards includes the numerals 1 through 9 printed on individual cards of the same size; the numerals 10, 20, 30, . . . , 90 are printed on cards that are twice as wide; the numerals 100, 200, . . . , 900 are printed on cards that are three times as wide; and the numerals 1000, 2000, . . . , 9000 are printed on cards that are four times as wide.

The teacher introduces, for example, the 3-ten (30) card by pointing to the 3 and saying, "three," then pointing to the 0 and saying, "ten." Likewise, for 400, the teacher points to the 4 and says, "four," then points to the two 0s in succession, saying "hundred." To construct 37, the child places the 7 card on top of the 0 of the 30 card. Using

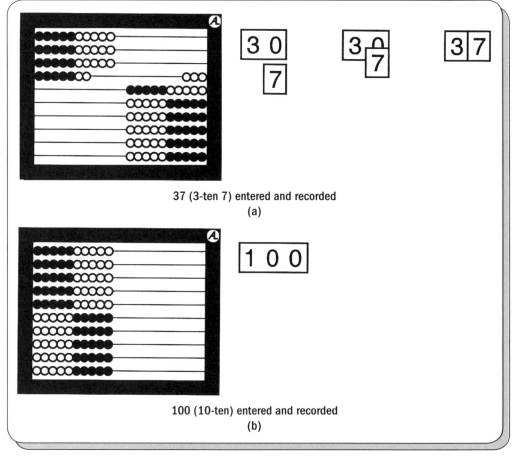

37 (3-ten 7) entered and recorded
(a)

100 (10-ten) entered and recorded
(b)

Fig. 4. Representations with the abacus and place-value cards

these cards, a child learns to recognize the tens digit by the single digit following it, not by its placement in a particular column.

Using these place-value cards has some interesting advantages over using the column model:

- The child sees a number, such as 37, as 30 and 7, not merely as 3 joined to 7, thereby avoiding a common error. Reversals are all but impossible.
- The child reads the numbers in the normal left-to-right pattern, not backward, as in the column model, in which a child starts at the right and says, "ones, tens, hundreds."
- The child can read the number 100 as "one hundred" or as "10-ten," which indeed it is. Also, a number such as 1200 makes sense read as "12 hundred."

Using these hands-on tools, children begin to visualize and construct for themselves the patterns of our number system. These skills enable them to understand computation and develop efficient strategies for learning the facts.

Visualization and Computation

In the United States, counting is considered the cornerstone of arithmetic; children engage in various counting strategies: counting all, counting on, and counting back. Japanese teachers have a different view of counting. Starting in first grade, students in Japan are discouraged from using one-by-one counting procedures. According to Hatano (1982), Japanese researchers found that mere practice in counting did not help children advance in the conservation tasks identified by Piaget. The Japanese Council of Mathematics Education states that children who are taught counting procedures have more difficulties in solving story problems, although they do well in computation (Hatano 1982, p. 215).

Counting, which is slow and unreliable until six years of age, creates other difficulties, as well. Children who use counting to add and subtract develop a unitary concept of number. That is, they think of 14 as 14 ones, not as 1 ten and 4 ones. Such thinking interferes with their understanding of place-value concepts and often results in rote learning of algorithms. To understand that our number system is based on tens, children must experience the pattern of trading: 10 ones for 1 ten, 10 tens for 1 hundred, and 10 hundreds for 1 thousand.

Visualization strategies offer efficient techniques for learning the facts. In the "complete the ten" strategy, for example, to add 9 + 4, the child enters 9 and 4 on the top two wires of the AL abacus; as the next step, he or she takes 1 from the 4 and combines it with the 9 to get 10 and 3, or 13 (see **fig. 5**). The child progresses from entering and rearranging the quantities physically to entering the quantities physically but rearranging them mentally, then to performing the entire process mentally. This strategy also works for higher numbers; for example, this same technique could be used for 59 + 4.

Another strategy that lends itself to visualization is the "2 fives" strategy, which works when both addends are 5 or more. For example, to add 8 + 6, the child enters 8 and 6 on the top two wires of the abacus, as shown in **figure 6**. The 2 fives formed by the dark-colored beads make a ten, and the amounts over 5, which are 3 and 1, total 4, to yield a sum of 14.

A Classroom Study

During the 1994–1995 school year, I conducted a study in two first-grade classrooms, each with sixteen children, in a rural Minnesota community (Cotter 1996). The

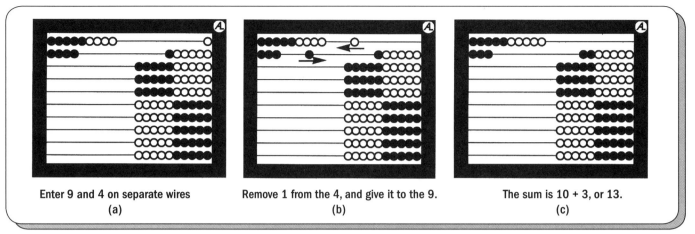

| Enter 9 and 4 on separate wires (a) | Remove 1 from the 4, and give it to the 9. (b) | The sum is 10 + 3, or 13. (c) |

Fig. 5. Using the "complete the ten" strategy to add 9 + 4

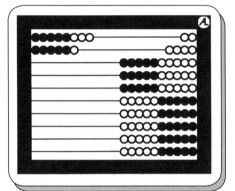

Fig. 6. Using the "2 fives" strategy to add 8 + 6 (the dark-colored beads equal 10, and the light-colored beads equal 4, giving a sum of 14)

experimental class used lesson materials that I supplied, whereas the matched control class used a traditional workbook.

For the first three months of the school year, the experimental class used the "Asian" method of counting with a slight modification for the numbers in the teens. I had the children say, "1-ten 3" rather than "ten 3." Parents had no complaints about this method. When introduced to 2-ten, one boy said that the number was twenty. The teacher told him that for now the class would call the number 2-ten, and the student was satisfied with that explanation.

The children learned the finger combinations to 10. They practiced both holding up the requisite number of fingers when asked and recognizing the quantity on the teacher's hands. They also practiced laying out and recognizing tiles or counters, but always being asked to change color or some other attribute after five to facilitate subitizing and visualizing.

The children easily learned to enter quantities from one to ten on the abacus without counting. They added two quantities with a sum of less than 10 by entering the two quantities in tandem and reading the sum immediately.

Around the fourth week of first grade, the teacher introduced tens. The children practiced entering various tens up to 10-ten. They soon discovered that adding 2-ten and 2-ten followed the pattern of adding 2 and 2. They added two quantities, such as 8 + 6, by entering 8 and 2 from the 6 to fill the first row, then entering the remaining 4 from the 6 in the second row. The sum was apparent.

The children constructed quantities using the abacuses and the place-value cards. When the stacks of abacuses became cumbersome for large numbers, the children used four-centimeter-square cards with drawings of thousands, hundreds, tens, and ones (see **fig.** 7).

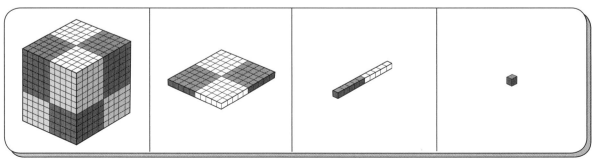

Fig. 7. Base-ten picture cards representing thousand, hundred, ten, and one

The children also used these base-ten cards for adding with trading. Each child in a group of three constructed a four-digit number with the cards. Then the group added its numbers by combining its cards and trading 10 of any denomination for 1 of the next higher denomination. They displayed their sums using place-value cards and recorded the sums on paper.

Along one edge of the reverse side of the AL abacus is a label that designates two wires each for the thousands, hundreds, tens, and ones; the third and seventh wires are not used (see **fig. 8**). Placing the abacus so that the label is at the top, the children entered four-digit numbers by moving up the requisite number of beads on the vertical wires. Using two wires allows an ample number of beads for entering both addends before trading.

For example, to add 8 + 6, the child enters 8 beads, followed by 6 beads (see fig. 8). The child can see the sum without counting, because the light-colored beads form a ten. Because no more than 9 can be recorded in any denomination, a trade is necessary. To trade, the child moves up 1 tens bead with the left hand while moving down 10 ones beads with the right hand.

The children enjoyed the bead-trading activity and benefited from the trading practice. They also added the numbers on cards labeled 1–10 by entering the numbers in the ones columns and trading when necessary. When the children mastered trading, they began entering and adding four-digit numbers. A few days later, they figured out for themselves how to perform addition on paper without the abacus.

During the fourth month of first grade, the children learned the traditional number names. First, the teacher explained that 4-ten has another name, *forty*, emphasizing that *ty* means *ten*. In the same way, she taught the numbers 60, 70, 80, and 90. For 30 and 50, she also used the words *third* and *fifth* to remind the students of the new names. *Twin-ten* helped them remember *twenty*.

Before introducing the teen names, the teacher played a word game in which the children reversed the syllables in such words as *sunset* and *bedroom*, changing them to *setsun* and *roombed*. Then the teacher told the students that *teen* means *ten*. Ten-4 becomes "teen four," and reversing the syllables completes the transformation to "fourteen." The children learned the names for 16–19 in the same way, then 13 and 15, which need the *thir* and *fif* modification.

The teacher explained the word *eleven* by laying out eleven objects with ten in a row and one below it. People once referred to this quantity by noticing the one object left over: "a one left," or with the words reversed, "left one." Eventually, this number came to be known as "eleven." The word *twelve* was derived in the same way: "two left."

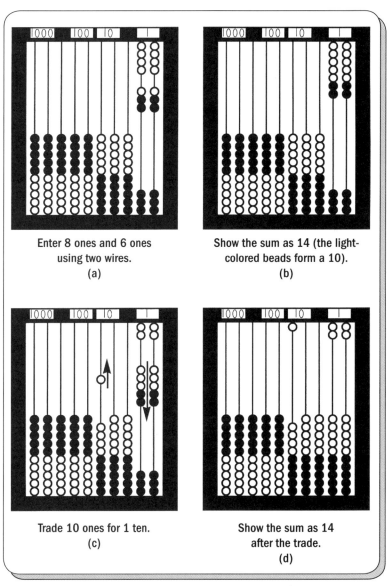

Enter 8 ones and 6 ones using two wires.
(a)

Show the sum as 14 (the light-colored beads form a 10).
(b)

Trade 10 ones for 1 ten.
(c)

Show the sum as 14 after the trade.
(d)

Fig. 8. Trading on the reverse side of the AL abacus

Results

After the experimental lessons, the teacher remarked that the children had advanced much farther than she had ever expected. She thought that the children who had learning difficulties learned much more than they would have achieved in a traditional program. Zachery, a first grader with learning difficulties, was able to add nines mentally using the "complete the ten" strategy.

At the end of the eighth month, I interviewed all the children in both classes. Using a procedure from Miura and others (1993), I gave the first-grade children tens and ones from base-ten blocks and showed them that the ten was equal to 10 ones. Then I asked the children to use the blocks to make 11, 13, 28, 30, and 42. In the interviews, 21 percent of the students in the experimental class and 44 percent of the students in the control class made their constructions without any tens blocks. Zachery made all five of his constructions with tens blocks and nine or fewer ones.

During the interviews, I also asked the children to construct 48 with the tens and ones blocks, then to subtract 14. Note that no "borrowing" is involved. Children who had a unitary concept of number removed 14 ones, that is, 8 separate ones and 6 more from a ten. In the experimental class, 81 percent removed a ten and 4 ones, but only 33 percent of the control class did so.

Some other results of the interviews are as follows:

- In the experimental class, 94 percent knew the sum of 10 + 3 and 88 percent knew 6 + 10; in the control class, 47 percent knew 10 + 3 and 33 percent knew 6 + 10.
- When asked to circle the tens place in the number 3924, 44 percent of the experimental class and 7 percent of the control class did so correctly.
- When asked to compute 85 – 70 mentally, 31 percent of the experimental class did so correctly; none of the students in the control class did so.
- Forty percent of the students in the control class wrote 512 for the sum of 38 + 24 or 812 for 57 + 35; none of the experimental class did so, even without using the abacus.

Concluding Comments

In the United States, we give our children numbers in bulk and they spend years learning to package the numbers efficiently. Asians give their children prepackaged numbers. Teaching children to count initially with consistent counting words follows good teaching practice; introducing exceptions should occur only after the students understand the general rule.

Some teachers and parents are concerned that the abacus may become a crutch. The best answer to this issue was given by five-year-old Stan from my class after I asked him how he knew that 11 plus 6 is 17. He replied by saying, "because I've got the abacus in my mind." Once the children visualize a concept, they do not use the abacus. A child who can visualize quantities and understand the patterns in the number system has developed good number sense and will search for more patterns, a quest that leads to an increase in abstract thought.

Young children are capable of adding and subtracting and performing other mathematical tasks before they develop accurate counting skills. Counting need not be the basis of arithmetic. These Minnesota children and subsequent classes have developed an understanding of our number system and efficient strategies for learning the facts. Language patterns and visualization are two components of learning that help young children construct mathematical knowledge.

BIBLIOGRAPHY

Activities for Learning. AL Abacus, n.d. Manipulative. (Available from Activities for Learning, 129 S.E. Second St., Linton, ND 58552.)

Clements, Douglas H. "Subitizing: What Is It? Why Teach It?" *Teaching Children Mathematics* 5 (March 1999): 400–404.

Cotter, Joan A. *Activities for the AL Abacus: A Hands-on Approach to Arithmetic.* 2nd ed. Hutchinson, Minn.: Activities for Learning, 1988.

———. "Constructing a Multidigit Concept of Numbers: A Teaching Experiment in the First Grade." Ph.D. diss., University of Minnesota, 1996. Abstract in *Dissertation Abstracts International* 9626354, DAI-A 57/04, p. 1465.

Hatano, Giyoo. "Learning to Add and Subtract: A Japanese Perspective." In *Addition and Subtraction: A Cognitive Perspective,* edited by Thomas P. Carpenter, James M. Moser, and Thomas A. Romberg, pp. 211–23. Hillsdale, N.J.: Lawrence Erlbaum Associates, 1982.

Miura, Irene T., Yukari Okamoto, Chungsoon C. Kim, Marcia Steere, and Michel Fayol. "First Graders' Cognitive Representation of Number and Understanding of Place Value: Cross-National Comparisons—France, Japan, Korea, Sweden, and the United States." *Journal of Education Psychology* 85 (January 1993): 24–30.

National Council of Teachers of Mathematics (NCTM). *Curriculum and Evaluation Standards for School Mathematics.* Reston, Va.: NCTM, 1989.

Piaget, Jean. *The Child's Conception of Number.* New York: W. W. Norton & Co., 1966.

Song, Myung J., and Herbert P. Ginsburg. "The Effect of the Korean Number System on Young Children's Counting: A Natural Experiment in Numerical Bilingualism." *International Journal of Psychology* 23 (1988): 319–32.

Strauss, Mark S., and Lynne E. Curtis. "Infant Perception of Numerosity." *Child Development* 52 (December 1981): 1146–52.

Wirtz, Robert. *New Beginnings.* Monterey, Calif.: Curriculum Development Associates, 1980.

Wynn, Karen. "Addition and Subtraction by Human Infants." *Nature* 358 (27 August 1992): 749–50.

This article has a striking impact on teachers at the conclusion of a course or institute. I ask teachers to reflect on the article in light of the assignments and tasks we have completed and then share the date of the original article.

–Frances (Skip) Fennell.

MEANING AND SKILL— MAINTAINING THE BALANCE*

William A. Brownell

The subject given me poses a question. It is no academic question arising out of purely theoretical considerations. It is assumed that both meaning and computational competence are proper ends of instruction in arithmetic. It is implied that somehow or other both ends are not always achieved and that there is evidence that this is so.

Indeed there is such evidence. More than one school system has embarked upon a program of so-called meaningful arithmetic, only to discover that on standardized tests of computation and "problem solving" pupils do none too well. In such schools officials and teachers are likely to believe that they have made a bad bargain; and school patrons are likely to support them vehemently in this belief.

We may try to convince all concerned that the instruments used to evaluate learning are inappropriate, or at least imperfect and incomplete. True, standardized tests rarely if ever provide means to assess understanding of arithmetical ideas and procedures. Hence, the program of meaningful instruction, even if well managed, has no chance to reveal directly and explicitly its contribution to this aspect of learning. On the other hand, can we deny that the learning outcomes that *are* measured are of no significance? To do so is to say in effect that computational skill is of negligible importance, and we can hardly justify this position.

Why is there now the necessity to talk about establishing and maintaining the desirable kind of balance between meaning on the one hand and computational competence on the other?

SOURCES OF THE DILEMMA

1. INCOMPLETE EXPOSITION OF "MEANINGFUL ARITHMETIC"

Perhaps those of us who have advocated meaningful learning in arithmetic are at fault. In objecting to the drill conception of the subject prevalent not so long ago, we may have failed to point out that practice for proficiency in skills has its place, too. It is questionable whether any who have spoken for meaningful instruction ever proposed that children be allowed to leave our schools unable to compute accurately, quickly, and confidently. I am sure that all, if asked, would have rejected this notion completely. But we may not have said so, or said so often enough or vigorously

* Paper read before the Elementary Section of the N.C.T.M. at the Milwaukee meeting of April 14, 1956.

Arithmetic Teacher 3 (October 1956): 129–36, 142; reprint *Teaching Children Mathematics* 9 (February 2003): 310–16

enough. Our comparative silence on this score may easily have been misconstrued to imply indifference about proficiency in computation.

If this has actually happened, we can scarcely blame classroom teachers if they have neglected computational skill as a learning outcome. It is characteristic of educational movements to behave pendulum-wise. When we correct, we tend to over-correct. Just this sort of thing seems to have happened in the teaching of arithmetic. In fleeing from over-reliance on one kind of practice, we may have fled too far. It is a curious state of affairs that those of us who deplored the limitations of this kind of practice must now speak out in its behalf and stress its positive usefulness.

2. Misunderstood Learning Theory

A second possible explanation for our dilemma may be found, as is frequently the case in the practical business of education, in misinterpretations or misapplications of psychological theories of learning. Over-simplification of certain generalizations in learning theories as widely apart as are those of conditioning and of field theory could lead, and may have led, to the lessening of emphasis on practice in arithmetic.

a. Conditioning theory. According to the learning theory of one influential exponent of conditioning, once a response has been made to a stimulus a connection has been established. Thus, the child who says "Seven" in replying to the question, "How many are two and five?", sets up a connection between 2 and 5 as stimulus and 7 as response. If this is so—if the connection is actually formed by the one response—then at first glance practice might seem to be utterly purposeless.

But this inference is quite unfounded, as the psychologist in question makes abundantly clear. In saying that a connection is "established" he means no more than that a new neuromuscular pattern is available. He does *not* mean that our hypothetical child after the one experience will always, only, and instantly respond with "seven" when asked the sum of 2 and 5. If the response "seven" is to be the invariable one, moreover if the response is to be made in situations differing ever so slightly from the original one, then practice is required. In this theory of learning as conditioning, therefore, there is no comfort for those who would abandon practice as a means to promote the learning of arithmetical facts and skills.

b. Field theory. No more comfort is to be found in field theory of learning. It is often said that one experience of "insight" or "hindsight"—before, during, or after success—is enough; but enough for *what?* It may be all that is needed to understand a situation and the method of dealing with it. Yet, it is one thing to know the general rationale for solving a complicated mechanical puzzle but quite another thing to be able to manipulate the parts correctly with facility, ease, and speed.

So in arithmetic it is one thing to comprehend the mathematical principles governing decomposition in subtraction—something that can come from a single insightful experience—but another thing to be able to subtract quickly and correctly. Indeed if in examples like 73 – 47 and 52 – 19 a child who possesses this understanding always thinks through the complete logical explanation, his performance will be impaired, at least from the standpoint of speed. Understanding and skill are not identical. A single instance of insight may lead to understanding but will hardly produce skill. For skill, practice is necessary.

3. Influence of General Educational Theory

A third explanation for failure to stress computational competence is to be found in the recent history of educational theory. It must be remembered that the place of meaning and understanding in arithmetic has been generally recognized for not more than fifteen or twenty years. In 1935 we were still under the influence of somewhat sentimental and unrealistic notions both about children and about the course of their development. In extreme form these notions led to a kind of teaching that was anything but systematic. Indeed, what children learned, when they learned it, and how they learned it was left pretty much to the children themselves. Attempts to guide and direct learning and to organize learning experiences were frowned upon as "violating child nature" and as almost certainly productive of serious derangements of child personality. To those who held these views practice was anathema.

In the public schools, as contrasted with college departments of Education, this conception of the processes of learning and teaching did not gain much of a foothold. Nevertheless, it was in this climate of thought that stress on meaning in arithmetic put in its appearance. Those who were committed to the educational theory I have mentioned welcomed the new emphasis as confirming both what they did and what they did not do. If, deliberately or unwittingly, they accepted only the part of the emerging view most congenial to them, they committed an error that is very common and that is altogether human. Be that as it may, the consequences were none too good. True, these teachers may have been largely responsible for the quick and general endorsement of one aspect of meaningful arithmetic—learning with understanding. On the other hand, they could not themselves absorb the whole of it, and they remained hostile to practice in learning. Pupils taught by such teachers cannot be expected to make high scores on standardized arithmetic tests of skill in computation and "problem solving."

4. Inadequate Instruction on Meanings

I have suggested three hypotheses as explaining why we may not be obtaining balance between understanding and computational competence in arithmetic, on the assumption that our shortcomings relate to the latter (computational competence) rather than to the former. These three are: the possible failure of advocates of meaningful arithmetic to emphasize sufficiently the importance of practice in acquiring arithmetical skills; misinterpretations of psychological theories of learning which have had the effect of minimizing the place of practice; and the unwillingness of some teachers, who believe completely that arithmetic must be made intelligible to children, to provide the practice necessary for computational proficiency. May I add a fourth? It is that we may not as yet be doing a very good job in teaching arithmetical meanings as they should be taught.

There is ample evidence in psychological research on learning that the effects of understanding are cumulative. There is also ample evidence, if not in arithmetic, then in other types of learning, that the greater the degree of understanding, the less the amount of practice necessary to promote and to fix learning. If these truths are sound—and I think they are—then they should hold in the field of arithmetical learning. It follows that computational skills among school children would be greater than they are if we *really* taught them to understand what they learn.

Again I remind you that meaningful arithmetic, as this phrase is commonly used, is a newcomer in educational thought. Many teachers, trained in instructional procedures suitable, say, to a view of arithmetic as a tool or a drill subject, find it difficult to comprehend fully what meaningful arithmetic is and what it implies for the direction of learning. Others than myself, I am sure, have, in conferences with teachers, been somewhat surprised to note that some of them are unfamiliar with major ideas in this conception and with methods of instruction adapted thereto.

Perhaps the commonest instructional error is, in a different context, the same one that has always distorted learning in arithmetic, namely, the acceptance of memorized responses in place of insistence upon understanding. Mathematical relationships, principles, and generalizations are couched in language. For example, the relationship between a given set of addends and their sum is expressed verbally in some such way as: "The order of the numbers to be added does not change the sum." It is about as easy for a child to master this statement by rote memorization as to master the number fact, 8 – 7 = 1, and the temptation is to be satisfied when children can repeat the words of the generalization *verbatim*. Similarly, the rationale of computation in examples such as: 33 + 48 and 71 – 16, makes use of concepts deriving from our number system and our notions of place value. But many a child glibly uses the language of "tens" and "ones" with no real comprehension of what he is saying. Such learning is a waste of time. To use an Irish bull, the meanings have no meaning.

I intend no criticism of teachers. Until recently there have been few professional books of high quality to set forth the mathematics of arithmetic and to describe the kind of instruction needed. Moreover, many teachers have had no access to these few books. Again, until recently not many courses of study and teachers' manuals for textbook series have been of much help. It is not strange, therefore, that though meaningful arithmetic is adopted in a given school system, not all members of the teaching staff are well equipped to teach it. As a result, their pupils, denied a full and intelligent treatment of arithmetic as a body of rational ideas and procedures, have been unable to bring to computation all the aid that could come through understanding.

Toward a Solution

So much for possible explanations—explanations of a general character—for our failure to keep meaning and skill in balance. I have no way of knowing the reality of any of the four hypotheses suggested or of the extent of its validity, to say nothing of the degree to which, taken together, the four account adequately for the situation. The fact remains that something needs to be done. What is the remedy?

Certainly we shall not get very far as long as we think of understanding and practice in absolute terms. I have deliberately done this so far in order to examine the issue in simple terms. Actually, it is erroneous to conceive of understanding as if it were either totally present or totally absent. Instead, there are degrees or levels of understanding. Likewise, not all forms of practice are alike. Rather, there are different types, and they have varying effects in learning.

Levels of Understanding

Consider the example 26 + 7. The child who first lays out twenty-six separate objects, next seven more objects, and then determines the total by counting the objects one by one has a meaning for the operation. So has the child who counts silently, starting with 26. So has the child who breaks the computation into two steps, 26 + 4 = 30 and 30 + 3 = 33. So has the child who employs the principle of adding by endings —6 + 7 = 13, so 26 + 7 = 33. So has the child who, capable of all these types of procedure, nevertheless recognizes 33 at once as the sum of 26 and 7. All these children "understand," but their understandings may be said to represent different points in the learning curve. The counter is at the bottom, and the child who through understanding has habituated his response "thirty-three" so that it comes automatically is at the top of a series of progressively higher levels of performance.

All these levels of performance or of understanding are good, depending upon the stage of learning when they are used. For instance, finding the sum of 26 and 7 by counting objects is a perfectly proper way of meeting the demand at first; but it is not the kind of performance we want of a child in grade 4. At some time in that grade, or earlier, he should arrive at the stage when he can announce the sum correctly, quickly, and confidently, with a maximum of understanding.

It is a mistake to believe that this last stage can be achieved at once, by command as it were. When a child is required to perform at a level higher than he has achieved, he can do only one of three things. (a) He can refuse to learn, and his refusal may take the form; "I won't," or "I can't," or "I don't care," the last named signifying frustration and indifference which we should seek to prevent at all costs. (b) Or, he can acquire such proficiency at the level he *has* attained that he will be credited for thinking at the level desired. Many children develop such expertness in silent counting that, in the absence of close observation and questioning, they are believed to have procedures much beyond those they do have. (c) Or, third, he may try to do what the teacher seems to want. If an immediate answer is apparently expected, he will supply one, by guessing or by recalling a memorized answer devoid of meaning. If he guesses, obviously he makes no progress at all in learning. If he memorizes, only unremitting practice will keep the association alive; and if he forgets, he is helpless or must drop back to a very immature level of performance such as counting objects or marks by 1's.

For the stage of performance we should aim for ultimately, as in the case of the simple number facts, higher-decade facts, and computational skills, we have no standard term. We may use the word "memorization" to refer to what a child does when he learns to say "Four and two are six" without understanding much about the numbers involved, about the process of addition, or about the idea of equivalence. If we employ "memorization" in this sense, then that word is inappropriate for the last step in the kind of learning we should foster. Hence for myself I have adopted the phrase "meaningful habituation." "Habituation" describes the almost automatic way in which the required response is invariably made; "meaningful" implies that the seemingly simple behavior has a firm basis in understanding. The particular word or phrase for this last step in meaningful learning is unimportant; but the *idea*, and its difference from "memorization," *are* important.

Teaching meaningfully consists in directing learning in such a way that children ascend, as it were, a stairway of levels of thinking arithmetically to the level of meaningful habituation in those aspects of arithmetic which should be thoroughly mastered, among them the basic computational skills. Too many pupils, even some supposedly taught through understanding, do not reach this last stage. Instead, they stop short thereof; and even if they are intelligent about what they do when they compute, they acquire little real proficiency. In instances of this kind both learning and teaching have been incomplete.

How are teachers to know the status of their pupils with respect to progress toward meaningful habituation? Little accurate information is to be had from their written work, for both correct and incorrect answers can be obtained in many ways, and inference is dangerous. Insightful observation and pupils' oral reports volunteered or elicited through questioning are more fruitful sources of authentic data. Since children differ so much in their thought procedures, a good deal of this probing must be done individually. One of the most fruitful devices I can suggest for this probing consists in noting what children do in the presence of error.

I recall a conversation with a fourth grade girl whom I knew very well and who was having difficulty in learning—in her case, in memorizing—the multiplication facts. I asked her—her name was June—"How many are five times nine?" (This form of expression was used in her school instead of the better "How many are five nines?") Immediately she responded, "Forty-five." When I shook my head and said, "No, forty-six," she was clearly upset. Her reply, after some hesitation, was, "No, it's forty-five." When I insisted that the correct product is 46, June said, "Well, that isn't the way I learned it." I suggested that perhaps she had learned the wrong answer. Her next statement was, "Well, that's what my teacher told me." This time I told her that she may have misunderstood her teacher or that her teacher was wrong. June was obviously puzzled; then she resorted to whispering the table, "One times nine is nine; two times nine is eighteen," and so on, until she reached "five times nine is forty-five." Again I shook my head and said, "Forty-six." When she was unable to reconcile my product with what she had become accustomed to say, I asked her, "June, have you no way of finding out whether forty-five or forty-six is the correct answer?" Her response was, "No, I just learned it as forty-five." Of course I did not leave her in her state of confusion; but the point of the illustration is, I hope, quite apparent: Her inability to deal with error was convincing evidence of the superficiality of her "learning" and of its worthlessness.

Compare my conversation with June with that I had with Anne, another fourth grade girl whom also I knew well and who, like the first girl, was learning the multiplication facts. When I asked Anne, "How many are five times nine?", the correct answer came at once, just as in June's case; but from here on, mark the difference. I introduced the error, saying that the product is 46, not 45. Anne looked at me in disgust and said, "Are you kidding?" I maintained my position that 5 × 9 = 46. Immediately she said, "Do you want me to prove it's forty-five?" I told her to go ahead if she thought she could. She answered, "Well, I can. Go to the blackboard." There I was instructed to write a column of five 9's and, not taking any chances with me, Anne told me to count the 9's to make sure I had 5. Next came the command, "Add them." When I deliberately made mistakes in addition, she corrected me, each time saying, "Do you want me to prove that, too?" Obviously, I had to arrive at a total of 45. Having done so, I said, "Oh, that's just a trick," to which she replied, "Do you want me to prove it another way?" Exposure to error held no terrors for this child; she did not become confused or fall back upon repetition of the multiplication table; nor did she cite her teacher as an authority. Instead, Anne had useful resources in the form of understandings that were quite lacking in the case of June whose discomfiture I have described.

Probing for understanding need not depend wholly on opportunities to work with individual children. On the contrary, there are possibilities also under conditions of group instruction when questions beginning with "How" and "Why" supplement the commoner questions starting with "What." The worth of valid knowledge concerning level of understanding is inestimable for the guidance of learning. The demands upon time are not inconsiderable, but no one should expect to get full knowledge concerning every pupil. The prospects are not hopeless if ingenuity is exercised and if the goal is set, not at 100% of knowledge, but more realistically at perhaps 20% more than is now ordinarily obtained.

TYPES OF PRACTICE

In crude terms, practice consists in doing the same thing over and over again. Actually, an individual never does the same thing twice, nor does he face the same situation twice, for the first reaction to a given situation alters both the organism and the situation. A changed being responds the second and the third time, and the situation is modified accordingly.

We must concede the truth of these facts. At the same time, for our purposes we may violate them a bit. Let us conceive of practice of whatever kind as falling somewhere along a continuum. At the one end of the continuum is practice in which the learner tries as best he can to repeat just what he has been doing. At the other end is practice in which the learner modifies his attack in dealing with what is objectively the same situation (or what to him are similar situations). We may call these extremes "repetitive practice" and "varied practice," respectively.

An instance of repetitive practice is memorizing the serial order of number names by rote or applying the series in the enumeration of groups of objects. An instance of varied practice is the attempt, by trying different approaches, to find steadily better ways of computing in such examples as $43 + 39$, $75 - 38$, $136 \div 4$, and 32×48. Between the two extreme types of practice are innumerable others, differing by degree in the extent to which either repetition or variation is employed. But again, for our purposes, we may disregard all the intervening sorts of practice: we could not possibly name them all, or describe them, or show their special contributions of learning.

Both repetitive and varied practice affect learning, but in quite unlike ways. For illustration we may choose an instance of learning outside of arithmetic, for example, the motor activity of swimming. Suppose the beginner engages in repetitive practice: What does he do, and what will happen? Well, he will continue to use as nearly as he can exactly the movements he employed the first time he was in deep water, and the result will be that he may become highly proficient in making just those movements. He will hardly become an expert swimmer, but he will become an expert in doing what he does, whether it be swimming or not.

On the other hand, suppose that the beginner engages in varied practice. In this case he will seek to *avoid* doing precisely what he did at the outset. He will discard uneconomical movements; he will try out other movements, select those that are most promising, and seek a final coordination that makes him a good swimmer. Then what will he do? He will change to repetitive practice, for, having the effective combination of movements he wants, he will seek to perfect it in order to become more proficient in it.

The differences between repetitive practice and varied practice, both in what the learner does and in what his practice produces, are clearly discernible when we think of motor activities. They are less easily identified when we think of ideational learning tasks like the number facts and computational skills. But the differences are there none the less. The child who counts and only counts in dealing with examples like $36 + 37$ and $24 + 69$ is employing repetitive practice. The more he counts, the more expert he becomes in counting; but the counting will not, and cannot, move him to a higher level of understanding and of performance. In contrast is the child who, through self-discovery or through instruction, tries different ways to add in such examples. Under guidance he can be led to adopt higher and higher levels of procedures until he is ready for meaningful habituation. If then he does not himself fix his automatic method of adding, he can be led to do so through repetitive practice. In any case it is safer to provide the repetitive practice in order to increase proficiency and make it permanent.

The distinction between repetitive and varied practice, in their nature and in their consequence, is not always recognized in teaching. If repetitive practice is introduced too soon, before understanding has been achieved, the result, for one thing, may be blind effort and frustration on the part of the learner. Or, it may fix his performance at a low level, the level he has attained. No new and better procedure can emerge from repetitive practice though it may appear under conditions of drill when a child, tired of repetition or disappointed in its result, abandons it in search of something new.

There may be an instructional error of another kind, one already alluded to. This error is to insist, to quote some, that "there is no place for drill in the modern conception of teaching." True, there is no place for unmotivated drill on ill-understood skills; but the statement goes too far in saying that there is no place at all for repetitive practice. How else, one may ask, is the final step of meaningful habituation to be made permanent; how else is real proficiency at this level of learning to be assured?

The kind of practice most beneficial at any time, then, is the kind best adapted to accomplish a given end. For illustration, let us return to June and Anne and their learning of the multiplication facts.

June was trying to master these facts by repetitive practice, by saying over and over and over again the special grouping of words for each separate fact. Her level of understanding of the numbers and relationships involved was close to zero. Unless engaged in almost ceaselessly, her repetition of verbalizations could give her little more than temporary control, and control, be it noted, of the verbalizations alone. Lapses of memory would be nearly fatal and would subject her to the hazards of guessing. In no way could her memorization of senseless phrases contribute much to sound learning of the facts themselves, not to mention the deficiencies of its results for more advanced forms of computation and functional use. What June needed was not repetitive, but varied practice. By contrast, Anne, who was able to "prove" her announced products and who thereby demonstrated her full understanding of the relationship of the numbers, no longer needed varied practice and could safely and properly be encouraged to engage in repetitive practice.

We employ varied practice, then, if we wish the child to move upward from where he is toward meaningful habituation, and repetitive practice if we are endeavoring to produce true competence, economy, and permanence in this last stage of learning (or at any earlier stage which represents a type of performance of worth in itself). Practice has to be designed to fit the learner's needs, a fact which brings us back again to the individual and to the critical importance of accurate knowledge concerning his learning status.

In Conclusion

To sum up, the balance between meaning and skill has been upset, if indeed it ever was properly established. The reasons are many, some of them relating to educational theory in general, others to misconceptions of psychological theories of learning, others to failure to teach arithmetical meanings thoroughly, and still others to carry learning in the case of computation to the level I have denoted meaningful habituation, and then to fix learning at that level. I have discussed these matters at length, perhaps at unwarranted extent in view of the fact that the remedy for the situation can be stated briefly. The remedy I propose is as follows:

1. Accord to competence in computation its rightful place among the outcomes to be achieved through arithmetic;

2. Continue to teach essential arithmetical meanings, but make sure that these meanings are just that and that they contribute as they should to greater computational skill;

3. Base instruction on as complete data as are reasonably possible concerning the status of children as they progress toward meaningful habituation;

4. Hold repetitive practice to a minimum until this ultimate stage has been achieved; then provide it in sufficient amount to assure real mastery of skills, real competence in computing accurately, quickly, and confidently.

EDITOR'S NOTE [1956]. Dr. Brownell is talking particularly to teachers in school systems that have initiated programs of meaningful arithmetic but his message is for all teachers. He indicates that it is difficult to comprehend fully what meaningful arithmetic is. He warns against teaching for memorized generalizations under the guise of teaching meanings. A good sound program of arithmetic will develop meanings and understandings as basic and necessary to a functional program which does not slight computation and problem solving. In this country we study arithmetic primarily for its contribution to intelligent and successful living in social, economic, and cultural situations. We do not thereby neglect those mathematical elements which give substance to arithmetic as a science; we are not primarily concerned with arithmetic as a puzzle for the amusement of the few. In our society the abilities to read and to form mathematical conclusions are very important.

Multiplying Fractions

Zhijun Wu

MULTIPLYING fractions challenges students to examine many of the ideas that they have developed about multiplication from their work with whole numbers. The challenge is not one of computation, but rather is one of conceptualization. As Kennedy and Tipps (1997) state,

> Algorithms for multiplication of common fractions are deceptively easy for teachers to teach and for children to use, but their meanings are elusive. Children who are taught rules for performing computation with these algorithms can multiply pairs of fractional numbers with ease. However, if they compute by rules alone, they will understand little of the meanings behind the computation.

Children who know only rules for computing have limited ability to generalize the information to other situations, especially when facing complex problems.

In the early elementary school years, multiplication with whole numbers is introduced as repeated addition. The repeated-addition model, although a useful link between multiplication and addition, is limited if it is the students' only concept of multiplication. Certain real-life situations represented by multiplication with mixed numbers, common fractions, or decimal fractions cannot easily be interpreted with the repeated-addition model (Graeber and Campbell 1993). Sometimes children have difficulty making sense of such situations.

Teachers should be able to extend multiplication from whole numbers to fractions in a meaningful way. Conceptual models for multiplying fractional numbers, such as the unified approach to multiplying fractions (Ott 1990) and the area model (Graeber and Tanenhaus 1993), can be used to teach the meaning of multiplication of fractions. According to Graeber and Tanenhaus, the area model is a familiar way for students to interpret the multiplication of whole numbers and can be used as a means of conceptualizing the product. This model may help students make sense of the idea of "multiplication making smaller" for such expressions as $9 \times 2/3$ and $1/2 \times 1/10$. These researchers also point out, however, that their expectation about the magnitude of a computational result is likely to interfere with students' ability to make sense of multiplying fractions (Graeber and Tanenhaus 1993). According to the unified approach, repeated addition can be generalized to a product of fractions; in other words, $1/4 \times 3/4$ means one-quarter of a group with a size of three-quarters (the size of one whole group) (Ott 1990). One might wonder how a student would interpret the idea of adding 3/4 to itself 1/4 times. How can a concrete or pictorial representation of the grouping be produced to communicate the abstract idea to students in a meaningful way and allow them to transfer the idea to a variety of problem-solving situations? The multiplication of rational numbers is open to many new and different interpretations compared with whole-number multiplication (Hiebert and Behr 1988). This article attempts to present one alternative model to make the extension of multiplication from whole numbers to fractional numbers meaningful. Before the model is presented and explained, two points are worth mentioning.

First, students' development of the meaning of multiplication with fractional numbers should emerge from experience with genuine problems (Graeber and Campbell 1993). The reasonable way to start a lesson on the meaning of multiplication with fractional numbers is to allow students to explore situations and make conjectures (NCTM 2000). Mathematical concepts for all operations are rooted in situations and problems.

Teaching Children Mathematics 8 (November 2001): 174–77

Problem situations should be appropriately designed to provide contexts in which students can solidify their existing knowledge, extend what they know, and further develop generalized ideas about the operation.

Second, ideas that must be generalized are abstract ideas. Educators usually believe that the meanings of mathematical concepts can be communicated to children with the help of manipulatives and pictures. Some mathematical ideas, however, such as the "fractional part of a whole," are abstract concepts that may have no single clear, concrete referent. As such, they are watershed concepts, moving mathematics beyond the concrete world. Students must encounter and reflect on these concepts at a relatively abstract level (Hiebert and Behr 1988). Sometimes, drawing a picture can be thought of as an intermediate step between a mental representation and a physical representation (van Essen and Hamaker 1990).

In the following discussion, real-life problem-solving situations are used to explore the meaning of multiplication of fractions, and pictorial representations are constructed to guide teachers and students to reason about situations, discover patterns, seek ways to validate their own thinking, and convince others that their thinking is correct.

Problem 1

At the supermarket, potatoes were bagged in 3/4-pound bags. Mom bought 3 bags of potatoes. How many pounds of potatoes did Mom buy?

Pictorially, this situation can be represented as shown in **figure 1**. To find the total amount of potatoes, we need to add 3/4 three times. Symbolically, this situation can be represented as 3/4 + 3/4 + 3/4, which is 3 × 3/4. This problem is similar to some of the multiplication situations that students encounter with whole numbers. The link between multiplication and addition is clearly shown. The repeated-addition model offers a satisfying interpretation; that is, the problem deals with three equal-sized groups wherein each group consists of one 3/4-pound bag of potatoes. Situations such as this one can be modeled by the repeated-addition interpretation. We can, however, find other situations that can be represented by multiplication but that do not involve direct repeated-addition situations. Let us look at another problem.

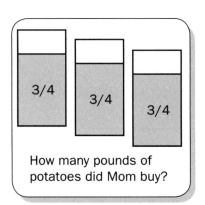

Fig. 1. Pictorial representation of problem 1

Problem 2

Mrs. Smith has 120 books in her fourth-grade classroom; 4/5 of the books are fiction. How many books are fiction?

Figure 2 shows a pictorial representation of this situation. The bar is used to represent a unit, or a whole, which in this problem is 120 books. Of this whole, 4/5 are fiction books. We divide the whole into five equal parts and shade four parts to represent 4/5 of the whole. In this representation, three components are identified: (1) a whole (unit), (2) a fractional part of the whole, and (3) this fractional part of the whole. For 4/5, the denominator of the fraction, 5, indicates the number of equal-sized groups into which the set of books (the whole) is subdivided: 120 ÷ 5 = 24. Twenty-four books represents one-fifth of the original set of 120 books,

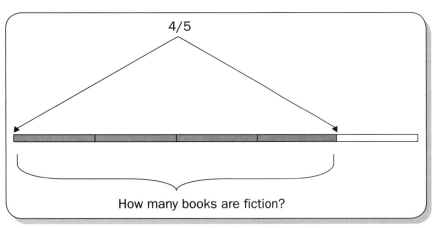

Fig. 2. Pictorial representation of problem 2

or the number of books in one group. The numerator of 4/5, 4, indicates the number of groups being considered: $4 \times 24 = 96$. Ninety-six is the number of books in four groups. For this problem, one sentence, $4/5 \times 120 = 96$, describes both of the operations used. The meaning of the sentence is that multiplication is used to represent the idea of taking a fractional part (4/5) of a whole (120). Problem 3 illustrates a similar situation.

Problem 3

Before the new semester, all the notebooks at the local store are discounted by 1/4. A notebook originally costs $0.96. How much do you save on one notebook if you buy it today?

In **figure 3**, we again use a bar to represent the original price for a notebook, which is $0.96. The $0.96 is considered to be the whole, or the unit, and 1/4 of the whole is discounted. In other words, the purchaser saves 1/4 of the $0.96 when buying one notebook. The whole $0.96 is partitioned into sections of equal size. When $0.96 is subdivided into four equal-sized sections, the mathematical sentence, $0.96 \div 4 = 0.24$, represents the action that has taken place. In this problem, we consider only one section, one-fourth of the original whole, which is $0.24.

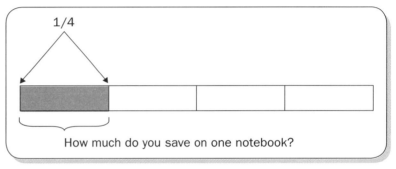

Fig. 3. Pictorial representation of problem 3

The sentence $1/4 \times 0.96 = 0.24$ describes the same procedure. In this situation, we again use multiplication to represent a fractional part of a whole. Problem 4 shows another situation that is similar to those in problems 2 and 3.

Problem 4

Julie bought 4/5 of a yard of material for her class project. Later, she found that she needed only 3/4 of the material. How much material did Julie use for her project?

The entire bar in **figure 4a** represents a yard of material. The shaded portion is 4/5 of a yard, which now becomes the whole in this problem. We need to find 3/4 of this whole. The whole, 4/5 of a yard, is divided into four equal parts. Each part is 1/5 of a yard. Three parts is 3/5 of a yard. The number sentence $3/4 \times 4/5 = 3/5$ indicates the action that we have taken to find the amount of material used, as shown in **figure 4b**. Again, multiplication is used to consider a fractional part of a whole. This situation is more complex than those in problems 2 and 3 because in this problem, students must realize that the whole is itself part of a unit. The fractions 4/5 and 3/5 refer to two

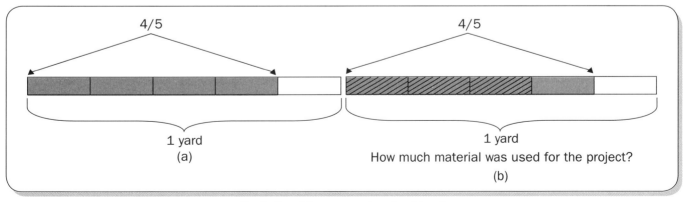

Fig. 4. Pictorial representation of problem 4

different wholes. As long as the correct whole units are identified, determining what fractional part of which whole to consider becomes obvious.

What about a situation where mixed fractions are involved, such as in this problem:

Problem 5

Cabbage costs $0.39 a pound. Julie bought 3 1/3 pounds of cabbage to prepare her dish. How much did she pay for the cabbage?

Figure 5 shows that this situation calls for a combination of both interpretations, repeated addition and part of a whole. The whole is $0.39. The problem involves three wholes (or equal-sized groups) and a fractional part (1/3) of the whole; the situation can be represented by the number sentence 3 1/3 × $0.39. According to the distributive property of multiplication over addition, this sentence is equivalent to (3 + 1/3) × 0.39 = 3 × 0.39 + 1/3 × 0.39. The interpretation of multiplication with fractional numbers both as repeated addition and as taking a fractional part of a whole provides a satisfactory explanation of this situation.

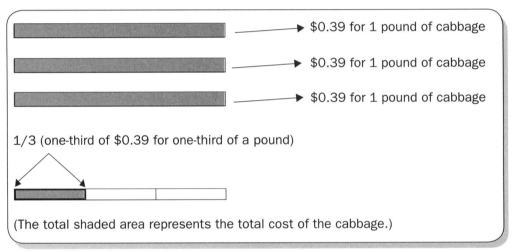

Fig. 5. Pictorial representation of problem 5

Conclusion

Problems 2, 3, and 4 illustrate why the repeated-addition model is inadequate for solving certain multiplication problems. In none of the three situations does the pictorial representation present a clear picture of repeated addition. All three situations have one feature in common, that is, each one considers a fractional part of a whole. The whole can be a set of objects or a fractional part of another whole.

Clearly, many multiplicative situations require an appropriate conceptualization of the whole (unit) before the situation can be understood and a solution procedure can be implemented (Hiebert and Behr 1988). In elementary school mathematics, a fraction is typically interpreted as a part-whole relationship, that is, partitioning an object or a set of objects into equal parts. A whole is cut into n slices; each slice is encoded as $1/n$; and if we refer to several slices (k), that idea is encoded as k/n. The idea of one whole is a basic feature in this representation.

Although the repeated-addition interpretation of multiplication is ordinarily the first one that students encounter, by the time that operations on whole numbers are expanded to include rational numbers, the meaning of multiplication must also be expanded. The models representing multiplication can be expanded from repeated addition to include a fractional part of a whole. Using real-life contexts, we can categorize many multiplicative situations that students encounter in the middle grades into those that involve repeated addition or those that require taking a fractional part of a whole, depending on the specific situation. To figure out what situation a problem is describing, students must understand the entire problem and be able to develop a coherent representation of it.

REFERENCES

Graeber, Anna O., and Patricia F. Campbell. "Misconceptions about Multiplication and Division." *Arithmetic Teacher* 40 (March 1993): 408–11.

Graeber, Anna O., and Elaine Tanenhaus. "Multiplication and Division: From Whole Numbers to Rational Numbers." In *Research Ideas for the Classroom: Middle Grades Mathematics,* edited by Douglas T. Owens, pp. 99–117. New York: Macmillian Publishing Co., 1993.

Hiebert, James, and Merlyn Behr. "Introduction: Capturing the Major Themes." In *Research Agenda for Mathematics Education: Number Concepts and Operations in the Middle Grades,* 2nd ed., edited by James Hiebert and Merlyn Behr, pp. 1–18. Hillsdale, N.J.: Lawrence Erlbaum Associates; Reston, Va.: National Council of Teachers of Mathematics, 1988.

Kennedy, Leonard M., and Steve Tipps. *Guiding Children's Learning of Mathematics.* 8th ed. Belmont, Calif.: Wadsworth/Thomson Learning, 1997.

National Council of Teachers of Mathematics (NCTM). *Principles and Standards for School Mathematics.* Reston, Va.: NCTM, 2000.

Ott, Jack M. "A Unified Approach to Multiplying Fractions." *Arithmetic Teacher* 37 (March 1990): 47–49.

van Essen, Gerard, and Christiaan Hamaker. "Using Self-Generated Drawings to Solve Arithmetic Word Problems." *Journal of Educational Research* 83 (July/August 1990): 301–12.

Developing Algebraic Reasoning through Generalization

John K. Lannin

NCTM's (2000) recommendations for algebra in the middle grades strive to assist students' transition to formal algebra by developing meaning for the algebraic symbols that students use. Further, students are expected to have opportunities to develop understanding of patterns and functions, represent and analyze mathematical situations, develop mathematical models, and analyze change. By helping students move from specific numeric situations to develop general rules that model all situations of that type, teachers in fact begin to address the NCTM's recommendations for algebra. Generalizing numeric situations can create strong connections between the mathematical content strands of number and operation and algebra (as well as with other content strands). In addition, these generalizing activities build on what students already know about number and operation and can help students develop a deeper understanding of formal algebraic symbols.

My research has examined the various strategies that students use as they attempt to generalize numeric situations and articulate corresponding justifications. Knowledge of students' algebraic reasoning can assist teachers in identifying common errors and guiding students toward an understanding of what constitutes a valid and a useful algebraic generalization.

Background on Generalizing Mathematical Situations

An excellent introductory situation for middle school students to generalize is the Cube Sticker problem (see **fig. 1**). This situation can be generalized using a variety of rules, each providing insight into the connections between the arithmetic and geometric relationships that exist in the situation. The following paragraphs describe how I present such situations to students to give them the opportunity to create and discuss their own rules as they attempt to construct generalizations.

A company makes colored rods by joining cubes in a row and using a sticker machine to place "smiley" stickers on the rods. The machine places exactly 1 sticker on each exposed face of each cube. Every exposed face of each cube has to have a sticker; this rod of length 2, then, would need 10 stickers.

1. How many stickers would you need for rods of lengths 1–10? Explain how you determined these values.
2. How many stickers would you need for a rod of length 20? Of length 56? Explain how you determined these values.
3. How many stickers would you need for a rod of length 137? Of length 213? Explain how you determined these values.
4. Write a rule that would allow you to find the number of stickers needed for a rod of any length. Explain your rule.

Fig. 1. The Cube Sticker problem (adapted from NCTM 2000)

Mathematics Teaching in the Middle School 8 (March 2003): 342–48

Growing Professionally

When introducing this problem, I spend a few minutes discussing the problem situation (what we are trying to find, what information is useful, and so on), to help students understand the nature of the problem. Then, I allow students to work individually on the problem for about ten minutes to enable each student to develop at least one strategy for finding the number of stickers.

After working individually, students share their thinking in small groups and discuss the validity of each strategy, along with its advantages and disadvantages. A whole-class discussion about each student's strategy is helpful to demonstrate the variety of possible solution strategies and to develop classroom expectations for what constitutes an efficient generalization and what constitutes a valid generalization.

Two important characteristics of the Cube Sticker problem facilitate generalization. First, the problem requires students to find the number of stickers for rods of different lengths before asking them to construct a general rule. This progression helps students identify which factors vary and which remain the same when calculating the number of stickers on a rod. Second, by requiring students to find the number of stickers for relatively short rods, followed by the number of stickers for much longer rods, the problem forces students to move beyond using drawing and counting strategies toward identifying a general relationship that exists in the situation.

Students' Strategies for Developing Generalizations

When students try to formulate generalizations for situations, such as the Cube Sticker problem, they bring both powerful reasoning and misconceptions regarding the application of mathematical operations to the problem. I invite you to think about how you would find the number of stickers for rods of various lengths before reading the descriptions that follow.

Figure 2 lists students' strategies that other researchers (Stacey 1989; Swafford and Langrall 2000) and I have observed. Note, however, that students often use more than one of these strategies as they attempt to generalize this situation. The following sections elaborate on these strategies.

Strategy	Description
Counting	Drawing a picture or constructing a model to represent the situation and counting the desired attribute
Recursion	Building on a previous term or terms in the sequence to construct the next term
Whole-object	Using a portion as a unit to construct a larger unit using multiples of the unit. This strategy may or may not require an adjustment for over- or under-counting.
Contextual	Constructing a rule on the basis of a relationship that is determined from the problem situation
Guess and check	Guessing a rule without regard to why the rule may work
Rate-adjust	Using the constant rate of change as a multiplying factor. An adjustment is then made by adding or subtracting a constant to attain a particular value of the dependent variable.

Fig. 2. Student strategies for generalizing numerical situations

Counting

A possible statement given by a student who uses the counting strategy is, "Make a rod of that length and count the number of stickers that you would need." This statement describes the level of problem solving at which every student should begin, that is, by building a rod and counting the number of stickers. As teachers, however, we must encourage students to move beyond counting by asking such questions as "Building and counting would be difficult for a rod of length 137. Can you use what you know about shorter rods to find a way to calculate the number of stickers on a rod of length 137?"

Recursion

A student using a recursive strategy has constructed a relationship for building a rod of a given length from a rod with a length that is 1 cube shorter than the desired rod. When using this strategy for the Cube Sticker problem, a student might state, "To find the number of stickers, start with 6 stickers for the first one and add 4 stickers for every cube that you add to the rod because you can peel the sticker off the end of the old rod and place it on the new cube. Then, you need only 4 more stickers for the new rod" (see **fig. 3**).

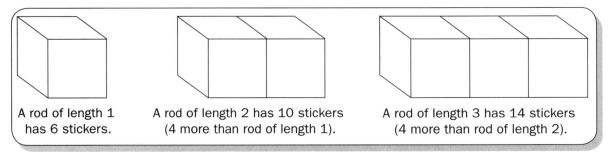

A rod of length 1 has 6 stickers.

A rod of length 2 has 10 stickers (4 more than rod of length 1).

A rod of length 3 has 14 stickers (4 more than rod of length 2).

Fig. 3 A recursive strategy involves building on the previous rod

A recursive strategy is a powerful means for finding the number of stickers if we know (or can find) the number of stickers on the previous rod. This rule is relatively simple to demonstrate on a computer spreadsheet (see **fig. 4**). The rule also provides a strong connection to the concept of slope (i.e., it describes the increase in the number of stickers when the length of the rod increases by 1), but for longer rods, this strategy is not as efficient as an explicit rule. This strategy can lead to the development of an explicit rule if the student can connect the number of times that he or she adds 4 (one less time than the length of the rod) and the number of stickers for a rod of length 1, which is 6. The resulting rule is $6 + 4(n - 1)$, where n is the length of the rod. Some middle school students have difficulty moving beyond the use of a recursive strategy in this situation, possibly because they do not have a strong understanding of the connection between addition and multiplication. I have witnessed middle school students enter 6 into their calculators and add 4 repeatedly to find the number of stickers for rods of length 137.

Length of Rod	Number of Stickers
1	6
2	=B2+4
3	=B3+4
4	=B4+4
5	=B5+4
6	=B6+4
7	=B7+4
8	=B8+4
9	=B9+4
10	=B10+4
11	=B11+4
12	=B12+4
13	=B13+4
14	=B14+4
15	=B15+4

Fig. 4. Use of a recursive formula on a spreadsheet

Whole-object

The use of the whole-object method is often a mistaken attempt to directly apply proportional reasoning to this situation. A student using this strategy might state, "Because a length-10 rod has 42 stickers, you could multiply 42 by the number of times that 10 goes into the length of the rod. For example, for a rod of length 20, you could multiply 42 (the number of stickers in a rod of length 10) by 2 (the number of rods of length 10 that are in a rod of length 20)." We need to encourage students to count the number of stickers for a rod of length 20 and examine why this strategy results in the incorrect number of stickers. Often, students are unaware that when they use this method, they are counting the extra stickers where the two length-10 rods are joined.

Some students who use this strategy will also adjust for overcounting of the stickers by subtracting the two extra stickers that would be counted when the two rods of length 10 are connected. This method correctly counts the number of stickers for a rod of length 20, but students may have difficulty using this strategy to create a rule that would find the number of stickers for a rod of any length; the strategy is confusing to apply to rods with lengths that are not multiples of 10.

Contextual

The contextual strategy is useful because it links the student's rule to the situation and allows for the immediate calculation of the number of stickers for a rod of any length. A student who uses this strategy might say, "All of the middle blocks have only 4 stickers, and the number of middle blocks is 2 less than the length of the rod [see **fig. 5**]. To find the number of middle blocks, subtract 2 from the length of the rod and

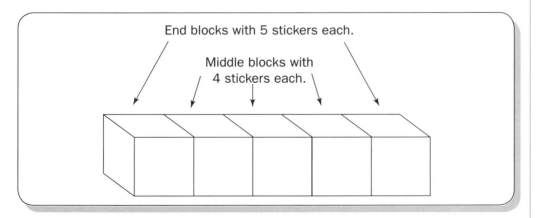

End blocks with 5 stickers each.

Middle blocks with 4 stickers each.

Fig. 5. A contextual strategy for the Cube Sticker problem

multiply that number by 4. Then, add 10 to that total because the 2 blocks on the end have 5 stickers each." (This generalization is one of many such explicit rules that could be constructed for this situation.) This strategy helps students make a connection to a general relationship that exists for all rods of length 2 or more. (Note that explaining how this rule can be related to a rod of length 1 is difficult, although the rule results in the correct number of stickers for that situation, as well.) The blocks on the ends of the rod will always have 5 stickers, and the blocks connecting them (if any) will always have 4 stickers.

Rate-adjust

The rate-adjust strategy is related to the contextual method but uses abstract reasoning regarding the number sequence generated for the number of stickers. A student using this strategy could state, "Because the number of stickers increases by 4 each time [from a rod of length n to a rod of length $n + 1$], you know that you have to multiply the length of the rod by 4. Then, to get 6 stickers for the rod of length 1, you multiply 1 by 4, but you have to add 2 [to make 6 stickers]. You should multiply the length of the rod by 4 and add 2 to find the number of stickers." This strategy requires the understanding that the expression $4n$ increases by 4 as n increases by 1, an important connection to the concept of slope.

Once the rate of change (four stickers added per cube) is identified, it is used to find the correct number of stickers for a rod of length 1, recognizing that the same rule will also result in the correct number of stickers for a rod of any length. This strategy is relatively easy to apply to linear situations, but it is difficult to apply to nonlinear situations; therefore, I encourage students who use this strategy to also develop rules using the contextual strategy.

Guess and check

The final strategy offers no connection to the context or the number sequence generated for the number of stickers. A student who uses this strategy might state, "I tried a few rules and came up with multiply by 4 and add 2." Although the rule is correct for this situation, it provides no insight into the relationship between the rule and the context and, therefore, is difficult to justify.

Student justifications

The following paragraphs discuss the types of justifications that students provide and how teachers can help students recognize what constitutes a valid justification for a generalization.

Avoiding the pitfalls of proof by example

Trying to justify a general situation by demonstrating that the rule results in the correct values for a few individual cases is a common error that occurs throughout grades K–12 (Hoyles 1997). Often, teachers model this type of reasoning. For example, when teaching a traditional algebra course, I remember illustrating the distributive property by calculating $4(5 + 7)$ and $4(5) + 4(7)$, noting that these two expressions resulted in the same value, and moving on. Such a misuse of examples can lead to student misunderstanding about what constitutes a valid justification for a general statement. What can we do to counter this type of misconception? An excerpt from a lesson that I taught demonstrates my attempt to deal with this situation:

Teacher. Jason, please explain your rule.

Jason. I multiplied by 4, and I added 2. It was like a guess and check.

Teacher. OK, so you guessed and it worked for the first [example.]

Jason. Yeah, and I checked it for others and it worked.

Teacher. And it worked for the second [example], and it worked for the third [example]. Is it always going to work?

[I then asked for a show of hands of students who agreed with Jason's reasoning and those who did not. The class was divided almost evenly between these two groups.]

Teacher. Lisa, what do you think?

Lisa. Well, I don't really understand what he is doing. I don't think that it will work every time. It may work for a few, but it may not work for some numbers.

My purpose in polling the class was to see how prevalent this misconception was. Then, I solicited further input from particular students who disagreed, and Lisa provided a counterargument (that we do not know whether this rule really works for all values) for Jason's faulty justification. Jason had no means of supporting his argument further, and for subsequent problems, no student offered this type of argument during a whole-class discussion.

Despite this discussion, many students continued to use proof by example when questioned individually. Further discussion about this type of justification is necessary to help students understand the need to establish a general relationship. The resilience of this error demonstrates that teachers should proceed with caution when faced with this type of argument.

Linking rules to the problem context

Linking the rule to a general relationship that exists in the problem situation, such as the explanations provided for the recursive and contextual strategies, is an acceptable means of justifying a generalization. In classroom discourse, teachers must consistently follow each student's valid or invalid justification by questioning the entire class to see whether others accept the student's justification. In doing so, we can establish the class expectation that all generalizations must be adequately justified. Similarly, we can continue to question students' attempts to provide justification by example by stating, "This rule appears to work for the examples that have been provided. Now we need to understand why this rule will always work." Many students will also reinforce this idea by asking others to explain why a rule works. The emphasis must be placed on understanding why the rule results in the correct values, as well as on finding a correct rule.

Teachers must also require students to explain each component of their rules. Students tend to provide rules without any explanation of what the various parts of their rules represent in relation to the context of the situation. For the Cube Sticker problem, students often state that their rule is to "Multiply the length of a rod by 4 and add 2." Although this rule results in the correct number of stickers when given the length of the rod, it supplies little insight into the situation. Such questions as "Why are you multiplying by 4?" "What does multiplying the length of the rod by 4 calculate?" and "Why are you adding 2 to the length of the rod?" require students to connect each part of the rule to the situation and, thus, help others achieve a deeper understanding of the generalization.

Using proof by induction

A third type of justification, an informal proof by induction, is sometimes offered by students who use the rate-adjust strategy. For the Cube Sticker problem, a student using this justification might say, "I know my rule works for a rod of length 1 because I tested it, and it gave me 6. I know that it will work for all other rods because if I increase the length of the rod by 1, the number of stickers increases by 4." A student providing this type of argument should also explain how he or she knows that the increase will be 4, establishing a strong connection to the concept of slope.

Formalizing and Extending Students' Understanding of Generalization

One of the advantages of generalizing numeric situations is that the variables actually represent varying quantities (as opposed to the view often promoted of a variable as a single unknown value in an equation, such as $2x + 3 = 11$). The introduction of formal notation can help students see algebraic symbols as representing varying quantities. For example, the contextual description provided earlier could easily be translated into the explicit rule $S = 4(n - 2) + 10$, where S is the number of stickers and n is the length of a rod.

Teachers can ask questions about what n represents in this situation and what values are appropriate for substituting for n, such as "Can n be 3/4?" Also, we can ask students whether we could construct a rod that contains exactly 40 stickers, for example, to encourage them to think about the output values of this function. By asking these types of questions, we can begin to address the concepts of domain and range that will be introduced later in formal algebra.

The recursive rule could also be written informally using notation that some middle and secondary school textbooks use (see for example, Coxford et al. [1998]; Romberg et al. [1998]). Students could write the rule as follows: "Start: 6, NEXT = NOW + 4." This notation describes both the initial value in the sequence and the relationship between the current term and the next term in the sequence. Further discussion about when the recursive rule is easier to use and when an explicit rule is easier to use helps students develop a deeper understanding of the advantages and disadvantages of reasoning recursively and explicitly.

Extending the Cube Sticker situation to related situations involving the joining of other prisms, such as triangular or pentagonal prisms, can encourage students to further reflect on the mathematical power of various strategies. For example, the recursive rule could be modified to generate a new rule if the triangular prisms were joined at their bases. The starting value would change to 5 for a rod of length 1. The NOW-NEXT rule would be changed to NEXT = NOW + 3, because 3 more stickers are needed each time the rod is increased in length by 1. The contextual rule described earlier could also be quickly reapplied to this situation, resulting in the explicit rule $S = 3(n - 2) + 8$. Extending the situation in this manner can help students extend their thinking about how their generalizations can be applied to related situations.

Conclusion

Generalizing numeric situations gives students an opportunity to engage in discussions about important mathematical ideas. As seen in the Cube Sticker problem, these situations often connect many mathematical strands and emphasize the need to formulate and validate conjectures. They encourage students to view variables as dynamic quantities that can be used to make sense of their environment.

One of the difficulties students face is that they may not have been asked to justify general statements in the past, and they often resort to justification through the use of examples. We must help students recognize the importance of linking their rules to the context of the situation and not promote the proof-by-example justification commonly used by students.

REFERENCES

Coxford, Arthur F., James T. Fey, Christian R. Hirsch, Harold L. Schoen, Gail Burrill, Eric W. Hart, and Ann E. Watkins. *Contemporary Mathematics in Context*. Chicago, Ill.: Everyday Learning, 1998.

Hoyles, Celia. "The Curricular Shaping of Students' Approaches to Proof." *For the Learning of Mathematics* 17 (February 1997): 7–16.

National Council of Teachers of Mathematics (NCTM). *Principles and Standards for School Mathematics*. Reston, Va.: NCTM, 2000.

Romberg, Thomas A., Gail Burrill, Mary A. Fix, James A. Middleton, Joan D. Pedro, Margaret R. Meyer, Sherian Foster, and Margaret A. Pligge. *Mathematics in Context*. Chicago, Ill.: Encyclopaedia Britannica, 1998.

Stacey, Kaye. "Finding and Using Patterns in Linear Generalizing Problems." *Educational Studies in Mathematics* 20 (1989): 147–64.

Swafford, Jane O., and Cynthia W. Langrall. "Grade 6 Students' Preinstructional Use of Equations to Describe and Represent Problem Situations." *Journal for Research in Mathematics Education* 31 (January 2000): 89–112.

Part-part-whole

Mrs. Jones put her students into groups of 5. Each group had 3 girls. If she has 25 students, how many girls and how many boys does she have in her class?

Associated sets

Ellen, Jim, and Steve bought 3 helium-filled balloons and paid $2 for all 3 balloons. They decided to go back to the store and buy enough balloons for everyone in the class. How much did they pay for 24 balloons?

Well-known measures

Dr. Day drove 156 miles and used 6 gallons of gasoline. At this rate, can he drive 561 miles on a full tank of 21 gallons of gasoline?

Growth (stretching and shrinking situations)

A 6" × 8" photograph was enlarged so that the width changed from 8" to 12". What is the height of the new photograph?

Fig. 1. Types of proportion problems

Mathematics Teaching in the Middle School 6 (December 2000): 254–61

Three Balloons for Two Dollars: Developing Proportional Reasoning

Cynthia W. Langrall
Jane Swafford

Ellen, Jim, and Steve bought three helium-filled balloons and paid $2 for all three. They decided to go back to the store and buy enough balloons for everyone in the class. How much did they pay for 24 balloons? (Lamon 1993b)

WHEN this proportion problem was given to a group of middle school students, some correctly answered $16, but others answered $8, $12, $24, and $26. What kind of reasoning would lead to such diverse responses?

Proportional reasoning is one of the most important abilities to be developed during the middle grades. Using proportional reasoning, students consolidate their knowledge of elementary school mathematics and build a foundation for high school mathematics and algebraic reasoning. Students who fail to develop proportional reasoning are likely to encounter obstacles in understanding higher-level mathematics, particularly algebra.

Through the Peoria Urban Mathematics Plan (PUMP) for Algebra project, a teacher-enhancement project funded by the National Science Foundation, we have been working with forty-four middle school teachers in a midsize urban city to increase the number of students enrolled in algebra. To help teachers understand why many of their students are not successful with proportions, we searched the literature on proportional reasoning to see whether it offered any insights. We drew from the work of Lamon (1993a, 1993b, 1994, 1995) and the Rational Number Project (Cramer, Post, and Currier 1993), incorporating their findings into our work in PUMP classrooms. This article shares the insights that we gained from studying the literature and our experiences using many of the problems found in the literature with our urban middle school students.

Overview of Proportion

A proportion is the statement that two ratios are equal in the sense that both convey the same relationship. For example, the ratio of three balloons for $2 is the same as that of twenty-four balloons for $16. Hence, the statement 3/2 = 24/16 is a proportion. A proportion expresses a multiplicative relationship between two quantities, in this example, balloons and dollars. The ratio 3/2 conveys this multiplicative relationship, meaning that for every 3 balloons, the cost is $2, or that for each dollar, 1.5 balloons can be purchased. To find how many balloons you can buy for a given amount of money, you must multiply the amount by 3/2 or 1.5.

Problem Types

Lamon (1993b) identifies four different types of proportion problems (see **fig. 1**). In a part-part-whole problem, a subset of a whole is compared with its complement (e.g., boys with girls, correct answers with incorrect answers) or with the whole itself (e.g., 12 boys out of 20 students, 80 correct answers out of 100 questions). Problems that involve associated sets relate two quantities, which are not ordinarily associated, through a problem context, such as balloons and dollars, people and pizza, or cookies and boxes.

Problems involving well-known measures express relationships that are well-known entities or rates, such as speed, which is the ratio of miles and hours, or unit price, which is the ratio of items and dollars. Growth problems express a relationship between two continuous quantities, such as height, length, width, or circumference, and involve either *scaling up*, that is, enlarging or stretching, or *scaling down*, that is, reducing or shrinking.

Different problem types elicit different solution strategies regardless of a student's level of understanding of proportional reasoning (Lamon 1993b). Research indicates that students tend to use a higher level of proportional reasoning strategies in solving problems of associated sets. The language of ratio is elicited more naturally when students are forced to think about two sets, not typically associated, as a composite that relates one to the other in the context of the problem. With part-part-whole problems, students are inclined to use informal methods of reasoning, even when they have demonstrated higher-level thinking in other problems, for the reason that part-part-whole problems lend themselves to counting, matching, and building-up strategies that do not require advanced proportional reasoning.

The literature recommends using problems of well-known measures. For some students, familiarity with such well-known measures as speed and price may facilitate proportional thinking; but for others, the familiar language may allow them to mask their lack of understanding (Lamon 1993b). Students who have learned formulas for working "miles per hour" problems, for example, may be able to solve these problems. However, they do not necessarily understand the multiplicative, or proportional, relationship involved in "miles per hour," that is, that a particular number of miles is traveled each hour. In all the literature we reviewed, growth problems were identified as the most difficult type (e.g., Cramer, Post, and Currier [1993]; Lamon [1993b]). Unlike the part-part-whole and associated-sets problems, which involve discrete quantities, growth problems involve continuous quantities, which are more difficult for students to represent with objects or pictures.

Student Solution Strategies

Using problems from each of the four categories, we interviewed sixteen middle school students from PUMP classrooms to discover what kind of strategies they would use to solve the different types of problems. Four students each from fifth through eighth grade were selected by their teachers for the interviews on the basis of their general levels of mathematics understanding. One student from each of the grades had a low level of understanding, two had average levels, and one had a high level.

Level 0

From our interpretation of the literature and the results of the interviews, we identified four different levels of strategies for proportional reasoning (see **fig. 2**). Strategies at level 0 involve no proportional reasoning. These strategies are characterized by additive rather than multiplicative comparisons or random use of numbers or operations in the problems. They do not lead to correct solutions or the development of more mature proportional reasoning. Kerry, for example, randomly selected numbers to divide in the Balloon Problem and obtained an answer of $8, as revealed in the following dialogue:

I. How did you determine your answer?

K. Because if they say they paid $2 for all three balloons, then they decided to go back and pay for twenty-four balloons. I took 3 divided by 24.

I. Why did you divide 24 by 3?

Level 0: Nonproportional reasoning

- Guesses or uses visual clues ("It looks like . . . ")
- Is unable to recognize multiplicative relationships
- Randomly uses numbers, operations, or strategies
- Is unable to link the two measures

Level 1: Informal reasoning about proportional situations

- Uses pictures, models, or manipulatives to make sense of situations
- Makes qualitative comparisons

Level 2: Quantitative reasoning

- Unitizes or uses composite units
- Finds and uses unit rate
- Identifies or uses scalar factor or table
- Uses equivalent fractions
- Builds up both measures

Level 3: Formal proportional reasoning

- Sets up proportion using variables and solves using cross-product rule or equivalent fractions
- Fully understands the invariant and covariant relationships

Fig. 2. Proportional reasoning strategies

D. Because when they said how much did they pay for twenty-four balloons, I thought of something you could divide by 24.

Although Kerry may have had a reason for choosing division as her operation, her choice of the numbers to divide was arbitrary and not justified by the context of the proportional situation.

Earl responded to the Balloon Problem as follows:

E. So they paid $2 for three balloons. You take 2 + 24.

I. Why are you doing 2 + 24?

E. To add up to see the number, because the balloons cost $2 and they want to buy twenty-four balloons for the whole class. So they need to figure how much it will cost, $26.

Clearly, Earl did not understand the multiplicative relationship that eight times as many balloons would cost eight times as much. In fact, throughout the interview, he interpreted many of the problems in terms of addition, whether or not the quantities he chose to add, such as dollars to balloons, were appropriate. For the photograph-enlargement problem (see **fig. 1**), because the 8-inch side had been enlarged by 4 inches, Earl added 4 inches to the 6-inch side instead of multiplying the length of the side by 1.5. To Earl, the enlargement was a question of adding 4 inches to each side of the photograph rather than stretching the whole photograph to 1.5 times its original size. Students like Earl and Kerry do not understand many of the fundamental components that underlie proportional reasoning.

Level 1

Level 1 strategies used by students in the interviews represent informal reasoning about proportional situations. At this level, students can think productively about problems, using manipulatives, pictures, or other models to make sense of the situations. For example, Belinda initially used a nonproportional reasoning strategy to solve the Balloon Problem by adding $2 (200 cents) and 24. She became confused when asked to explain her reasoning, and the interviewer suggested that she sketch out what she was thinking. She drew twenty-four circles on her paper, crossed out three, and wrote $2.00 (see **fig. 3a**). She continued to cross out circles in groups of three, keeping track of the $2.00 amounts on her paper. She then added the column of $2.00 amounts. Similarly, Tonya used a picture to make sense of the groups-of-students problem (see fig. 1). She first made five groups of five boxes (see **fig. 3b**) because "five groups with 5 divided by 25 puts five people in each group." She then labeled three boxes in each group as "G" and two boxes as "B" because "three girls were in the groups, and 2 + 3 = 5. So there are two boys in each group." She next counted the "B" boxes and correctly concluded that the class had ten boys.

(a)   (b)

Fig. 3. Students' informal reasoning strategies

Manipulatives, unit cubes in this instance, also helped students make sense of the Balloon Problem. Joycie first gave $12 as the answer. When asked by the interviewer to try the unit cubes, she made eight groups of three. She then counted by threes to verify that she had twenty-four cubes.

J. So, if three balloons cost $2, for twenty-four balloons, you need $8.

I. How did you use the cubes to determine the cost of the balloons?

J. Well, these [pointing to one group of three] cost $2, and another three will cost $4, and then another three will cost $6, $8, $10, $12, $14, $16. So it must be $16.

I. You gave a number of different answers. How can you determine which answer is correct?

J. I think $16 because I had a model show me. For $8 I was kind of guessing.

Other students started with twenty-four cubes. For example, Jody made stacks of three with her twenty-four cubes, counting by twos as she worked. Corey made similar stacks of three, then put two cubes in front of each stack to represent the $2. Then he counted the number of cubes in the two-cube stacks.

Both associated-sets and part-part-whole problems lend themselves to modeling with pictures, diagrams, and manipulatives. Students should be allowed to develop such strategies before being taught how to set up proportions and use the cross-product rule. This modeling helps students build on their informal reasoning to develop a better understanding of how the two measures in each of the ratios in a proportion vary together.

Level 2

At the more sophisticated strategy level 2, students can use quantitative reasoning without manipulatives or can link their models with numerical calculations. Diamond first used a nonproportional reasoning strategy for the Balloon Problem; she divided 24 by 3 to arrive at $8, then divided 24 by 2 to arrive at $12. Unable to reconcile her two answers, Diamond resorted to using the cubes and made eight groups of three.

I. So how much would it cost for twenty-four?

D. $16 because it's $2 for each pack and there is three balloons in each pack. Put each group in three, and they cost $2, so that would be eight packs that cost $16. . . . I take three in each package times eight groups equals twenty-four, and then $2 \times 8 = 16$. (See **fig. 4a**.)

By using cubes, Diamond understood that she would need $2 eight times. This realization enabled her to move meaningfully to the numerical calculation 2×8, thereby demonstrating quantitative reasoning at level 2. Other students could build up the two measures without pictures or cubes, as Tonya's solution shows (see **fig. 4b**). Marti calculated the unit price for one balloon by dividing $2 by 3 on her calculator. She got 0.66667, which she called "a wacky number," then multiplied this result by 24. Cramer and others (1993) reported that using a unit rate was students' most popular strategy and the one that was responsible for the largest number of correct answers. These researchers also found, however, that problems involving nonintegral rates, such as $0.67, were more difficult.

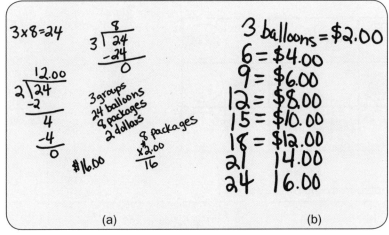

(a)   (b)

Fig. 4. Students' quantitative reasoning strategies

Level 3

At the level of formal proportional reasoning, level 3, students can set up a proportion using a variable and solve for the variable using the cross-product rule or equivalent fractions, with full understanding of the structural relationships that exist. Students must understand that the relationship between the two measures, here balloons and dollars, remains the same, that is, is *invariant*, while the two measures in each ratio vary together, that is, *covary*. In these interviews, conducted at the beginning of our work on proportional reasoning, no student demonstrated formal proportional reasoning by setting up a symbolic proportion, although setting up proportions was presented in the textbook beginning in grade 5.

Four Essential Components of Proportional Reasoning

Four essential prerequisite components of formal proportional reasoning might help explain the limitations in our students' proportional reasoning.

Component 1

Students must recognize the difference between *absolute*, or additive, and *relative*, or multiplicative, change. Absolute change alters the original amount by an absolute, or fixed, amount, such as $10. Relative change alters the original amount by a quantity relative to the original amount, such as 10 percent. Relative change is multiplicative because the amount of the alteration is found by multiplying the original quantity by the rate, again, such as 10 percent. In the photograph-enlargement problem, Earl did not think about the change in the 8-inch side of the photograph using relative thinking. He thought in terms of the absolute change of adding 4 inches to go from 8 to 12 inches, rather than the relative change of adding one-half of the photograph's original length. Thus, he incorrectly added 4 inches to the 6-inch side, rather than 3 inches.

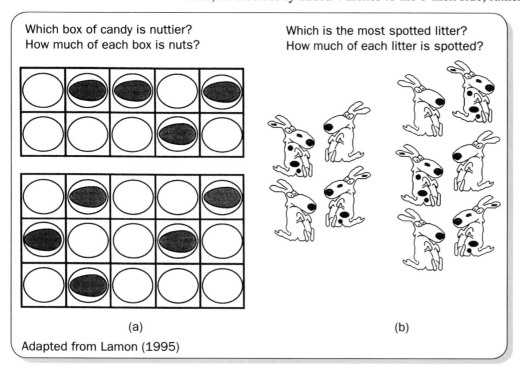

Which box of candy is nuttier?
How much of each box is nuts?

Which is the most spotted litter?
How much of each litter is spotted?

(a)

(b)

Adapted from Lamon (1995)

Fig. 5. How much?

Teachers can help students develop relative thinking by asking, "How much?" rather than "How many?" as the problems adapted from Lamon (1995) in **figure 5** illustrate. The question "How much of each?" focuses students' attention on the part in relation to the whole rather than on an absolute quantity in and of itself.

Component 2

Closely related to component 1 is the need to recognize situations in which using a ratio is reasonable or appropriate. Before students begin to solve problems involving missing values in proportions, they must be able to recognize whether a ratio is the appropriate comparison. Problems such as those in **figure 6** provide this experience.

Component 3

Another essential component in proportional reasoning is understanding that the quantities that make up a ratio covary in such a way that the relationship between them remains unchanged, or is *invariant*. Students tend to see problems in terms of either-or relationships; that is, either the quantities are the same or they are different. However, many different ratios can be proportional because the relationship between the two pairs of numbers is the same. Even students who can generate sets of equivalent fractions often have difficulty recognizing the invariance in equivalent ratios. The problems in **figure 7** give students the opportunity to focus on what has changed and what has remained the same. Students should also start with problems that they can solve with the help of manipulatives or pictures before proceeding to work with problems that are more difficult to model (Lamon 1995).

During mathematics class, the fifth-grade students were grouped at three tables with 2 girls and 4 boys at each table.

During science class, they were arranged into two groups with 3 girls and 6 boys in each group.

What has changed? What has not changed?

Fig. 7. What has changed?

Component 4

The ability to build increasingly complex unit structures is essential; this approach is called *unitizing*. Students engage in qualitative proportional reasoning from level 2 when they choose one ratio as a unit and use that unit to build up to or measure the other. Diamond's use of three-packs of balloons for $2 is an example of unitizing. She used her three-pack as a unit to build up to twenty-four balloons, or eight packs. In the well-known-measures problem in **figure 1**, several students used a unit of 156 miles per 6 gallons of gasoline to determine whether some number of these mile/gallon units would equal 561 miles and 21 gallons of gasoline. Students should be presented with situations that encourage the unitizing process and prompt them to reconceptualize a whole in terms of as many different units as possible.

PUMP Instructional Changes

After our review of the literature and assessment of students' strategies, we shared our information with PUMP teachers through a series of seminars. We examined videotapes of the student interviews and discussed the diversity of student responses to introduce PUMP teachers to the four different problem types, students' solution strategies, and the four essential components of proportional reasoning. After gaining an understanding of the different levels of proportional reasoning, the teachers presented an associated-sets problem to their students and reported to other seminar participants the types of solution strategies exhibited in their classrooms.

These PUMP teachers refrained from introducing formal methods of setting up proportions using variables and instead asked students to represent and solve proportions informally and quantitatively using objects or pictures. We encouraged the teachers to allow their students to present different solution strategies to the class and to require their students to explain their thinking.

Together, we also examined the instructional approach to ratio and proportion that appeared in the textbook series that the PUMP teachers were using. We found that part-

Discuss the statements below. Do they make sense? What distinguishes those that make sense from those that do not?

1. If one girl can walk to school in 10 minutes, two girls can walk to school in 20 minutes.

2. If one box of cereal costs $2.80, two boxes of cereal cost $5.60.

3. If one boy makes one model car in 2 hours, then he can make three models in 6 hours.

4. If Huck can paint the fence in 2 days, then Huck, Tom, and a third boy can paint the fence in 6 days.

5. If one girl has 2 cats, then four girls have 8 cats.

When does it make sense to use a ratio?

Adapted from Lamon (1995)

Fig. 6. Making sense

part-whole problems appeared in the fifth-grade textbook almost twice as often as problems involving associated sets. In the sixth-grade textbook, the problem mix between part-part-whole and associated sets was about equal. Most of the problems in seventh- and eighth-grade textbooks were growth problems, the most difficult type. Well-known-measures problems appeared at all grade levels and dominated none. Regardless of the grade level or problem type, students were asked to calculate a missing value in a proportion but never asked to make comparisons between ratios. To compensate for the deficiencies that they found, PUMP teachers supplemented the textbook with more problems of associated sets and added comparison problems to the curriculum.

Adding different problem types and encouraging and supporting students' informal and quantitative reasoning about proportion had a positive impact on student achievement. The state assessment test results showed an increase of ten points that year, although proportion problems were not specifically identified as such. Greater achievement gains are expected to occur as students spend more time over the span of the middle grades in developing and building on informal and quantitative reasoning strategies.

Recommendations for Instruction

From our work with PUMP teachers and our review of a variety of literature, we have identified several recommendations for classroom instruction.

The emphasis in textbooks is often on developing procedural skills rather than conceptual understanding. The cross-product rule is introduced early in the curriculum, without giving students an opportunity to model proportional relationships with objects or pictures. Instruction with proportional reasoning should begin with situations that can be visualized or modeled. To help students think about situations in which two measures change in relationship with each other, qualitative comparisons should be introduced before numerical comparisons are made or missing values are found. These comparisons can take the form of such statements as "If the speed is faster, you cover the same distance in less time." Beginning instruction should also emphasize informal reasoning with associated-sets and part-part-whole problems. After students can solve proportion problems using informal reasoning, the quantitative reasoning strategies of unit rates and scalar factors can be developed. Well-known-measures problems and growth problems can be introduced to those students who no longer require models. Gradually, a full range of quantitative strategies should be encouraged for solving missing-value problems. Formally setting up proportions using variables and applying the cross-product rule should be delayed until after students have had an opportunity to build on their informal knowledge and develop an understanding of the essential components of proportional reasoning.

Conclusion

Proportional reasoning is complex, both in terms of the underlying mathematics and of the developmental experiences that it requires. Proportional reasoning must be developed over a long period of time, not in a single unit or chapter. Because proportional reasoning is used in geometry, rational numbers, and many other mathematical subject areas and because it appears to be foundational to the development of algebraic reasoning, it should be a unifying theme throughout the middle grades.

REFERENCES

Cramer, Kathleen, Thomas Post, and Sarah Currier. "Learning and Teaching Ratio and Proportion: Research Implications." In *Research Ideas for the Classroom: Middle Grades Mathematics*, edited by Douglas T. Owens, 159–78. New York: Macmillan Publishing Co., 1993.

Lamon, Susan J. "Ratio and Proportion: Children's Cognitive and Metacognitive Processes." In *Rational Numbers: An Integration of Research*, edited by Thomas P. Carpenter, Elizabeth Fennema, and Thomas A. Romberg, 131–56. Hillsdale, N.J.: Lawrence Erlbaum Associates, 1993a.

———. "Ratio and Proportion: Connecting Content and Children's Thinking." *Journal for Research in Mathematics Education* 24 (January 1993b): 41–61.

———. "Ratio and Proportion: Cognitive Foundations in Unitizing and Norming." In *The Development of Multiplicative Reasoning in the Learning of Mathematics*, edited by Guershon Harel and Jere Confrey, 89–120. Albany, N.Y.: State University of New York Press, 1994.

———. "Ratio and Proportion: Elementary Didactical Phenomenology." In *Providing a Foundation for Teaching Mathematics in the Middle Grades*, edited by Judith T. Sowder and Bonnie P. Schappell, 167–98. Albany, N.Y.: State University of New York Press, 1995.

National Council of Teachers of Mathematics (NCTM). *Curriculum and Evaluation Standards for School Mathematics*. Reston, Va.: NCTM, 1989. C

COMMENTARY

Part of developing a deeper understanding is addressing misconceptions. This collection of articles emphasizes what Salvatori suggests are ways to turn "moments of difficulty" into opportunities for learning (2000). Many things cause misconceptions in mathematics, from instruction that focuses on superficial aspects of the content to terminology that is misleading, such as "reducing" fractions. Students' misconceptions can be very subtle and can require careful analysis to uncover. This chapter offers some exemplars of ways teachers respond to students' misconceptions. These articles afford teachers the opportunity to observe the types of misconceptions that students might have, identify possible causes of such misconceptions, and learn to focus instruction on the identification and effective use of misconceptions as opportunities for learning.

© NEWSPAPER ENTERPRISE ASSOCIATION, INC. REPRODUCED BY PERMISSION.

Synopses of Articles

In "Balancing Act: The Truth behind the Equals Sign," Mann (2004) points out how students resist the algebraic function of the equals sign. Through a series of dialogue clips and detailed figures, teachers can gain insights into the instructional challenges of making sense of the equals sign as well as increase their awareness of the crucial importance of doing so.

Baroody's (2006) "Why Children Have Difficulties Mastering the Basic Number Combinations and How to Help Them" adds another dimension to the way in which we view children's development of computational mastery. He starts with a series of myths about learning the basic facts and then proceeds to contrast those with research-based approaches. Baroody provides justification for several strategies that emphasize number sense and fluency while pointing out common missteps and areas of confusion.

The article recommended by more respondents than any other on this topic was "The Harmful Effects of Algorithms in Grades 1–4," by Kamii and Dominick (1998). This provocative piece questions the use of traditional algorithms and is a sure-fire conversation starter. By using examples of procedures invented by children, teachers have opportunities to get inside the thinking of children as they approach a variety of computational problems.

"Mean and Median: Are They Really So Easy?" by Zawojewski and Shaughnessy (2000), uses NAEP data to point out that students are not good at finding the mean and median, nor are they good at selecting and using different statistics appropriately. The article poses three NAEP tasks to use with students (or in professional development with teachers) and discusses how students performed on the national assessment, as well as shares implications for teaching.

Balancing Act:
The Truth behind the Equals Sign

Rebecca L. Mann

I walked into the classroom, wrote " = " on the board, and asked the third graders, "What does this mean?" You can probably anticipate the response—the students all thought it meant "The answer is." Only after some gentle nudging did the students agree that it could also mean "is the same as."

Thinking of the equals sign operationally is a common interpretation of many elementary school students (Kilpatrick, Swafford, and Findell 2001; NCTM 2000). Falkner, Levi, and Carpenter (1999) state, "Children in the elementary grades generally think that the equals sign means that they should carry out the calculation that precedes it and that the number after the equals sign is the answer to the calculation. Elementary school children generally do not see the equals sign as a symbol that expresses the relationship 'is the same as' " (p. 233). The equals sign is a symbol that indicates that a state of equality exists and that the two values on either side of the equals sign are the same. It does not mean that the answer is coming or that the answer is on the other side of the sign.

An understanding of the concept of equality is vital to successful algebraic thinking and is one of the big ideas of algebra about which students should reason. The concept of balance, or equivalence, is the basis for the comprehension of equations and inequalities (Greenes and Findell 1999). Exposing students to this important algebraic concept in the lower grades is essential to develop an understanding of equality (NCTM 2000). Instead of waiting to introduce the concept during the middle school years, teachers should help students in elementary school come to recognize the equals sign as a symbol that represents equivalence and balance.

To help my students transition from "The answer is" to the "is the same as" mode of thinking, I initiated a discussion about seesaws:

Me. What happens when you go on a seesaw?

Sara. You go up and down and if the other person is big, like your dad, you stay up in the air.

Me. What do you think Sara meant when she said if the other person is big, you stay up in the air?

Matthew. She means that if the other person is heavier, then you sometimes only go up and can't come down no matter how hard you try.

Steven. Yeah, but if I go on with my little brother, I go down and I can't stay up in the air. Even if I try, I keep coming back down.

Me. Why do you suppose that happens?

Maria. It's because you weigh different. I mean, because Sara and her dad or Steven and his brother don't weigh the same. See, when Arcelia (*her twin sister*) and I go on the seesaw, we can balance it. That's because we are the same size.

Me. So it is easier to balance the seesaw if you are both the same size? (*The class agrees*.)

The students then became seesaws. With elbows bent and palms at shoulder level

Exposing students to the concept of balance in the lower grades is essential

Teaching Children Mathematics 11 (September 2004): 65–69

Fig. 1. A student simulates a balanced seesaw.

facing the ceiling, each student simulated a balanced seesaw (see **fig. 1**). I told the class that I had an imaginary basket of oranges, stressing that each big, fat, juicy orange weighed exactly the same as every other orange. I also told them about my imaginary basket of apples. These poor apples were the scrawny leftovers. Each apple weighed the same as the others, but each apple weighed much less than each orange. Next, I took the students through a series of scenarios. "I just put an orange in your right hand," I said. "What happened to your seesaw?" The students leaned to the right. "Now I am coming around and putting an orange in your left hand," I continued. "Remember that the oranges weigh the same. Now what happens to your seesaw?" The students straightened up, bringing their seesaws into balance. "The oranges are still in your hands and I am going to add an apple to your left hand," I said. Without waiting for the question, the class quickly tipped to the left. "What should we do to bring you into balance again?" I asked. "Add an apple," Jamal replied. "OK, I am coming around and adding an apple to your left hand," I said.

Some of the students stood up straight, a few leaned even farther to their left, and the majority looked confused. A discussion began about the placement of the last apple. The class came to the consensus that I had made a mistake. "So what you are telling me is that if I had wanted to balance the seesaw, it makes a difference as to on which side I place the apple?" I asked. The students responded, "Yes!"

"OK!" I said. "I'll take the apple I just added to your left hand and move it to your right hand."

The students looked relieved to be back in balance. The lesson continued as I "removed" apples and oranges in different arrangements from the students' seesaws. I made a point during the activity to add one orange to each hand at the same time, to demonstrate how adding equal weight to both sides of the seesaw does not change the tilt of the seesaw.

After shaking out their tired arms, the students were asked to create a set of "Seesaw Rules." Students worked in small groups to discuss their ideas and develop a series of statements about balancing seesaws as I moved from group to group, encouraging the students to extend their thinking. When I had determined that all the essential ideas were incorporated into at least one of the lists, the groups shared their lists and came to consensus as a class on the following set of rules:

- For a seesaw to be balanced, it must have the same amount of weight on each end.
- If one end weighs more or less, the seesaw will not be balanced.
- If you have a balanced seesaw and add something to one end, it will not be balanced anymore.
- If you have a balanced seesaw and take away something from one end, it will not be balanced anymore.
- If you have a balanced seesaw and add the same amount of weight to both ends, it will still be balanced.
- If you have a balanced seesaw and take away the same amount of weight from both ends, it will still be balanced.

Next, I directed the students' attention to the blackboard, on which I had written the equals sign. I encouraged students to make a connection between the equals sign and their seesaws.

Me. What does this equals sign on the board have to do with your seesaws?

Debbie. If you have two oranges on one side and two oranges on the other side, it's the same.

Theo. Yeah, you have to have the same thing on both sides of the seesaw and you have to have the same thing on both sides of the equals sign.

Blake. Look at our rules! If you want the seesaw to be balanced, the same weight must be on each side and if you take something away from one side, it won't be balanced. It's the same for the equals sign.

Me. Can anyone explain what Blake is talking about?

Sari. Well … if you have 3 oranges and 2 apples on one side of the equals sign, just like we had on the seesaw, and 3 oranges and 2 apples on the other side of the equals sign, you have the same on both sides. If you took 2 apples away from one side, it wouldn't be equal anymore; you wouldn't have the same on both sides.

Nancy. Yeah! And it wouldn't balance. See? To be equal, both sides have to balance. Cool!

The third graders were beginning to see the connections between the equals sign, equivalence, and balance.

After this initial "seesaw" lesson, I encouraged students to revisit the balance idea on a regular basis. Over the course of the school year, I placed emphasis on viewing the equals sign not as an indication that it was time to execute an operation but as an indicator of the presence of a relationship. Falkner, Levi, and Carpenter (1999) state,

> A concerted effort over an extended period of time is required to establish appropriate notions of equality. Teachers should also be concerned about children's conceptions of equality as soon as symbols for representing number operations are introduced. Otherwise, misconceptions about equality can become more firmly entrenched. (p. 233)

Additional activities to which the students were exposed that reinforced the concepts of balance and equivalence were pan-balance scales, missing-addend problems, and open-number sentences such as $5 + 6 = \triangle + 2$ or $7 - 3 = 1 + \diamondsuit$. If students were in doubt about their solutions, they were encouraged to act out the problem on their own created seesaw (see **fig. 2**). Students also had opportunities to experiment with balance-scale activities that encourage the development of number sense; in these activities, balance is maintained without assigning numerical values to different shapes. **Figure 3** illustrates activities designed to encourage the development of number sense. After

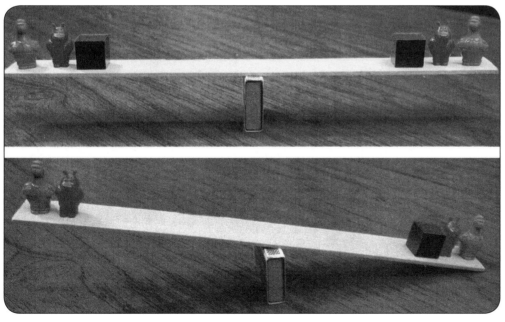

Fig. 2. Students' seesaws

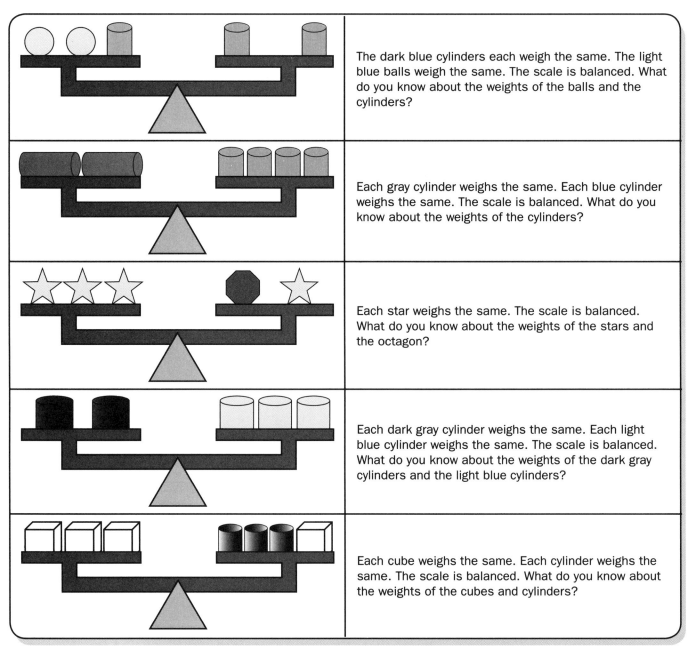

The dark blue cylinders each weigh the same. The light blue balls weigh the same. The scale is balanced. What do you know about the weights of the balls and the cylinders?

Each gray cylinder weighs the same. Each blue cylinder weighs the same. The scale is balanced. What do you know about the weights of the cylinders?

Each star weighs the same. The scale is balanced. What do you know about the weights of the stars and the octagon?

Each dark gray cylinder weighs the same. Each light blue cylinder weighs the same. The scale is balanced. What do you know about the weights of the dark gray cylinders and the light blue cylinders?

Each cube weighs the same. Each cylinder weighs the same. The scale is balanced. What do you know about the weights of the cubes and cylinders?

Fig. 3. Activities designed to encourage the development of number sense

analyzing the contents of the two pans, students compare the two sides and use their deductive-reasoning skills to determine relationships between the objects, as Latisha did in **figure 4**.

The concepts of equivalence and balance are an essential first step in algebraic thinking. In preparation for manipulating equations in secondary school, students should learn how to create and maintain balance (equality) in the lower grades (Greenes et al. 2001). Thinking algebraically should not be the domain of middle school or high school students. The foundation can be, and should be, set during the primary years.

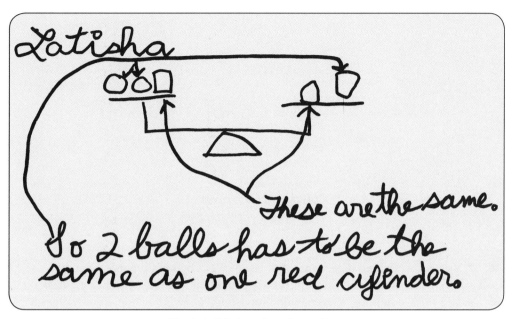

These are the same.

So 2 balls has to be the same as one red cylinder.

Fig. 4. Latisha's work on the problem

Resources for Teaching Equivalency

The resources listed below contain balance-scale activities appropriate for students in kindergarten through fifth grade.

Cuevas, Gilbert J., and Karol Yeatts. *Navigating through Algebra in Grades 3–5*. Reston, Va.: National Council of Teachers of Mathematics, 2001.

Gavin, M. Katherine, Carol R. Findell, Carole E. Greenes, and Linda Jensen Sheffield. *Awesome Math Problems for Creative Thinking*. Series of six books. Mountain View, Calif.: Creative Publications, 2000.

Greenes, Carole, and Carol Findell. *Groundworks: Algebraic Thinking*. Palo Alto, Calif.: Creative Publications, 1999.

Greenes, Carole, Mary Cavanagh, Linda Dacey, Carol Findell, and Marian Small. *Navigating through Algebra in Prekindergarten–Grade 2*. Reston, Va.: National Council of Teachers of Mathematics, 2001.

Hoogeboom, Shirley, and Judy Goodnow. *Beginning Algebra Thinking for Grades 3–4*. Alsip, Ill.: Ideal School Supply Company, 1994.

REFERENCES

Falkner, Karen P., Linda Levi, and Thomas P. Carpenter. "Children's Understanding of Equality: A Foundation for Algebra." *Teaching Children Mathematics* 6 (4) (December 1999): 232–36.

Greenes, Carole, Mary Cavanagh, Linda Dacey, Carol Findell, and Marian Small. *Navigating through Algebra in Prekindergarten–Grade 2*. Reston, Va.: National Council of Teachers of Mathematics, 2001.

Greenes, Carole, and Carol Findell. "Developing Students' Algebraic Reasoning Abilities." In *Developing Mathematical Reasoning in Grades K–12*, 1999 Yearbook of the National Council of Teachers of Mathematics, edited by Lee V. Stiff. Reston, Va.: National Council of Teachers of Mathematics, 1999.

Kilpatrick, Jeremy, Jane Swafford, and Bradford Findell, eds. *Adding It Up: Helping Children Learn Mathematics*. Washington, D.C.: National Academy Press, 2001.

National Council of Teachers of Mathematics (NCTM). *Principles and Standards for School Mathematics*. Reston, Va.: NCTM, 2000.

Why Children Have Difficulties Mastering the Basic Number Combinations and How to Help Them

Arthur J. Baroody

XENA, a first grader, determines the sum of 6 + 5 by saying almost inaudibly, "Six," and then, while surreptitiously extending five fingers under her desk one at a time, counting, "Seven, eight, nine, ten, *eleven*." Yolanda, a second grader, tackles 6 + 5 by mentally reasoning that if 5 + 5 is 10 and 6 is 1 more than 5, then 6 + 5 must be 1 more than 10, or *11*. Zenith, a third grader, immediately and reliably answers, "Six plus five is eleven."

The three approaches just described illustrate the three phases through which children typically progress in mastering the basic number combinations—the single-digit addition and multiplication combinations and their complementary subtraction and division combinations:

- Phase 1: Counting strategies—using object counting (e.g., with blocks, fingers, marks) or verbal counting to determine an answer
- Phase 2: Reasoning strategies—using known information (e.g., known facts and relationships) to logically determine (deduce) the answer of an unknown combination
- Phase 3: Mastery—efficient (fast and accurate) production of answers

Educators generally agree that children should master the basic number combinations—that is, should achieve phase 3 as stated above (e.g., NCTM 2000). For example, in *Adding It Up: Helping Children Learn Mathematics* (Kilpatrick, Swafford, and Findell 2001), the National Research Council (NRC) concluded that attaining computational fluency—the efficient, appropriate, and flexible application of single-digit and multidigit calculation skills—is an essential aspect of mathematical proficiency.

Considerable disagreement is found, however, about how basic number combinations are learned, the causes of learning difficulties, and how best to help children achieve mastery. Although exaggerated to illustrate the point, the vignettes that follow, all based on actual people and events, illustrate the *conventional wisdom* on these issues or its practical consequences. (The names have been changed to protect the long-suffering.) This view is then contrasted with a radically different view (*the number-sense view*), originally advanced by William Brownell (1935) but only recently supported by substantial research.

Vignette 1: Normal children can master the basic number combinations quickly; those who cannot are mentally impaired, lazy, or otherwise at fault.

Alan, a third grader, appeared at my office extremely anxious and cautious. His apprehension was not surprising, as he had just been classified as "learning disabled" and, at his worried parents' insistence, had come to see the "doctor whose specialty was learning problems." His mother had informed me that Alan's biggest problem was that

Teaching Children Mathematics 13 (August 2006): 22–31

he could not master the basic facts. After an initial discussion to help him feel comfortable, we played a series of mathematics games designed to create an enjoyable experience and to provide diagnostic information. The testing revealed, among other things, that Alan had mastered some of the single-digit multiplication combinations, namely, the n × 0, n × 1, n × 2, *and* n × 5 *combinations. Although not at the level expected by his teacher, his performance was not seriously abnormal. (Alan's class had spent just one day each on the 10-fact families* n × 0 *to* n × 9, *and the teacher had expected everyone to master all 100 basic multiplication combinations in this time. When it was pointed out that mastering such combinations typically takes children considerably more time than ten days, the teacher revised her approach, saying, "Well, then, we'll spend two days on the hard facts like the 9-fact family.")*

Vignette 2: Children are naturally unmindful of mathematics and need strong incentives to learn it.

Bridget's fourth-grade teacher was dismayed and frustrated that her new students had apparently forgotten most of the basic multiplication and division facts that they had studied the previous year. In an effort to motivate her students, Mrs. Burnside lit a blowtorch and said menacingly, "You will learn the basic multiplication and division facts, or you will get burned in my classroom!" Given the prop, Bridget and her classmates took the threat literally, not figuratively, as presumably the teacher intended.

Vignette 3: Informal strategies are bad habits that interfere with achieving mastery and must be prevented or overcome.

Carol, a second grader, consistently won the weekly 'Round the World game. (The game entails having the students stand and form a line and then asking each student in turn a question—for example, "What is 8 + 5?" or "How much is 9 take away 4?" Participants sit down if they respond incorrectly or too slowly, until only one child—the winner—is left.) Hoping to instruct and motivate the other students, Carol's teacher asked her, "What is your secret to winning? How do you respond accurately so quickly?" Carol responded honestly, "I can count really fast." Disappointed and dismayed by this response, the teacher wrote a note to Carol's parents explaining that the girl insisted on using "immature strategies" and was proud of it. After reading the note, Carol's mother was furious with her and demanded an explanation. The baffled girl responded, "But, Mom, I'm a really, really fast counter. I am so fast, no one can beat me."

Vignette 4: Memorizing basic facts by rote through extensive drill and practice is the most efficient way to help children achieve mastery.

Darrell, a college senior, can still recall the answers to the first row of basic division combinations on a fifth-grade worksheet that he had to complete each day for a week until he could complete the whole worksheet in one minute. Unfortunately, he cannot recall what the combinations themselves were.

How Children Learn Basic Combinations

The conventional wisdom and the number-sense view differ dramatically about the role of phases 1 and 2 (counting and reasoning strategies) in achieving mastery and about the nature of phase 3 (mastery) itself.

Conventional wisdom: Mastery grows out of memorizing individual facts by rote through repeated practice and reinforcement.

Although many proponents of the conventional wisdom see little or no need for the counting and reasoning phases, other proponents of this perspective at least view these preliminary phases as opportunities to practice basic combinations or to imbue the basic combinations with meaning before they are memorized. Even so, all proponents of the conventional wisdom view agree that phases 1 and 2 are not necessary for achieving the storehouse of facts that is the basis of combination mastery. This conclusion is the logical consequence of the following common assumptions about mastering the number combinations and mental-arithmetic expertise:

- Learning a basic number combination is a simple process of forming an association or bond between an expression, such as $7 + 6$ or "seven plus six," and its answer, 13 or "thirteen." This basic process requires neither conceptual understanding nor taking into account a child's developmental readiness—his or her existing everyday or informal knowledge. As the teachers in vignettes 1 and 4 assumed, forming a bond merely requires practice, a process that can be accomplished directly and in fairly short order without counting or reasoning, through flash-card drills and timed tests, for example.

- Children in general and those with learning difficulties in particular have little or no interest in learning mathematics. Therefore, teachers must overcome this reluctance either by profusely rewarding progress (e.g., with a sticker, smile, candy bar, extra playtime, or a good grade) or, if necessary, by resorting to punishment (e.g., a frown, extra work, reduced playtime, or a failing grade) or the threat of it (as the teacher in vignette 2 did).

- Mastery consists of a single process, namely, fact recall. (This assumption is made by the teacher and the mother in vignette 3.) Fact recall entails the automatic retrieval of the associated answer to an expression. This fact-retrieval component of the brain is independent of the conceptual and reasoning components of the brain.

Number-sense view: Mastery *that underlies computational fluency* grows out of discovering the numerous patterns and relationships that interconnect the basic combinations.

According to the number-sense view, phases 1 and 2 play an integral and necessary role in achieving phase 3; mastery of basic number combinations is viewed as an outgrowth or consequence of number sense, which is defined as well-interconnected knowledge about numbers and how they operate or interact. This perspective is based on the following assumptions for which research support is growing:

- Achieving mastery of the basic number combinations efficiently and in a manner that promotes computational fluency is probably more complicated than the simple associative-learning process suggested by conventional wisdom, for the reason that learning any large body of factual knowledge meaningfully is easier than learning it by rote. Consider, for example, the task of memorizing the eleven-digit number 25811141720. Memorizing this number by rote, even if done in chunks (e.g., 258-111-417-20), requires more time and effort than memorizing it in a meaningful manner—recognizing a pattern or relationship (start with 2 and repeatedly add 3). Put differently, psychologists have long known that people more easily learn a body of knowledge by focusing on its structure (i.e., underlying patterns and relationships) than by memorizing individual facts by rote. Furthermore, psychologists have long known that well-connected factual knowledge is easier to retain in memory and to transfer to learning other new but related facts than are isolated facts. As with any worthwhile knowledge, meaningful memorization of the basic combinations entails discovering patterns or relationships. For example, children who understand the "big idea" of *composition*—that a whole, such as a number, can be composed from its parts, often in different ways and with different parts (e.g., $1 + 7$, $2 + 6$, $3 + 5$, and $4 + 4 = 8$)—can recognize $1 + 7$, $2 + 6$, $3 + 5$, and $4 + 4$ as related facts, as a family of facts that "sum to eight." This recognition can help them understand the related big idea of *decomposition*—that a whole, such as a number, can be decomposed into its constitute parts, often in different ways (e.g., $8 = 1 + 7$, $2 + 6$, $3 + 5$, $4 + 4$, . . .). Children who understand the big ideas of composition and decomposition are more likely to invent reasoning strategies, such as translating combinations into easier or known expressions (e.g., $7 + 8 = 7 + [7 + 1] = [7 + 7] + 1 = 14 + 1$ or $9 + 7 = 9 + [1 + 6] = [9 + 1] + 6 = 10 + 6 = 16$). That is, children with a rich grasp of number and arithmetic patterns and relationships are more likely to achieve level 2.

- Children are intrinsically motivated to make sense of the world and, thus, look for regularities. Exploration and discovery are exciting to them.

- Combination mastery that ensures computational fluency may be more complicated than suggested by the conventional wisdom. Typically, with practice, many of the reasoning strategies devised in phase 2 become semiautomatic or automatic. Even adults use a variety of methods, including efficient reasoning strategies or—as Carol did in vignette 3—fast counting, to accurately and quickly determine answers to basic combinations. For example, children may first memorize by rote a few $n + 1$ combinations. However, once they realize that such combinations are related to their existing counting knowledge—specifically their already efficient number-after knowledge (e.g., "after 8 comes 9")—they do not have to repeatedly practice the remaining $n + 1$ combinations to produce them. That is, they discover the number-after rule for such combinations: "The sum of $n + 1$ is the number after n in the counting sequence." This reasoning process can be applied efficiently to any $n + 1$ combination for which a child knows the counting sequence, even those counting numbers that a child has not previously practiced, including large combinations, such as $1,000,128 + 1$. (Note: The application of the number-after rule with multidigit numbers builds on previously learned and automatic rules for generating the counting sequence.) In time, the number-after rule for $n + 1$ combinations becomes automatic and can be applied quickly, efficiently, and without thought.

Recent research supports the view that the basic number-combination knowledge of mental-arithmetic experts is not merely a collection of isolated or discrete facts but rather a web of richly interconnected ideas. For example, evidence indicates not only that an understanding of commutativity enables children to learn all basic multiplication combinations by practicing only half of them but also that this conceptual knowledge may also enable a person's memory to store both combinations as a single representation. This view is supported by the observation that the calculation prowess of arithmetic savants does not stem from a rich store of isolated facts but from a rich number sense (Heavey 2003). In brief, phases 1 and 2 are essential for laying the conceptual groundwork—the discovery of patterns and relationships—and providing the reasoning strategies that underlie the attainment of computational fluency with the basic combinations in phase 3.

Reasons for Children's Difficulties

According to the conventional wisdom, learning difficulties are due largely to defects in the learner. According to the number-sense view, they are due largely to inadequate or inappropriate instruction.

Conventional wisdom: Difficulties are due to deficits inherent in the learner.

All too often, children's learning difficulties, such as Alan's as described in vignette 1, are attributed largely or solely to *their* cognitive limitations. Indeed, children labeled "learning disabled" are often characterized as inattentive, forgetful, prone to confusion, and unable to apply knowledge to even moderately new problems or tasks. As vignette 1 illustrates, these cognitive characteristics are presumed to be the result of mental-processing deficits and to account for the following nearly universal symptoms of children labeled learning disabled:

- A heavy reliance on counting strategies
- The capacity to learn reasoning strategies but an apparent inability to spontaneously invent such strategies
- An inability to learn or retain basic number combinations, particularly those involving numbers greater than 5 (e.g., sums over 10)
- A high error rate in recalling facts (e.g., "associative confusions," such as responding to $8 + 7$ with "16"—the sum of $8 + 8$—or with "56"—the product of 8×7)

In other words, children with learning difficulties, particularly those labeled learning disabled, seem to get stuck in phase 1 of number-combination development. They can sometimes achieve phase 2, at least temporarily, if they are taught reasoning strategies directly. Many, however, never achieve phase 3.

Number-sense view: Difficulties are due to defects inherent in conventional instruction.

Although some children labeled learning disabled certainly have impairments of cognitive processes, many or even most such children and other struggling students have difficulties mastering the basic combinations for two reasons. One is that, unlike their more successful peers, they lack adequate informal knowledge, which is a critical basis for understanding and successfully learning formal mathematics and devising effective problem-solving and reasoning strategies. For example, they may lack the informal experiences that allow them to construct a robust understanding of composition and decomposition; such understanding is foundational to developing many reasoning strategies.

A second reason is that the conventional approach makes learning the basic number combinations unduly difficult and anxiety provoking. The focus on memorizing individual combinations robs children of mathematical proficiency. For example, it discourages looking for patterns and relationships (conceptual learning), deflects efforts to reason out answers (strategic mathematical thinking), and undermines interest in mathematics and confidence in mathematical ability (a productive disposition). Indeed, such an approach even subverts computational fluency and creates the very symptoms of learning difficulties often attributed to children with learning disabilities and seen in other struggling children:

- *Inefficiency*. Because memorizing combinations by rote is far more challenging than meaningful memorization, many children give up on learning all the basic combinations; they may appear inattentive or unmotivated or otherwise fail to learn the combinations (as vignettes 1 and 4 illustrate). Because isolated facts are far more difficult to remember than interrelated ones, many children forget many facts (as vignette 2 illustrates). Put differently, as vignette 4 illustrates, a common consequence of memorizing basic combinations or other information by rote is forgetfulness. If they do not understand teacher-imposed rules, students may be prone to associative confusion. If a child does not understand why any number times 0 is 0 or why any number times 1 is the number itself, for instance, they may well confuse these rules with those for adding 0 and 1 (e.g., respond to

7×0 with "7" and to 7×1 with "8"). Because they are forced to rely on counting strategies and use these informal strategies surreptitiously and quickly, they are prone to errors (e.g., in an effort to use skip-counting by 7s to determine the product 4×7, or four groups of seven, a child might lose track of the number of groups counted, count "7, 14, 21," and respond "21" instead of "28").

- *Inappropriate applications*. When children focus on memorizing facts by rote instead of making sense of school mathematics or connecting it with their existing knowledge, they are more prone to misapply this knowledge because they make no effort to check themselves or they miss opportunities for applying what they do know (e.g., they fail to recognize that the answer "three" for $2 + 5$ does not make sense). For example, Darrell's rote and unconnected knowledge in vignette 4 temporarily satisfied his teacher's demands but was virtually useless in the long run.

- *Inflexibility*. When instruction does not help or encourage children to construct concepts or look for patterns or relationships, they are less likely to spontaneously invent reasoning strategies, and thus they continue to rely on counting strategies. For example, children who do not have the opportunity to become familiar with composing and decomposing numbers up to 18 are unlikely to invent reasoning strategies for sums greater than 10.

Helping Children Master Basic Combinations

Proponents of the conventional wisdom recommend focusing on a short-term, direct approach, whereas those of the number-sense view recommend a long-term, indirect approach.

Conventional wisdom: Mastery can best be achieved by well-designed drill.

According to the conventional wisdom, the best approach for ensuring mastery of basic number combinations is extensive drill and practice. Because children labeled learning disabled are assumed to have learning or memory deficits, "over-learning" (i.e., massive practice) is often recommended so that such children retain these basic facts.

In recent years, some concern has arisen about the brute-force approach of requiring children, particularly those labeled learning disabled, to memorize all the basic combinations of an operation in relatively short order (e.g., Gersten and Chard [1999]). That is, concerns have been raised about the conventional approach of practicing and timed-testing many basic combinations at once. Some researchers have recommended limiting the number of combinations to be learned to a few at a time, ensuring that these are mastered before introducing a new set of combinations to be learned. A controlled- or constant-response-time procedure entails giving children only a few seconds to answer and providing them the correct answer if they either respond incorrectly or do not respond within the prescribed time frame. These procedures are recommended to minimize associative confusions during learning and to avoid reinforcing incorrect associations and "immature" (counting and reasoning) strategies. In this updated version of the conventional wisdom, then, phases 1 and 2 of number-combination development are still seen as largely unnecessary steps for, or even a barrier to, achieving phase [3].

Number-sense view: Mastery can best be achieved by purposeful, meaningful, inquiry-based instruction—instruction that promotes number sense.

A focus on promoting mastery of individual basic number combinations by rote does not make sense. Even if a teacher focuses on small groups of combinations at a time and uses other constant-response-time procedures, the limitations and difficulties of a rote approach largely remain. For this reason, the NRC recommends in *Adding It Up* that efforts to promote computational fluency be intertwined with efforts to foster conceptual understanding, strategic mathematical thinking (e.g., reasoning and problem-solving abilities), and a productive disposition. Four instructional implications of this recommendation and current research follow.

1. *Patiently help children construct number sense by encouraging them to invent, share, and refine informal strategies* (e.g., see phase 1 of **fig. 1**). Keep in mind that number sense is not something that adults can easily impose. Help children gradually build up big ideas, such as composition and decomposition. (See **fig. 2** and **fig. 3** for examples of games involving these big ideas.) Children typically adopt more efficient strategies as their number sense expands or when they have a real need to do so (e.g., to determine an outcome of a dice roll in an interesting game, such as the additive composition version of Road Hog, described in **fig. 3**).

Phase 1. Encourage children to summarize the results of their informal multiplication computations in a table. For example, suppose that a child needs to multiply 7 × 7. She could hold up seven fingers, count the fingers once (to represent one group of seven), and record the result (7) on the first line below the ◯. She could repeat this process a second time (to represent two groups of seven) and write 14 on the next line. The child could continue this process until she has counted her fingers a seventh time to represent seven groups of seven, then record the answer, 49, on the seventh line below ◯. This written record could then be used later to compute the product of 8 × 7 (eight groups of seven). The child would just count down the list until she comes to the product for 7 × 7 (1 seven is 7, 2 sevens is 14, . . . , 7 sevens is 49) and count on seven more (50, 51, 52, 53, 54, 55, 56). In time, children will have created their own multiplication table.

×	①	②	③	④	⑤	⑥	⑦	⑧	⑨
1	1	2	3	4	5	6	7	8	9
2	2	4	6	8	10	12	14	16	18
3	3	6	9	12	15	18	21	24	27
4	4	8	12	16	20	24	28	32	36
5	5	10	15	20	25	30	35	40	45
6	6	12	18	24	30	36	42	48	
7	7	14	21	28	35	42	49	56	
8	8	16	24	32	40	48			
9	9	18	27		45				

Phase 2. Once the table is completed, children can be encouraged to find patterns or relationships within and between families. See, for example, part III ("Product Patterns") of probe 5.5 on page 5–25 and the "Multiplication and Division" section in box 5.6 on page 5–32, Baroody with Coslick (1998).

Fig. 1. Vertical keeping-track method (based on Wynroth [1986])

2. *Promote meaningful memorization or mastery of basic combinations by encouraging children to focus on looking for patterns and relationships;* to use these discoveries to construct reasoning strategies; and to share, justify, and discuss their strategies (see, e.g., phase 2 of **fig. 1**). Three major implications stem from this guideline:

• Instruction should concentrate on "fact families," not individual facts, and how these combinations are related (see box 5.6 on pp. 5–31 to 5–33, Baroody with Coslick [1998] for a thorough discussion of the developmental bases and learning of these fact families).

- Encourage children to build on what they already know. For example, mastering subtraction combinations is easier if children understand that such combinations are related to complementary and previously learned addition combinations (e.g., $5 - 3$ can be thought of as $3 + ? = 5$). Children who have already learned the addition doubles by discovering, for example, that their sums are the even numbers from 2 to 18, can use this existing knowledge to readily master $2 \times n$ combinations by recognizing that the latter is equal to the former (e.g., $2 \times 7 = 7 + 7 = 14$). Relating unknown combinations to previously learned ones can greatly reduce the amount of practice needed to master a family of combinations.
- Different reasoning strategies may require different approaches. Research indicates that patterns and relationships differ in their salience. Unguided discovery learning might be appropriate for highly salient patterns or relationships, such as additive commutativity. More structured discovery learning activities may be needed for less obvious ones, such as the complementary relationships between addition and subtraction (see, e.g., **fig. 4**).

3. *Practice is important, but use it wisely.*
- Use practice as an opportunity to discover patterns and relationships.
- Practice should focus on making reasoning strategies more automatic, not on drilling isolated facts.
- The learning and practice of number combinations should be done purposefully. Purposeful practice is more effective than drill and practice.
- Practice to ensure that efficiency not be done prematurely—that is, before children have constructed a conceptual understanding of written arithmetic and had the chance to go through the counting and reasoning phases.

4. *Just as "experts" use a variety of strategies, including automatic or semi-automatic rules and reasoning processes, number-combination proficiency or mastery should be defined broadly as including any efficient strategy, not narrowly as fact retrieval.* Thus, students should be encouraged in, not discouraged from, flexibly using a variety of strategies.

The Number Goal game

Two to six children can play this game. A large, square center card is placed in the middle with a number, such as 13, printed on it. From a pile of small squares, all facing down and having a number from 1 to 10, each player draws six squares. The players turn over their squares. Taking turns, each player tries to combine two or more of his or her squares to yield a sum equal to the number on the center card.

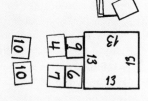
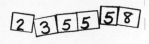

If a player had squares 2, 3, 5, 5, 5, and 8, she could combine 5 and 8 and also combine 3, 5, and 5 to make 13. Because each solution would be worth 1 point, the player would get 2 points for the round. If the player had chosen to combine 2 + 3 + 8, no other possible combinations of 13 would be left, and the player would have scored only 1 point for the round. An alternative way of playing (scoring) the game is to award points for both the number of parts used to compose the target number (e.g., the play 3 + 5 + 5 and 5 + 8 would be scored as 5 points, whereas the play 2 + 3 + 8 would be scored as 3 points).

Number Goal Tic-Tac-Toe (or Three in a Row)

This game is similar to the Number Goal game. Two children can play this game. From a pile of small squares, all facing down and having a number from 1 to 10, each player draws six squares. The players turn over their squares. Taking turns, each player tries to combine two or more of his or her squares to create a sum equal to one of the numbers in the 3 × 3 grid. If a player can do so, she or he places her or his marker on that sum in the 3 × 3 grid, discards the squares used, and draws replacement squares. The goal is the same as that for tic-tac-toe —that is, to get three markers in a row.

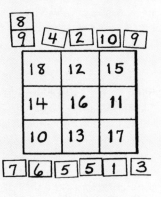

Fig. 2. Examples of a composition-decomposition activity (based on Baroody, Lai, and Mix [2006])

Additive composition version

In this version of Road Hog, each player has two race cars. The aim of the game is to be the first player to have both race cars reach the finish line. On a player's turn, he or she rolls two number cubes to determine how many spaces to move the car forward. (Cars may never move sideways or backward.) Play at the basic level involves two six-sided number cubes with 0 to 5 dots each. At the intermediate level, one number cube has 0 to 5 dots; the other, the numerals 0 to 5. This distinction may encourage counting on (e.g., for a roll of 4 and · · ·, a child might start with "four" and then count "five, six, seven" while successively pointing to the three dots). At the advanced level, both number cubes have the numerals 0 to 5. A similar progression of number cubes can be used for the super basic, super intermediate, and super advanced levels that involve sums up to 18 (i.e., played with six-sided number cubes, both having dots; or one number cube having dots and the other having numerals; or both number cubes having the numerals 5 to 9).

After rolling the number cubes, a player must decide whether to move each race car the number of spaces specified by one number cube in the pair (e.g., for a roll of 3 and 5, the player could move one car three spaces and the other five) or sum the two number cubes and move one car the distance specified by the sum (e.g., with a roll of 3 and 5, the player could move a single car the sum of 3 + 5, or 8, spaces). Note that in this version and all others, opponents must agree that the answer is correct. If an opponent catches the player in an error, the latter forfeits her or his turn.

Deciding which course to take depends on the circumstances of the game at the moment and a player's strategy. The racetrack, a portion of which is depicted below, consists of hexagons two or three wide. In the example depicted, by moving one car three spaces and the other car five spaces, the player could effectively block the road. The rules of the game specify that a car cannot go off the road or over another car. Thus, the cars of other players must stop at the roadblock created by the "road hog"—regardless of what number they roll.

Additive decomposition versions

The game has two decomposition versions. In both versions, the game can be played at three levels of difficulty. The basic level involves cards depicting whole numbers from 1 to 5; the intermediate level, 1 to 10; and the advanced level, 2 to 18.

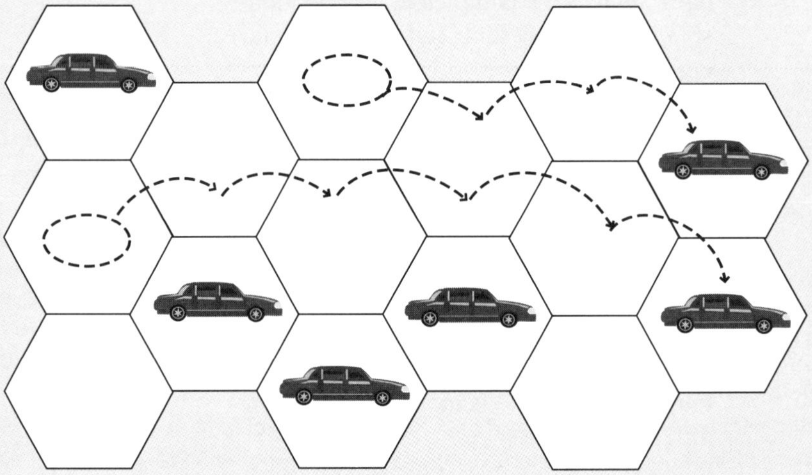

In single additive decomposition, a child draws a card, for example, 3+?, which depicts a part (3), and a missing part (□), and a second card, which depicts the whole (5). The child must then determine the missing part (2), move one car a number of spaces equal to the known part (3), and move the other car a number of spaces equal to the missing part (2).

In double additive decomposition, a child draws a number card, such as 5, and can decompose it into parts any way she or he wishes (e.g., moving one car five spaces and the other none or moving one car three spaces and the other two).

Teachers may wish to tailor the game to children's individual needs. For example, for highly advanced children, the teacher may set up a desk with the version of the game involving whole numbers from 10 to 18.

Multiplicative decomposition versions

The multiplicative decomposition versions would be analogous to those for additive decomposition. For example, at the basic level, a player would draw a card—for example, 3 + ?—and would have to determine the missing factor—6. At the advanced level, a player would draw a card—for example, 18—and would have to determine both nonunit factors—2 and 9 or 3 and 6 (1 and 18 would be illegal).

Fig. 3. Road Hog car-race games

Objectives: (a) Reinforce explicitly the addition-subtraction complement principle and (b) provide purposeful practice of the basic subtraction combinations with single-digit minuends (basic version) or teen minuends (advanced version)

Grade level: 1 or 2 (basic version); 2 or 3 (advanced version)

Participants: Two to six players

Materials: Deck of subtraction combinations with single-digit minuends (basic version) or teen minuends (advanced version) and a deck of related addition combinations

Procedure: From the subtraction deck, the dealer deals out three cards faceup to each player (see figure). The dealer places the addition deck in the middle of the table and turns over the top card. The player to the dealer's left begins play. If the player has a card with a subtraction combination that is related to the combination on the addition card, he or she may take the cards and place them in a discard pile. The dealer then flips over the next card in the addition deck, and play continues. The first player(s) to match (discard) all three subtraction cards wins the game (short version) or a point (long version). (Unless the dealer is the first to go out, a round should be completed so that all players have an equal number of chances to make a match.)

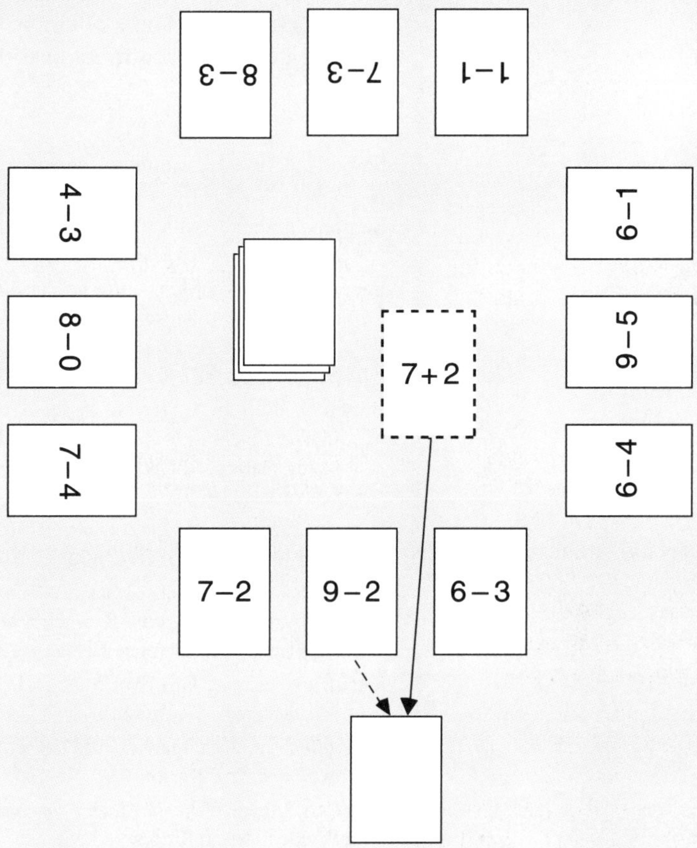

Fig. 4. What's Related (based on Baroody [1989])

Conclusion

An approach based on the conventional wisdom, including its modern hybrid (the constant-response-time procedure) can help children achieve mastery with the basic number combinations but often only with considerable effort and difficulty. Furthermore, such an approach may help children achieve efficiency but not other aspects of computational fluency—namely, appropriate and flexible application—or other aspects of mathematical proficiency—namely, conceptual understanding, strategic mathematical thinking, and a productive disposition. Indeed, an approach based on the conventional wisdom is likely to serve as a roadblock to mathematical proficiency (e.g., to create inflexibility and math anxiety).

Achieving computational fluency with the basic number combinations is more likely if teachers use the guidelines for meaningful, inquiry-based, and purposeful instruction discussed here. Children who learn the basic combinations in such a manner will have the ability to use this basic knowledge accurately and quickly (efficiently), thoughtfully in both familiar and unfamiliar situations (appropriately), and inventively in new situations (flexibly). Using the guidelines for meaningful, inquiry-based, and purposeful approach can also help students achieve the other aspects of mathematical proficiency: conceptual understanding, strategic mathematical thinking, and a productive disposition toward learning and using mathematics. Such an approach can help all children and may be particularly helpful for children who have been labeled learning disabled but who do not exhibit hard signs of cognitive dysfunction. Indeed, it may also help those with genuine genetic or acquired disabilities.

REFERENCES

Baroody, Arthur J. *A Guide to Teaching Mathematics in the Primary Grades.* Boston: Allyn & Bacon, 1989.

Baroody, Arthur J., with Ronald T. Coslick. *Fostering Children's Mathematical Power: An Investigative Approach to K–8 Mathematics Instruction.* Mahwah, N.J.: Lawrence Erlbaum Associates, 1998.

Baroody, Arthur J., Meng-lung Lai, and Kelly S. Mix. "The Development of Young Children's Number and Operation Sense and Its Implications for Early Childhood Education." In *Handbook of Research on the Education of Young Children,* 2d ed., edited by Bernard Spodek and Olivia Saracho, pp. 187–221. Mahwah, NJ: Lawrence Erlbaum Associates, 2006.

Brownell, William A. "Psychological Considerations in the Learning and the Teaching of Arithmetic." In T*he Teaching of Arithmetic,* Tenth Yearbook of the National Council of Teachers of Mathematics, edited by William D. Reeve, pp. 1–50. New York: Teachers College, Columbia University, 1935.

Gersten, Russell, and David Chard. "Number Sense: Rethinking Arithmetic Instruction for Students with Mathematical Disabilities." *The Journal of Special Education* 33, no. 1 (1999): 18–28.

Heavey, Lisa. "Arithmetical Savants." In *The Development of Arithmetic Concepts and Skills: Constructing Adaptive Expertise,* edited by Arthur J. Baroody and Ann Dowker, pp. 409–33. Mahwah, NJ: Lawrence Erlbaum Associates, 2003.

Kilpatrick, Jeremy, Jane Swafford, and Bradford Findell, eds. *Adding It Up: Helping Children Learn Mathematics.* Washington, DC: National Academy Press, 2001.

National Council of Teachers of Mathematics (NCTM). *Principles and Standards for School Mathematics.* Reston, VA: NCTM, 2000.

Wynroth, Lloyd. *Wynroth Math Program—The Natural Numbers Sequence.* 1975. Reprint, Ithaca, NY: Wynroth Math Program, 1986.

This article is based on talks given at the Math Forum, Mountain Plains Regional Resource Center, Denver, Colorado, March 28, 2003, and the EDCO Math Workshop, Lincoln, Massachusetts, July 28–30, 2003. The preparation of this manuscript was supported, in part, by National Science Foundation grant number BCS-0111829 ("Foundations of Number and Operation Sense") and a grant from the Spencer Foundation ("Key Transitions in Preschoolers' Number and Arithmetic Development: The Psychological Foundations of Early Childhood Mathematics Education"). The opinions expressed are solely those of the author and do not necessarily reflect the position, policy, or endorsement of the National Science Foundation or the Spencer Foundation.

The Harmful Effects of Algorithms in Grades 1–4

Constance Kamii and Ann Dominick

S TARTING in the 1970s, researchers such as Ashlock (1972, 1976, 1982) and Brown and Burton (1978) have documented the erroneous but consistent ways in which children inadvertently change the algorithms for multidigit computation. The rules children made up showed that their focus was on trying to remember the steps instead of on logically solving the problems.

Although some researchers studied children's unsuccessful efforts to use the conventional algorithms, others reported the surprising procedures invented by many children, adolescents, and adults. (Throughout this paper, the term *algorithm* is used to refer to the conventional rules of "carrying," "borrowing," etc.; child-invented procedures are referred to as *procedures*.) Cochran, Barson, and Davis (1970), for example, described how an eight-year-old solved $62 - 28$: first, $60 - 20 = 40$, then, $2 - 8 = -6$, and finally, $40 - 6 = 34$. Similar findings have been reported in Argentina (Ferreiro 1988, personal communication 1976), the Netherlands (ter Heege 1978), England (Plunkett 1979), and South Africa (Murray and Olivier 1989).

By the 1980s, some researchers were seriously questioning the wisdom of teaching conventional algorithms. In Brazil, Carraher, Carraher, and Schliemann (1985) and Carraher and Schliemann (1985) compared children who used algorithms with those who used their own procedures. Posing as a customer, for example, the researcher asked a child street vendor how much four coconuts would cost if one cost 35 cruzeiros. The vendor replied, "Three will be 105, plus … 35 … 140!" (Carraher, Carraher, and Schliemann 1985, p. 26). In a subsequent interview, however, the same child wrote the answer 200, as shown in **figure 1**, and explained, "Four times 5 is 20, carry the 2; 2 plus 3 is 5, times 4 is 20" (p. 26). The researchers concluded that children who use their own procedures are much more likely to produce correct answers than those who try to use algorithms. They thus began to think that algorithms were a hindrance rather than a help. Jones (1975) in England, Vakali (1984) in Greece, and Dominick (1991) in the United States reached the same conclusion.

Some investigators went further in the 1990s and concluded that algorithms are harmful to children. Narode, Board, and Davenport (1993) compared second graders before and after they were taught algorithms and concluded that children lose conceptual knowledge when they learn these rules. Kamii (1994) compared children in grades 2–4 who had been taught algorithms with those who had never been taught any algorithms and found that those who did their own thinking got more correct answers and had much better knowledge of place value. She also pointed out that the algorithms that are now conventional are the results of centuries of construction by adult mathematicians. Although it is not necessary for children to go through every historical step, it is unrealistic to expect them to skip the entire process of construction.

Many algorithms that were conventional centuries ago reveal a parallel between an individual's construction of numerical reasoning and humanity's construction of these rules. For example, some Hindus added 278 and 356 on a "dust" board in the following way (Groza 1968):

In my preservice methods class, I use this article along with two others related to teaching algorithms. I ask the students to answer and then discuss in small groups the following three questions: (1) According to the article, why is teaching algorithms (of any kind) harmful? (2) What evidence does this article provide? (3) What should we do instead? This activity is among those that help preservice teachers, especially those who have not necessarily been successful in mathematics, begin to understand why they do not understand why and how an algorithm works. They are actually mad that they weren't "allowed" to think in their elementary years. At the same time, those who have been successful play devil's advocate to criticize the article and say that they "did fine" using the algorithms and are at a loss as to what one would do differently. It makes for a very lively and thoughtful conversation about the way we should be teaching elementary mathematics.

—M. Lynn Breyfogle.

$$\overset{2}{35} \\ \underline{\times\ 4} \\ 200$$

Fig. 1. A Brazilian child's way of using the algorithm

The Teaching and Learning of Algorithms in School Mathematics: 1998 Yearbook, Lorna J. Morrow, ed. (Reston, Va.: NCTM, 1998, chapter 17

$$278 \text{ ------ } 578 \text{ ------ } 628 \text{ ------ } 634$$
$$356 \qquad 56 \qquad 6$$

In this algorithm, 200 and 300 of the 278 and 356 were added first and erased, and changed to 500 (the "5" of 578). The next step was to add 70 and 50, erase them and the 500, and change them to 620 (the "62" of 628). The 8 and 6 were then added and erased, as well as the 2, and changed to 34.

Some leaders in mathematics education also began to say that we must stop teaching algorithms because they make no sense to most children and discourage logical thinking. The most convincing arguments based on systematic study of children in classrooms were advanced by Madell (1985), Burns (1994), and Leinwand (1994).

The purpose of this paper is to present the evidence that led us to the conviction that algorithms not only are not helpful in learning arithmetic but also hinder children's development of numerical reasoning. We begin by discussing our data and go on to describe teachers' observations.

Research Based on Piaget's Constructivism

The distinction Piaget made among the three kinds of knowledge on the basis of their ultimate sources shows why teaching conventional algorithms does not foster children's learning of mathematics. The three kinds of knowledge he distinguished are physical, social, and logico-mathematical knowledge.

Physical knowledge is knowledge of objects in external reality. The color and weight of a block are examples of physical properties that are *in* objects in external reality and can be known empirically by observation.

Examples of *social (conventional) knowledge* are holidays, such as the Fourth of July, and written and spoken languages, such as the word block. Whereas an ultimate source of physical knowledge is in objects, an ultimate source of social knowledge is in conventions made by people.

Logico-mathematical knowledge consists of mental relationships, and the ultimate source of these relationships is each person's mental actions. For example, the child's knowing that one quantity combined with another gives a larger quantity results from his or her making a mental relationship. Someone else can explain this relationship, but this explanation does not become the child's knowledge until he or she makes the relationship. Likewise, an adult can explain to a child the algorithm for two-digit addition. However, listening to this explanation does not ensure the child's making the mental relationships about how to combine the two quantities.

A characteristic of logico-mathematical knowledge is that there is nothing arbitrary in it. For example, adding 356 to 278 results in 634 in every culture. The social (conventional) rule, or algorithm, stating that one *must* add the ones first, then the tens, and then the hundreds is arbitrary. The teaching of algorithms is based on the erroneous assumption that mathematics is a cultural heritage that must be *transmitted* to the next generation.

Piaget's constructivism and the more than sixty years of scientific research by him and others all over the world led Kamii to a compelling hypothesis: Children in the primary grades should be able to invent their own arithmetic without the instruction they are now receiving from textbooks and workbooks. This hypothesis was amply verified, as can be seen in Kamii (1985, 1989b, 1994).

A significant byproduct of this research was the finding that when children are encouraged to do their own thinking to add, subtract, and multiply multidigit numbers, they always deal with the large units first, such as the tens, and then with the ones. As can be seen in **figure 2** and Kamii (1989a, 1989b, 1994), this finding confirmed Madell's (1985) statement that when children are encouraged to think in their own ways, they "*universally* proceed from left to right" (p. 21).

18 +17	10 + 10 = 20 8 + 7 = 15 20 + 10 = 30 30 + 5 = 35	10 + 10 = 20 8 + 2 = another ten 20 + 10 = 30 30 + 5 = 35	10 + 10 = 20 7 + 7 = 14 14 + 1 = 15 20 + 10 = 30 30 + 5 = 35
44 −15	40 − 10 = 30 4 − 5 = 1 less than 0 30 − 1 = 29	40 − 10 = 30 30 − 5 = 25 25 + 4 = 29	40 − 10 = 30 30 + 4 = 34 34 − 5 = 29
135 × 4	4 × 100 = 400 4 × 30 = 120 4 × 5 = 20 400 + 120 + 20 = 540	4 × 100 = 400 4 × 35 = 70 + 70 = 140 400 + 140 = 540	

Fig. 2. Procedures invented by children for addition, subtraction, and multiplication

Another significant finding was that at the end of second and third grade, the children in "constructivist" classes consistently excel over those in traditional classrooms where algorithms are taught. Hypothesizing that algorithms are harmful to children, Kamii compared the performance of children who had never been taught these conventional rules with those who had.

In the school where she worked in 1989–91, some teachers taught algorithms whereas others did not, according to the following distribution:

Grade 1:   None of the four teachers taught algorithms.

Grade 2:   One of the three teachers taught algorithms; of the two who did not, one convinced parents that they should not teach algorithms at home, either.

Grade 3:   Two of the three teachers taught algorithms.

Grade 4:   All four teachers taught algorithms.

All the classes were heterogenous and comparable (the principal mixed up all the children at each grade level and divided them as randomly as possible before each school year). Students who transferred in from other schools were also distributed randomly among all the classes.

One of the problems Kamii asked each child to solve mentally in individual interviews was 7 + 52 + 186 (or 6 + 53 + 185). The answers given by the second, third, and fourth graders are presented in **tables 1–3**. It can be seen in **tables 1** and **2** that the "No algorithms" classes, both in second and in third grade, produced the highest percentages of correct answers (45 percent and 50 percent respectively). It is also evident that the "No algorithms" second- and third-grade classes produced more correct answers than all the fourth-grade classes **(table 3)**, who were all taught algorithms.

More significant are the incorrect answers listed in **tables 1–3**. All the wrong answers given in each class appear in these tables. The broken lines through the middle of each table indicate a range of answers that can be considered reasonable. It can be seen in table 1 that the incorrect answers of the "No algorithms" class were much more reasonable than the wrong answers of the "Algorithms" class. (The class that was exposed to some algorithms at home came out in between.) In third grade **(table 2)**, too, the incorrect answers of the "No algorithms" class were much more reasonable than those of the "Algorithms" classes. The fourth graders **(table 3)**, who had had an additional year of algorithms, gave incorrect answers that were more unreasonable than those of the third-grade "Algorithms" classes. In fourth grade, there were more answers in the 700s and 800s, and some four- and five-digit answers. A new symptom also emerged in fourth grade: Answers such as "4, 4, 4" consisting of single digits, demonstrating that children were thinking about three independent columns.

Table 1

Answers to 7 + 52 + 186 Given by Three Classes of Second Graders in May 1990

Algorithms $n = 17$	Some algorithms taught at home $n = 19$	No algorithms $n = 20$
9308		
1000		
989		
986		
938	989	
906	938	
838	810	
295	356	617
		255
		246
245 (12%)	245 (26%)	245 (45%)
		243
		236
	235	
200	213	138
198	213	—
30	199	—
29	133	—
29	125	—
—	114	
—	—	
	—	
	—	

Note. Blanks indicate that the child declined to try to work the problem.

Why Algorithms Are Harmful

We have two reasons for saying that algorithms are harmful: (1) They encourage children to give up their own thinking, and (2) they "unteach" place value, thereby preventing children from developing number sense.

As stated earlier, when children invent their own procedures, they proceed from left to right. Because there is no compromise possible between going toward the right and going toward the left as the algorithms require, children have to give up their own thinking to use the algorithms.

Table 2

Answers to 6 + 53 + 185 Given by Three Classes of Third Graders in May 1991		
Algorithms $n = 19$	Algorithms $n = 20$	No algorithms $n = 10$
	800 + 38	
838	800	
768	444	
533	344	284
246		245
244 (32%)	244 (20%)	244 (50%)
235	243	243
234	239	238
	238	
	234	
213	204	221
194	202	
194	190	
74	187	
29	144	
—	139	
—	—	
	—	

Note. Blanks indicate that the child declined to try to work the problem.

When we listen to children using the algorithm to do

$$\begin{array}{r} 89 \\ +34 \\ \hline \end{array},$$

for example, we can hear them say, "Nine and four is thirteen. Put down the three; carry the one. One and eight is nine, plus three is twelve...." The algorithm is convenient for adults, who already know that the "one," the "eight," and the "three" stand for 10, 80, and 30. However, for primary school children, who have a tendency to think that the "8" means eight, and so on, the algorithm serves to reinforce this error. The incorrect answers given by the "Algorithms" classes in **tables 1–3** demonstrate that the algorithms "untaught" place value and prevented the children from developing number sense. The children in all the "Algorithms" classes did not notice that their answers of 144, 783, and so on, were unreasonable for 6 + 53 + 185.

Most of the children in the "No algorithms" classes typically began by saying, "One hundred eighty and fifty is two hundred thirty." This is why their errors were reasonable even when they got an incorrect answer. Those in the "Algorithms" classes, however, typically said, "Six and three is nine, plus five is fourteen. Put down the four; carry the one...." Many of them then added 6 (the first addend) to the 1 in 185 (the third addend) and got an answer in the 700s or 800s.

Table 3

Answers to 6 + 53 + 185 Given by Four Classes of Fourth Graders in May 1991

Algorithms $n = 20$	Algorithms $n = 21$	Algorithms $n = 21$	Algorithms $n = 18$
	1215		
	848		
	844		
	783		
1300	783		10,099
814	783		838
744	718	791	835
715	713	738	745
713 + 8	445	721	274
	245		
244 (30%)	244 (24%)	244 (19%)	244 (17%)
243	234		234
	224		234
			234
194	194	144	225
177	127	138	"8, 3, 8"
144	—	134	"4, 3, 2"
143	—	"8, 3, 7"	"4, 3, 2"
134		"8, 1, 7"	—
"4, 4, 4"		—	—
"1, 3, 2"		—	—
		—	
		—	
		—	
		—	
		—	
		—	
		—	

Note. Blanks indicate that the child declined to try to work the problem.

Observations in Classrooms

The harmful effects of algorithms became even more evident when, in 1991–92, one of the fourth-grade teachers, Cheryl Ingram, decided to change her teaching to a constructivist approach. One of the ways in which she tried to wean the children away from algorithms was to write problems such as 876 + 359 horizontally on the board and to ask the class to invent many different ways of solving them without using a pencil. As the children volunteered to explain how they got the answer of 1235 by using the algorithm in their heads, she wrote exactly what students said for each column (6 + 9 = 15, 7 + 5 + 1 = 13, and 8 + 3 + 1 = 12) as follows:

Growing Professionally

$$
\begin{array}{r}
15 \\
13 \\
+12 \\
\hline
40
\end{array}
$$

After the child finished explaining how he or she got the answer of 1235, Ms. Ingram said, "But I followed your way, and when I put 15, 13, and 12 together, I got 40 as my answer. How did you get 1235?" Most of the children were stumped and became silent, until someone pointed out that the teacher's 13 was really 130 and that her 12 stood for 1200.

This kind of place-value problem was not too hard to correct. The persisting difficulty lay in the column-by-column, single-digit approach that prevented children from thinking about multidigit numbers. Presented with problems like 876 + 359, the children continued to give fragmented answers from right to left, such as "5, 130, 1200" (for 6 + 9, 10 + 70 + 50, and 100 + 800 + 300, respectively).

One day, to encourage the children to think about multidigit numbers, Ms. Ingram put on the board one problem after another that had 99 (or 98 or 95) in one of the addends, such as 366 + 199, 493 + 99, and 601 + 199. Only this kind of problem was presented during the entire hour, and the children were asked as usual to think about different ways of adding them.

Almost all the children in the class continued to use the algorithm during the entire hour. One child, however, whom we will call Joe, had been in "constructivist" classes since first grade and volunteered solutions like the following for each problem: "I changed '366 + 199' to '365 + 200,' and my answer is 565." After an entire hour of this kind of "interaction," the number of children imitating Joe by the end of the hour was only three! The rest of the class continued to deal with each column separately.

The school year proceeded with many ups and downs as Ms. Ingram continued her struggle to revive the children's own thinking (see Kamii [1994] for further detail). In May 1992, 6 + 53 + 185 was presented to her fourth graders, and the results were gratifying, as can be seen in **figure 3**.

In **figure 3**, the top row of each 2 × 2 matrix shows the number of children who gave the correct answer, and the bottom row shows the number of those who gave incorrect answers. The left-hand column of each matrix indicates the number who used the conventional algorithm, and the right-hand column shows the number who used their own invented procedures. By comparing these matrices, we can see that when Ms. Ingram taught algorithms in 1990–91, almost all her students used the algorithm, and

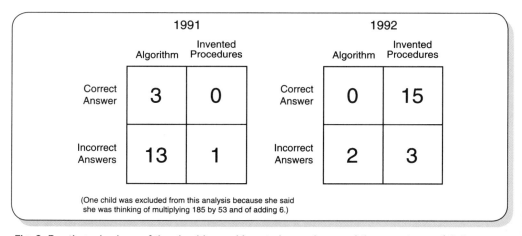

Fig. 3. Fourth graders' use of the algorithm and invented procedures and the correctness of their answer to 6 + 53 + 185 in May 1991 and May 1992

most of them got incorrect answers (shown in the last column of table 3). In 1991–92, by contrast, when Ms. Ingram encouraged her students to do their own thinking, most of her students used invented procedures and got the correct answer.

Ann Dominick, the second author, is a classroom teacher who has taught third and fourth graders for twelve years. When she worked with fourth graders in one school, almost every child in her class had been taught algorithms before coming to her. Now that she teaches third graders in another school, most of the children in her class have never been taught these algorithms. The difference in students' thinking is astounding.

The most striking differences are in students' confidence and their knowledge of place value. Those who have made sense of mathematics approach it with confidence rather than fear and hesitation. The students' intellectual pace is a gallop instead of a walk.

At the beginning of each year Ms. Dominick conducts individual interviews with each student to assess, among other things, their knowledge of place value. In the place-value task (Kamii 1989b), children are asked to show with sixteen counters what each digit in the numeral 16 means. When students came to her class using algorithms, approximately 20 percent of the fourth graders each year showed ten counters for the 1 in 16. (The other 80 percent showed only one counter.) In the third-grade classes where most of the children have not been taught any algorithms, about 85 percent show ten counters for the 1 in 16.

Ms. Dominick's thinking about teaching algorithms has changed over the years from (*a*) teaching arithmetic by teaching algorithms to (*b*) teaching the algorithms after "laying the groundwork for understanding" to (*c*) not teaching algorithms at all. The final shift came from reflecting on what happened when she "laid the groundwork for understanding" and then taught the algorithm.

The first rationale for teaching the algorithms was that it seemed to be the most efficient method. Once students started inventing their own methods, however, this argument no longer held true. For example, using the algorithm for multiplying by 25 often *slows* students' thinking. Frequently, children use their knowledge of $25 \times 4 = 100$ to reason that $25 \times 16 = 400$. Similarly, it takes much more time to use the algorithm to compute $502 - 304$. A more efficient way is to reason that $500 - 300 = 200$ and that $200 - 2 = 198$. An understanding of place value and reference points such as $25 \times 4 = 100$ and $250 \times 4 = 1000$ allow children the flexibility to determine for themselves the most efficient method for solving a problem in a given situation.

The second argument for teaching algorithms was to give struggling students a method for getting answers. It seemed that these students deserved to be given a method for at least getting an answer to achieve some degree of success. It later became evident, however, that when these students forgot a step or developed a "buggy" algorithm, they had nothing to fall back on. Teaching algorithms to these students also sent them the message that "the logic of this procedure is too much for you; so just follow these steps and you'll get the right answer." Some students need more time than others to develop the logic of mathematics. These children deserve the time they need to develop confidence in their ability to make sense of mathematics.

Conclusion

Children come to school with enormous potential for powerful thinking. Educators must try to develop this potential instead of continuing to "put the cart in front of the horse." Adults may pack the cart with treasures, but children need to go through their own constructive process and to proceed with confidence in their own ability to solve problems every step of the way.

REFERENCES

Ashlock, Robert B. *Error Patterns in Computation.* Columbus, Ohio: Charles E. Merrill Publishing Co., 1972, 1976, 1982.

Brown, John Seely, and Richard R. Burton. "Diagnostic Models for Procedural Bugs in Basic Mathematical Skills." *Cognitive Science* 2 (1978): 155–92.

Burns, Marilyn. "Arithmetic: The Last Holdout." *Phi Delta Kappan* 75 (1994): 471–76.

Carraher, Terezinha Nunes, David William Carraher, and Analucia Dias Schliemann. "Mathematics in the Streets and in Schools." *British Journal of Developmental Psychology* 3 (1985): 21–29.

Carraher, Terezinha Nunes, and Analucia Dias Schliemann. "Computation Routines Prescribed by Schools: Help or Hindrance?" *Journal for Research in Mathematics Education* 16 (1985): 37–44.

Cochran, Beryl S., Alan Barson, and Robert B. Davis. "Child-Created Mathematics." *Arithmetic Teacher* 17 (March 1970): 211–15.

Dominick, Ann McNamee. "Third Graders' Understanding of the Multidigit Subtraction Algorithm." Doctoral dissertation, Peabody College for Teachers, Vanderbilt University, 1991.

Ferreiro, Emilia. *Alfabetizacão em processo.* São Paulo: Cortez,1988.

Groza, Vivian S. *A Survey of Mathematics: Elementary Concepts and Their Historical Development.* New York: Holt, Rinehart & Winston, 1968.

Jones, D. A. "Don't Just Mark the Answer—Have a Look at the Method!" *Mathematics in School* 4 (May 1975): 29–31.

Kamii, Constance. *Double-Digit Addition: A Teacher Uses Piaget's Theory.* Videotape. New York: Teachers College Press, 1989a.

——. *Young Children Continue to Reinvent Arithmetic, 2nd Grade.* New York: Teachers College Press, 1989b.

——. *Young Children Continue to Reinvent Arithmetic, 3rd Grade.* New York: Teachers College Press, 1994.

——. *Young Children Reinvent Arithmetic.* New York: Teachers College Press, 1985.

Leinwand, Steven. "It's Time to Abandon Computational Algorithms." *Education Week*, 9 February 1994, p. 36.

Madell, Rob. "Children's Natural Processes." *Arithmetic Teacher* 32 (March 1985): 20–22.

Murray, Hanlie, and Alwyn Olivier. "A Model of Understanding Two-Digit Numeration and Computation." In *Proceedings of the Thirteenth Meeting of the International Group for the Psychology of Mathematics Education*, edited by Gerard Vergnaud, Janine Rogalski, and Michele Artigue, pp. 3–10. Paris: Laboratoire PSYDEE of the National Center of Scientific Research, 1989.

Narode, Ronald, Jill Board, and Linda Davenport. "Algorithms Supplant Understanding: Case Studies of Primary Students' Strategies for Double-Digit Addition and Subtraction." In *Proceedings of the Fifteenth Annual Meeting, North American Chapter of the International Group for the Psychology of Mathematics Education*, Vol. 1, edited by Joanne Rossi Becker and Barbara J. Pence, pp. 254–60. San Jose, Calif.: San Jose State University, Center for Mathematics and Computer Science Education, 1993.

Plunkett, Stuart. "Decomposition and All That Rot." *Mathematics in School* 8, no. 3 (1979): 2–7.

ter Heege, Hans. "Testing the Maturity for Learning the Algorithm of Multiplication." *Educational Studies in Mathematics* 9 (1978): 75–83.

Vakali, Mary. "Children's Thinking in Arithmetic Word Problem Solving." *Journal of Experimental Education* 53 (1984): 106–13.

Mean and Median:
Are They Really So Easy?

Judith S. Zawojewski and J. Michael Shaughnessy

WHEN are median and mean taught to students in your curriculum? Our first introduction to these terms as students was in college-level statistics courses in the mathematics department. Of course, we had learned to find arithmetic averages during elementary school as an application of long division and again in high school algebra when learning to use variables to represent relationships in equations. One of the authors taught middle school in the 1970s, and the other taught finite mathematics to college students in the 1970s. We were surprised to find these measures of central tendency in the books for middle school and equally surprised that college students had not previously seen median and mode. The procedure for finding the median is much easier than the one for finding the mean, so why not include it in the middle school curriculum? To teach the mean, all we had to say to students was that it was the same as the average that they had already learned in fifth- and sixth-grade mathematics.

Yet, are the concepts really so easy? Do students understand the mean and the median? Do they understand that the median and mean give us information about clustering in a distribution and about centering amid variation and that in some situations one is actually more appropriate to use than the other? Data from the National Assessment of Educational Progress (NAEP) (Brown and Silver 1989; Zawojewski and Heckman 1997; Zawojewski and Shaughnessy [2000]) over the past fifteen years indicate that middle school students have some difficulty finding the mean and median. Further, results indicate even greater problems in selecting and using the different statistics appropriately and that these difficulties persist into the high school years. Selected insights from different NAEP reports follow:

> Most students in the 7th and 11th grades appeared not to understand technical statistical terms such as *mean, median, mode,* and *range.* However, there is evidence that they could compute the mean when asked for the average (Brown and Silver 1989, 28).
>
> There is confusion about the meaning of the measures of central tendency, especially that of median, for eighth- and twelfth-grade students (Zawojewski and Heckman 1997, 196)
>
> There was significant growth from 1992 to 1996 in eighth- and twelfth-grade students' performance on NAEP items that required they find the mean and median for particular data sets. However, when given a choice about which statistic to use, students tend to select the mean over the median, regardless of the distribution of the data (Zawojewski and Shaughnessy in press).

Before reading on, have your students respond to the three released NAEP items in **figure 1**. How did your students do on the items? What questions emerge for you as you consider their performance? How can you find out more about what they know and can do? Examine the responses to item 3 in particular. How do students explain their choice of statistic? Do they use the distribution of the data in their explanations, or do they use other reasons?

Discussion of the Items

Item 1 in **figure 1** asks students to find the various measures of central tendency. This item is intended to assess whether students know and can distinguish among the

Mathematics Teaching in the Middle School 5 (March 2000): 436–40

procedures for finding each. Brown and Silver (1989) reported that although the high school students did better than the middle school students in 1985–1986, fewer than half, generally only about 40 percent in each grade, responded correctly with 15 for the mode and 16 for the median and mean, when they responded at all (see **table 1**). Another item administered at the same time used the term *average*, asking for the average age of six children ages 13, 10, 8, 5, 3, and 3. Almost all the students responded to the item, perhaps indicating a greater familiarity with the term *average*. Interestingly,

Item 1: Inches of Snow in January

Year	Inches of Snow	Year	Inches of Snow
1970	15	1978	15
1971	16	1979	17
1972	17	1980	15
1973	15	1981	17
1974	15	1982	16
1975	16	1983	17
1976	16	1984	15
1977	18		

a. What is the mode?   b. What is the median?
c. What is the mean?

Released item from the fourth mathematics assessment in 1985–1986 (Brown and Silver 1989, 28)

Item 2: Number of Sit-ups vs. Age in Years

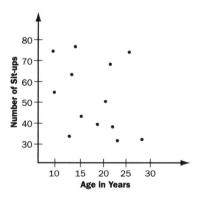

(Kenney and Silver 1997, 215)

In the graph above, each dot shows the number of sit-ups and the corresponding age for one of 13 people. According to this graph, what is the median number of sit-ups for these 13 people?

a. 15   b. 20   c. 45   d. 50   e. 55

Released item from the sixth mathematics assessment in 1992 (Zawojewski and Heckman 1997, 215)

Item 3: Movie Theater Attendance

This question requires you to show your work and explain your reasoning. You may use drawings, words, and numbers in your explanation. Your answer should be clear enough so that another person could read it and understand your thinking. It is important that you show all of your work.

The table below shows the daily attendance at two movie theaters for 5 days and the mean (average) and the median attendance.

	Theater A	Theater B
Day 1	100	72
Day 2	87	97
Day 3	90	70
Day 4	10	71
Day 5	91	100
Mean (average)	75.6	82
Median	90	72

(a) Which statistic, the mean or the median, would you use to describe the typical daily attendance for the 5 days at Theater A? Justify your answer.

(b) Which statistic, the mean or the median, would you use to describe the typical daily attendance for the 5 days at Theater B? Justify your answer.

Released item from the seventh mathematics assessment in 1996 (Zawojewski and Shaughnessy [2000])

Fig. 1. Released NAEP items to try with students

Table 1

Percent Correct and Response Rate for Students in Grades 7 and 11 on Item 1 (Fourth NAEP, 1985–1986)

	Percent Correct	Response Rate
Item	Grade 7	Grade 11
a. What is the mode?	26 [.65]	40 [.41]
b. What is the median?	38 [.65]	47 [.41]
c. What is the mean?	40 [.66]	41 [.72]

(Brown and Silver 1989, 29)

Table 2

Percent of Students in Grades 8 and 12 Responding to Choices on Item 2 (Sixth NAEP, 1992)

	Percent Responding	
Choice	Grade 8	Grade 12
A	5	2
B	11	6
C	32	27
D (correct)	23	31
E	26	32

(Zawojewski and Heckman 1997, 215)

the percent of correct answers from responding seventh graders was not much higher for the *average* item (46%) than for the *mean* item (40%); however, the high school students were more successful on the item that used the term *average*. You may be interested in determining whether your students respond differently when the word *average* is used instead of *mean*.

Item 2 in **figure 1** asks students to identify the median when the data are represented in a scatterplot. The item assesses the combined knowledge of interpreting a graph and using the procedure for finding the median. Zawojewski and Heckman (1997) reported that only a little more than a fifth of eighth-grade students and fewer than a third of twelfth-grade students in the NAEP samples for those grade levels responded correctly, as shown in **table 2**. One common wrong answer was 55 (choice E), which is disturbing because this response may indicate that some students may have added the labels on the *y*-axis ($30 + 40 + 50 + 60 + 70 + 80 = 330$), then divided by 6 to get 55. If so, these students are not only confused about the median and the mean but also unable to use and interpret information given in graphical form. If students in your class respond with choice E, ask students in a follow-up interview question or writing prompt why they chose this answer.

Item 3 in **figure 1** is different from the first two because it requires students to make a choice between mean and median rather than find the measures of central tendency. This type of item assesses students' conceptual understanding of mean and median, which is different than just knowing the procedures for finding them. The written responses indicated that a number of students had not made their choices on the basis of the mathematical characteristics of the two measures of central tendency. Instead, when faced with a choice of mean or median, some students selected the mean, apparently without regard for the shape of the distribution. Some claimed that the mean is the better choice because it is the typical value, or the average, as illustrated by one student's comment, "The mean. To make a generalization of 'typical' attendance, averages are used, not middle points." These types of responses seem to imply that students may think that the median is not representative of a typical value, whereas the mean is. Others claimed that the mean was better because it was superior to the median in some way, as illustrated by the student who wrote, "The mean. An average gives a more accurate # because it involves all the #s."

The idea that the mean is more precise, or more accurate, than the median may actually reveal some understanding that the procedure for the mean incorporates all the values, whereas the median reports just one value. These and other responses suggest that these students did not have an understanding of the trade-offs and relative advantages of each statistic. The prevalence of explanations indicating an "absolute" belief that the mean is better than the median, no matter what, may in part explain why only 4 percent of the grade-12 students in the 1996 NAEP responded with correct answers for item 3—that is, that the median is appropriate for Theater A and the mean, for Theater B—with a complete explanation for at least one measure (see **fig. 2**).

In examining students' difficulty with summary statistics, we must also think about how the problems, or tasks, and their scoring rubrics are designed. For example, item 3 in **figure 1** asked students to respond without knowing why they needed to decide between the two measures of central tendency. To get the highest score (called "extended" by NAEP) on this item, a response had to include a statement that attendance on Day 4 for Theater A (10 attendees) was much lower than for the other days and how this outlier can affect the mean. Because the NAEP scoring rubric accepted as "extended" only those responses that addressed the distribution of the data, the implication is that whenever a set of data contains an outlier, the median is the best statistic to use regardless of purpose. Imagine a situation in which the attendance figures of the two theaters are to be compared directly; in this instance, it could be argued that the identical statistic should be reported for both theaters. In fact, reporting both the mean and the median for each theater would be very effective for making direct comparisons. A statistician might assume that the theater attendances were to be compared using a specific statistical test, such as a t-test; if so, only the mean, not the median, would be required for Theaters A and B. Because the students taking the test were neither given nor asked to make up a reason for choosing the mean or the median, many may have simply chosen the statistic with which they were more familiar.

The NAEP results raise some questions that you may want to ask yourself as you review your students' performance:

- Do your students understand the procedures for finding the mean and median? For example, if they are able to find the mean or median of the years in item 1 rather than the inches of snow, you will be able to tell that your students know the procedure but do not understand when or where to apply it.

- Do your students understand the terminology of *mean* and *median?* For example, if you ask similar questions using *average* or *middle data point* instead of *mean* or *median*, you will be able to tell whether the students are connecting the words to known procedures.

- Do your students make mathematical connections between statistics and other branches of mathematics? For example, if they are able to interpret points on a scatterplot and apply their understanding of median, tasks such as item 2 will help you determine whether they can use their combined knowledge on a single task.

- Do your students understand the relationship between the distribution of the data set and the selection of mean and median? For example, when students explain their choices of mean or median in item 3, you can determine whether they are using information about outliers in their decision making.

a. "The median. Day 4's attendance of 10 obviously lowered the mean a good bit so the median would be more typical."

b. "I would use the mean, since the median gives an artificially low number—it does not reflect at all the two days of high attendance."

Fig. 2. Sample correct response for item 3

Implications for Teaching

The NAEP data we have shared here can help illuminate aspects of students' understanding of measures of central tendency that need attention, such as confusion about the procedures for finding mean and median, as well as difficulty selecting appropriate statistics. Furthermore, other summary statistics are equally important for developing students' conception of a distribution, such as measures of spread and variation (i.e., range, standard deviation, confidence interval, and so on). Unfortunately, past NAEP assessments had few items assessing students' understanding of variation and spread, and none of these items has been released yet.

You can, however, incorporate some of your own questions into items that are similar to these NAEP items to assess your students' understanding of spread, as well as of center. For example, you might want to implement the middle school activity on standard deviation described by Wilmot (1991) in *Dealing with Data and Chance*. Only through additional data gathering can you, as the teacher, understand student difficulties and use that insight to guide your instruction. As the classroom teacher, you are in a good position to probe students' knowledge by asking, orally and in writing, such follow-up questions as those suggested previously. You can also modify tasks to elicit more explanation from students or to provide a familiar context that may elicit better responses.

In addition, you can enhance your curriculum by selecting or creating additional worthwhile tasks to contribute to both teaching and assessment opportunities for data analysis. One of the reasons that students do not find the concepts of mean and median easy may be that they have not had sufficient opportunities to make connections between *centers* and *spreads;* that is, they have not made the link between the measures of central tendency and the distribution of the data set. As you take time for action, you can create teaching and assessment opportunities by selecting items from large-scale assessments, such as the NAEP, as well as tasks from supplementary curriculum materials, to guide your own data-driven instructional decisions.

REFERENCES

Brown, Catherine A., and Edward A. Silver. "Data Organization and Interpretation." In *Results from the Fourth Mathematics Assessment of the National Assessment of Educational Progress,* edited by Mary M. Lindquist, pp. 28–34. Reston, Va.: National Council of Teachers of Mathematics, 1989.

Wilmot, Barbara. "Exploring Standard Deviation." In *Dealing with Data and Chance, Curriculum and Evaluation Standards for School Mathematics* Addenda Series, Grades 5–8, pp. 29–32, 68–69. Reston, Va.: National Council of Teachers of Mathematics, 1991.

Zawojewski, Judith S., and David Heckman. "What Do Students Know about Data Analysis, Statistics, and Probability?" In *Results from the Sixth Mathematics Assessment of the National Assessment of Educational Progress,* edited by Patricia Ann Kenney and Edward A. Silver, pp. 195–224. Reston, Va.: National Council of Teachers of Mathematics, 1997.

Zawojewski, Judith, and J. Michael Shaughnessy. "Data and Chance." In *Results from the Seventh Mathematics Assessment of the National Assessment of Educational Progress,* edited by Edward A. Silver and Patricia Ann Kenney, pp. 235–68. Reston, Va.: National Council of Teachers of Mathematics, 2000.

Preparation of this article was supported in part by a grant to the National Council of Teachers of Mathematics from the National Science Foundation, grant no. RED-9453189, with additional support from the National Center for Education Statistics. Any opinions expressed herein are those of the authors and do not necessarily reflect the views of the National Council of Teachers of Mathematics, the National Science Foundation, or the National Center for Education Statistics.

REFERENCES

Bacon, Francis. "On Studies." In *Essays, Civil and Moral*. New York: Collier & Sons, 1909.

Bright, George W., and Rheta N. Rubenstein. *Professional Development Guidebook for Perspectives on the Teaching of Mathematics, Sixty-sixth Yearbook*. Reston, Va.: National Council of Teachers of Mathematics, 2004.

Burke, Carolyn. Conversation with Phyllis Whitin and David Whitin, April 10, 1995.

Cockcroft, Wilfred. *Mathematics Counts*. London: Her Majesty's Stationery Office, 1986.

Fosnot, Catherine Twomey, and Maarten Dolk. *Young Mathematicians at Work: Constructing Number Sense, Addition and Subtraction*. Portsmouth, N.H.: Heinemann, 2001.

Herrera, Terese. "An Interview with Liping Ma: Do Not Forget Yourself as a Teacher of Yourself." *ENC Focus* 9, no. 3 (July 2002):16–20.

Hiebert, James. "Signposts for Teaching Mathematics through Problem Solving." In *Teaching Mathematics through Problem Solving: Prekindergarten–Grade 6,* edited by Frank K. Lester, Jr., and Randall I. Charles, pp. 53–61. Reston, Va.: National Council of Teachers of Mathematics, 2003.

Holt, John. *How Children Learn*. Harmondsworth, England: Pelican, 1970.

Kamii, Constance, and Ann Dominick. "The Harmful Effects of Algorithms in Grades 1–4." In *The Teaching and Learning of Algorithms in School Mathematics: 1998 Yearbook*, edited by Lorna J. Morrow, pp. 130–40. Reston, Va.: National Council of Teachers of Mathematics, 1988.

Ma, Liping. *Knowing and Teaching Elementary Mathematics: Teachers' Understanding of Fundamental Mathematics in China and the United States*. Mahwah, N.J.: Lawrence Erlbaum Associates, 1999.

National School Reform Faculty (NSRF). "NSRF Protocols: Learning from Text." www.harmonyschool.org/nsrf/protocol/learning_texts.html (accessed July 15, 2007).

Salvatori, Mariolina. "Difficulty: The Great Educational Divide." In *Opening Lines: Approaches to the Scholarship of Teaching and Learning,* edited by Pat Hutchings, pp. 81–93. Menlo Park, Calif.: Carnegie Foundation for the Advancement of Teaching, 2000.

Schlechty, Phillip. Shaking Up the School House: *How to Support and Sustain Educational Innovation*. San Francisco, Calif.: Jossey-Bass, 2001.

Style, Emily. *Listening for All Voices*. Summit, N.J.: Oak Knoll School Monograph, 1988.

SUGGESTED READING

In our search for articles to support professional growth, we found many more than could be included in one book. The articles listed here are more favorites that may support your own professional growth and that of the teachers you support.

—Editors.

Bass, Hyman. "Computational Fluency, Algorithms, and Mathematical Proficiency: One Mathematician's Perspective." *Teaching Children Mathematics* 9 (February 2003): 322–27.

Baxter, Juliet A. "Some Reflections on Problem Posing: A Conversation with Marion Walter." *Teaching Children Mathematics* 13 (October 2005): 122–28.

Behrend, Jean L. "Are Rules Interfering with Children's Mathematical Understanding?" *Teaching Children Mathematics* 8 (September 2001): 36–40.

Berkman, Robert M. "One, Some, or None: Finding Beauty in Ambiguity." *Mathematics Teaching in the Middle School* 11 (March 2006): 324–27.

Boats, Jeff J., Nancy K. Dwyer, Sharon Laing, and Mark P. Fratella, "Geometric Conjectures: The Importance of Counterexamples." *Mathematics Teaching in the Middle School* 9 (December 2003): 210–15.

Bright, George W., Jeane M. Joyner, and Charles Wallis. "Assessing Proportional Thinking." *Mathematics Teaching in the Middle School* 9 (November 2003): 166–72.

Cady, JoAnn. "Implementing Reform Practices in a Middle School Classroom." *Mathematics Teaching in the Middle School* 11 (May 2006): 460–66.

Clements, Douglas H. "Mathematics in the Preschool." *Teaching Children Mathematics* 7 (January 2001): 270–75.

Day, Roger, and Graham A. Jones. "Building Bridges to Algebraic Thinking." *Mathematics Teaching in the Middle School* 2 (February 1997): 209–12.

Edwards, Thomas G. and Sarah M. Hensien. "Using Probability Experiments to Foster Discourse." *Teaching Children Mathematics* 6 (April 2000): 524–29.

Femiano, Robert B. "Algebraic Problem Solving in the Primary Grades." *Teaching Children Mathematics* 9 (April 2003): 444–49.

Flores, Alfinio, and Erika Klein. "From Students' Problem Solving Strategies to Connections with Fractions." *Teaching Children Mathematics* 11 (May 2005): 452–57.

Flowers, Judith, Angela S. Krebs, and Rheta N. Rubenstein. "Problems to Deepen Teachers Mathematical Understanding: Examples in Multiplication." *Teaching Children Mathematics* 12 (May 2006): 478–84.

Hedges, Melissa, DeAnn Huinker, and Meghan Steinmeyer. "Unpacking Division to Build Teachers' Mathematical Knowledge." *Teaching Children Mathematics* 11 (May 2005): 478–83.

Johnson, Art. "Krystal's Method." *Mathematics Teaching in the Middle School* 5 (November 1999): 148–50.

Knuth, Eric, and Dominic Peressini. "Unpacking the Nature of Discourse in Mathematics Classrooms." *Mathematics Teaching in the Middle School* 6 (January 2001): 320–25.

Langham, Belinda, Sue Sundberg, and Terry Goodman. "Developing Algebraic Thinking: An Academy Model for Professional Development." *Mathematics Teaching in the Middle School* 11 (March 2006): 318–23.

Lanius, Cynthia S., and Susan E. Williams. "Proportionality: A Unifying Theme for the Middle Grades." *Mathematics Teaching in the Middle School* 8 (April 2003): 392–96.

Lee, Hea-Jin, and Woo Sik Jung. "Limited-English Proficient (LEP) Students' Mathematical Understanding." *Mathematics Teaching in the Middle School* 9 (January 2001): 269–72.

Losq, Christine S. "Number Concepts and Special Needs Students: The Power of Ten-Frame Tiles." *Teaching Children Mathematics* 11 (February 2005): 310–15.

Lovin, LouAnn, Maggie Kyger, and David Allsopp. "Differentiation for Special Needs Learners." *Teaching Children Mathematics* 11 (October 2004): 158–67.

McGatha, Maggie B., and Linda J. Sheffield. "Mighty Mathematicians: Using Problem Posing and Problem Solving to Develop Mathematical Power." *Teaching Children Mathematics* 13 (September 2006): 79–85.

Neumann, Maureen D. "Preservice Teachers Examine Gender Equity in Teaching Mathematics." *Teaching Children Mathematics* 13 (March 2007): 388–95.

O'Donnell, Barbara D. "On Becoming a Better Problem-Solving Teacher." *Teaching Children Mathematics* 12 (March 2006): 346–51.

Russell, Susan J. "Developing Computational Fluency with Whole Numbers." *Teaching Children Mathematics* 7 (November 2000): 154–58.

Soto-Johnson, Hortensia, Michele Iiams, April Hoffmeister, Barbara Boschmans, and Todd Oberg. "Our Voyage with Knowing and Teaching Elementary Mathematics." *Teaching Children Mathematics* 13 (May 2007): 493–97.

Stein, Mary Kay. "Mathematical Argumentation: Putting Umph into Classroom Discussions." *Mathematics Teaching in the Middle School* 7 (October 2001): 110–12.

Stylianou, Despina A., Patricia A. Kenney, Edward A. Silver, and Cenzig Alacaci. "Gaining Insight into Students' Thinking through Assessment Tasks." *Mathematics Teaching in the Middle School* 6 (October 2000): 136–44.

Thatcher, Debra H. "The Tangram Conundrum." *Mathematics Teaching in the Middle School* 6 (March 2001): 394–99.

Van Zoest, Laura, and Ann Enyart. "Discourse, of Course: Encouraging Genuine Mathematical Conversations." *Mathematics Teaching in the Middle School* 4 (November–December 1998): 150–57.

Watson, Jane M. and J. Michael Shaughnessy. "Proportional Reasoning: Lessons from Research in Data and Chance." *Mathematics Teaching in the Middle School* 10 (September 2004). 104–9

Whitin, Phyllis. "Promoting Problem-Posing Explorations." *Teaching Children Mathematics* 11 (November 2004): 180–86.

Additional Resources from NCTM on Professional Development

Readers wishing to further explore the vital topic of professional development of mathematics educators may want to add one or more of these valuable NCTM publications to their repertoire of resources:

- *Learning from NAEP: Professional Development Materials for Teachers of Mathematics,* edited by Catherine A. Brown and Lynn V. Clark, 2006. This book-plus-CD resource contains activities and workshops that facilitators can use as is or tailor to the unique needs of participants in their school's or district's professional development program. The materials draw on the experiences of educators and the expertise of researchers to help users better understand national assessment data and how such data relate to student learning in mathematics classrooms. Stock #13023

- *Mathematics Teaching Today: Improving Practice, Improving Student Learning,* edited by Tami S. Martin, 2007. This update to the NCTM groundbreaking publication *Professional Standards for Teaching Mathematics* describes a vision for effective mathematics teaching and the support systems required to achieve that vision. It delineates Standards for various aspects of the teaching profession, including teachers' practice, professional supervision, collegial interaction, and career-long professional growth. It defines the roles of teachers and others in improving the teaching and learning of mathematics, expounding on the framework for professional practice found in *Principles and Standards for School Mathematics* in support of NCTM's vision of mathematics learning of the highest quality for all students. Stock #13201

- *Perspectives on the Teaching of Mathematics (Sixty-sixth Yearbook) with Professional Development Guidebook,* edited by Rheta N. Rubenstein, 2004. Teaching is a complex, ongoing endeavor that involves myriad decisions. This publication is organized around three aspects of teaching: foundations for teaching, the enactment of teaching, and the support of teaching nurtured in preservice education and strengthened throughout a teacher's career. The accompanying Professional Development Guidebook, edited by George W. Bright, contains activities that illustrate and extend some of the ideas in the yearbook to further teachers' growth and development in understanding and implementing effective mathematics instruction. Stock #12709

Gain new teaching ideas and engage in ongoing professional development. **Apply for NCTM membership,** including subscription to a **members-only journal** at one of three school levels—elementary, middle school, or secondary—or to the research journal. Apply quickly and easily at www.nctm.org/membership, or call (800)235-7566.

Please consult www.nctm.org/catalog for the availability of these titles and for a plethora of resources for teachers of mathematics at all grade levels.

For the most up-to-date listing of NCTM resources on topics of interest to mathematics educators, as well as on membership benefits, conferences, and workshops, visit the NCTM Web site at www.nctm.org.